EVOLUTION AND THE HUMANITIES

For

Doris Meyer

and in grateful and affectionate memory of

Meyer Fortes

in whose dining room this book began.

Evolution and the Humanities

DAVID HOLBROOK
Fellow of Downing College, Cambridge

St. Martin's Press New York

 For information write:
St. Martin's Press Inc.,
175 Fifth Avenue,
New York,
N.Y. 10010
Printed in Great Britain
First published in the United States of America in 1987

Library of Congress Cataloging-in-Publication Data

Holbrook, David
Evolution and the humanities

Bibliography: p.
Includes index.
1. Evolution–philosophy. 2. Humanities.
I. Title.
QH371.H64 1987 575′.001 86-22114
ISBN 0-312-00379-X

Contents

1 English and evolution

This study follows on from my previous book *Lost Bearings in English Poetry*, in which I examined the problems of poetry in an age in which people find it hard to believe in anything. I am also a teacher, and one meets the problem there, too: at the moment of writing a student reports in an essay that the existentialists see life as a 'scientific accident'. A great deal of error is contained in these words.

In such a phrase we find ourselves meeting a metaphysic. It has become a stereotype accepted almost without thought, without question - and remains unexamined. 'Science says', it seems, that life and man are the 'random' results of an 'accident' in 'chemistry'. And although this contradicts all that we know of the world and ourselves, it remains the only philosophy of life that seems to be upheld by the one remaining authority in the modern world, science.

I hope this explains how an English specialist came to adventure into debate on evolutionary theory - for the grounds of the myth are in Darwinism. Week after week one is reminded by chance remarks in publications that many people accept a certain attitude to life, a metaphysic, coming across to the humanities from science, which can only menacing to any sense that life can have a meaning. It is not only accepted by deliberate materialists and Marxists, but by agnostics, liberals, and even political conservatives.

For instances, I was told in a review of one of my own books, that man must

> see himself as no more than an evolved biological specimen, a surviving carrier of DNA, or, as a leading sociobiologist has put it, DNA'S way of making more DNA.
> (Anne Stevenson, reviewing *Sylvia Plath - Poetry and Existence*, *Times Literary Supplement*, 12 November, 1976).

Elsewhere I noticed that all values in literary criticism were challenged on a similar 'scientific' basis:

> But how can such values retain their credibility in the godless universe which most people now inhabit? Modern man is quite used to the idea that we are the temporary occupants of a cooling solar system; that human life is an accident of chemistry; that all the ages, from the first dawn on earth to its extinction, will amount to no more than a brief parenthesis in the endless night of space...
> (John Carey, writing about literary values in the *Times Literary Supplement*, 22 February, 1980)

When we are talking to scientists, most give the impression that we have no alternative to the view that life and man came into being by chance, and that all developments in life came only by fortuitous mutations, selected out by the negative process of natural selection - which operates on the principle of ruthless competition for survival. Even if we discount questions of religious belief, and reject such concepts as the *élan vital*, there must be no teleological talk, no recognition of higher and lower, or gradients: no 'goals', except of course the implicit goal of 'survival'. The origins of life, its development, and even the coming into being of man, we must believe, occured by processes analogous with the accidental dumping of a heap of bricks on a patch of waste land, which, in millions of millions of years, becomes a house.[1] As the quotation from Professor Carey shows, this kind of belief is taken to be the grounds for a deep scepticism about values, and the point of human life: it is formidably nihilistic. At the other end of the scale, however, as we shall see, there are those like the philosopher E.W.F. Tomlin who simply says that 'it is inconceivable that consciousness could have been produced by accident'. How can we find our way between these views?

The metaphysical dogma of mechanistic science is apparent in powerful popular forms. Not long ago, both in the publicity for the remarkable biology programmes on BBC television, *Life on Earth* and in the book which followed, what amounted to a philosophy of life was offered. For instance, in the *Radio Times* some charts were accompanied by a text which proclaimed that: 'evolution began when DNA... in copying itself, produced different life cells by mistake'. In his book, David Attenborough writes of Darwin that he:

> demonstrated that the driving force of evolution comes from the accumulation, over countless generations, of *chance* genetical changes sifted by the *rigours of natural selection.*
>
> (my italics).

Later, Attenborough makes the claim that:

> *mistakes in copying* are the source of the variations from which *natural selection can produce evolutionary change.*
>
> (my italics).

In accordance with this dogma, it seems that man's life was fortuitous - the consequence of a 'mistake in coding', and the conclusion from this must be that, rather than being seen as a higher being manifesting some astonishing achievement in the universe, man is a 'mistake' with all that follows - a view that somehow we humans are an absurd aberration.

The objection to the vague metaphysical implications of notions which seem to be coming over to us from science are twofold. First of all, they seem to be based on inescapable *truths* when these views should really be seen as *hypotheses*. Here it will be argued that the hypotheses are doubtful anyway, and often do not explain what they offer to explain: indeed, as we shall see, there are many critics and heretics. Secondly, some of the theories about the nature of things, even though they contradict the observable evident nature of the world, are taken to imply a certain kind of philosophy of life, and this is a false kind of extrapolation. Thus if, we consider the word 'accident' (let alone the word 'error'), we might admit that it could be that things might have been otherwise - life might have taken other forms. But there is another perspective to be taken into account: there does not seem to be a remarkable degree of order in the world; there does seem to be something we must call 'achievement', and something we may call 'directiveness'. So, we are faced with developments which cannot be explained without recognition of some other dynamic than mere 'chance' and 'accident', however much we may be suspicious of teleology: there is some kind of 'gradient', whatever that may mean, towards higher forms of being. So, the widespread belief that life and man were '*mistakes*' (a kind of absurd 'whoops!' in the Universe) is questionable, not least if this is taken to imply that nothing has any kind of impulse, organising power or that directiveness: as Polanyi puts it, there does seem to be a great deal of order and meaning in the world, and it is their fascination with that that impels scientists, even when they think they are being 'objective'. It could be that life and man are the outcome of some tendency or urge inherent in the Universe, some universal 'intelligence'.

While at large many believe they cannot escape from a bleak view of existence, closer acquaintance reveals many doubts in the very areas of science upon which modern beliefs are based. I leave aside here the very abstruse but perhaps fundamental arguments of John Wheeler based on Quantum Mechanics that since nothing exists unless it is observed, the very existence of things requires an intelligence to observe them.

1

> 'To suppose that the evolution of the wonderfully adapted biological mechanisms has depended only on a selection out of haphazard set of variations, each produced by blind chance, is like suggesting that if we went on throwing bricks together into heaps, we would eventually be able to choose ourselves the most desirable houses'.
>
> C.H. Waddington, *The Listener*, 13 November 1952

Eminent biologists like Pierre-Paul Grassé doubt whether small chance mutations can be the basis of evolution. Darwin himself said the use of the word 'chance' was 'incorrect'.[2] C.H. Waddington disputes the concept of chance altogether, and points to a kind of intelligent exploitation of possibilities by populations. The philosopher of science, Professor Marjorie Grene, suggests that Darwin does not account at all for the origin of species, or evolution; while Norman Macbeth, an acute lawyer with a logical mind, finds many serious faults in evolutionary argument. Recently scientists seem to be groping about, to try to find more satisfactory explanations, so many doubts have become apparent.

Yet, despite many falsities and errors in the prevailing view (Attenborough for instance was wrong to attribute to Darwin the theory of *genetical* changes) the dogma is constantly defended and upheld.[3] A debate broke out in the 1982 in *The Times*, but had inevitably to be concluded by an Editorial headed: 'Darwin cleared : official'. One of the most telling letters sent, from E.W.F. Tomlin, was not published. Tomlin pointed out that David Attenborough said

> that his method has been 'to try to perceive the single most significant thread in the history of a group and then concentrate on tracing that'. The remarks of Darwin that he 'demonstrated that the *driving force of evolution* comes from the accumulation, over countless generations, of chance genetic changes sifted by *the rigours of natural selection*', and (later) that 'mistakes in copying are the source of the variations from which *natural selection can produce evolutionary change*' (my italics).

The Times had said that 'the mechanism of evolution is the process of natural selection'. Tomlin made the point:

> that while natural selection may serve to serve to explain the disappearance of certain unwieldly species, as of certain primitive races, it is not by its very nature and operation *a positive agent of organo-formative development.* It cannot therefore be the 'driving force' of evolution. And one has only to read Darwin's correspondence, as between the lines of *The Origin of Species* itself, to perceive that he never ceased to entertain doubts about the positive role he had assigned to natural selection in his account of evolution.

Another relevant letter was from Professor H. Lipson, FRS, and this indicates the dilemma to which the present work is addressed: he said,

> 'I am one of the physical scientists who doubt Darwin's theory of evolution. My doubts arise not from any religious motive or desire to add fuel to either side of any controversy but merely because I think that Darwinism is scientifically indefensible.
>
> As Professor Thoday says, we have no option but to accept evolution; all the fossil evidence points to it. The contention is only about the cause. Darwin maintains that the cause was chance: as generation succeeded generation there would be minor variations at random; those that gave some advantage would persist and those that did not would disappear. Thus living beings would gradually improve, with, for example, enhanced powers of obtaining food or of destroying their enemies. This process Darwin called natural selection.
>
> As a physicist, I cannot accept this. It seems to me to be impossible that chance variation should have produced the remarkable machine that is the human body. Take only one example - the eye. Darwin admitted that this defeated him; he could not see how it could have evolved from a simple light-sensitive organ. He made a brave attempt in his book, The Origin of Species, to account for the evolution of an image-forming lens, but he made no suggestions concerning the light-controlled pupil; the eyelids or the tear ducts. And the eye is trivial compared with the brain!

---

[2] Darwin wrote: 'due to the selection of what may incorrectly be called *chance* variations...'. He does not say why it is incorrect to use the term 'chance', however (*Letters* ii. p.87.). I suppose it was because he believed in 'the ordering of the whole universe' (*More Letters*, i, 154).

[3] Yet that it is a dogma is even jokingly admitted by scientists in asides: Dr Gabriel Dover in *Bioscience*, June 1982, 'for Evolution has long acquired the robes of the new Theology'.

Darwin was aware of the difficulties and much of his book is concerned with trying to explain away the lack of evidence in favour of it. I myself can see no alternative to the hypothesis that living matter was designed. The origin of life is not explainable in terms of standard science nor is the wonderful succession of living creatures formed throughout the thousands of millions of years of this planet's existence.

But who was the Designer?'

(The University of Manchester Institute of Science and Technology).

While this leaves the question hanging in the air, it indicates that we need not simply accept the dogma thrust upon us, as by David Attenborough, so confidently: this honestly admits ignorance and bewilderment - and that, if we are to be honest, is the proper state of belief.

The reason for an English specialist to take up this issue is not only that the metaphysic in question reduces the world to meaninglessness but that it is in stark contradiction to the facts. The world that presents itself to us is one which exhibits directiveness - it clearly and patently manifests an urge to more complex, 'higher' and more orderly relationships. Everything we see is full of meaning. We know that our own consciousness, our own being-in-the-world, our own cultural achievements are primary to us. Yet such is the authority of science that it seems we must contradict what is true in our perception of the world for some abstract dogma that makes life seem meaningless, and excludes from its realities the very manifestations of human culture and consciousness with which we deal in the humanities. So, while our scientist colleagues may be blandly unaware there is a serious crisis in all this, we must try to expose it, and face it not least because, in the attitude to mind in the evolutionary process, the scepticism implicit in the false metaphysic threatens to undermine science itself.

Edmund Husserl, in his *Crisis of the European Sciences*, asks,

> But can the world, and human existence in it, truthfully have meaning if the Sciences recognise as true only what is objectively established.... and if history has nothing more the teach us than that all the shapes of the spiritual world, all the conditions of life, ideals, norms upon which man relies, form and dissolve themselves like fleeting waves, that it always was and ever will be so, that again and again reason must turn into nonsense, and well-being into misery? Can we console ourselves with that? Can we live in this world where historical occurence is nothing but an unending concatenation of illusory progress or bitter disappointment?
>
> *Crisis*, p.7.

Many modern literary and artistic works express the view to which Husserl is referring here. They believe their nihilism to be based on the unanswerable truths which science presents to us. The effect is to deprive man of faith in himself. Husserl again:

> If man loses his faith (ie. his capacity to secure rational meaning for his individual and common human existence), it means nothing less than the loss of faith in his true being...
>
> p. 13.

And, of course, if we lose faith in this way, we lose faith too in the pursuit of truth of which science itself is a part. Fortunately, the clue to our escape from this appalling position of philosophical impotence is within science itself : biology itself will save us, as will appear.

Within biology, there are signs that the fundamental belief in Darwinist Evolutionary theory is due for revision. If it can be revised, this may bring a new metaphysical freedom to our culture.

When examined closely, the Darwinist mechanistic view of the universe appears as only one hypothesis - not a fact at all - and a hypothesis that now seems very doubtful. Perhaps the strongest resistance to doubt is itself a fallacy - that evolutionary theory must be upheld because there is no alternative explanation. In face of this it is urgent for us all to admit our ignorance and live with it, as science should demand.

My critical approach does not deny that evolution has taken place : but it is to say that Darwinism evolutionary theory does not offer anything like an adequate explanation of the origins and nature of life. No religious arguments will be invoked here : the 'Creationist' alternative, and all the explanations based on God, and on the invocations of spiritual forces intervening in the world, do not seem to provide an adequate explanation. My conclusion is that we urgently need new forms of thinking, new paradigms, new modes of understanding : but until we have these, we must live with our awe and ignorance - as the poet does anyway, while there is no 'factual' justification for absurdism or nihilism.

Certainly we do not have simply to sit down under the metaphysical implications of conventional scientific theories about the origin and nature of life. Moreover, we must not surrender proper humanistic disciplines in the Universities, as some urge that we do, to reductionist and mechanistic theories which offer themselves as so exclusive that they must be taken to supplant other subjects. An example is the attempt of the socio-biologists like Hawkins or E.O. Wilson, to suggest that all questions of morality, human nature and society should be given over to reductionist biology. Such claims are based on confusions about the nature of thought and truth, which will be found to be in the background here. It is not true that such theories have made the old philosophical questions out-of-date. On the contrary, they have only made things more perplexing : and (as Marjorie Grene points out) even the simplest questions about the nature of knowing, like those discussed in Plato's *Meno*, are still unsolved.[4] We may pursue questions of the nature and meaning of human existence, and questions of morality and values in the subjective disciplines, without any need to feel that there is a greater, more adequate and more 'realistic' delineation of the truth about existence in the physical sciences to which we *must* defer.

Whether or not we can believe anything in a fundamental problem in the Humanities and especially in the arts. As Peter Mudford says, at the beginning of his book *The Art of Celebration* (Faber, 1975), 'The purpose of art is to celebrate the human world we all share'. The motto to his book is from Eliot's *Ash Wednesday*:

> Consequently I rejoice having to construct something
> Upon which to rejoice.

What Eliot constructed was a Christian faith. But he also said:

> I cannot see that poetry can ever be separated from something which I should call belief, and to which I cannot see any reason for refusing the name of belief, unless we are to reshuffle names altogether... it takes application and a kind of genius to believe anything, and to believe *anything* (I do *not* mean merely to believe in some 'religion') will probably become more and more difficult as time goes on....
>
> T.S. Eliot, 'A note on Poetry and Belief', *The Enemy*, January, 1927, p.16.

In his study Mudford discusses the impact upon writers of the scientific ideas of the nineteenth century. For some, like Tolstoy, the problem seemed at times insoluble. Even after writing *War and Peace* and *Anna Karenina*, Tolstoy found himself 'increasingly obsessed' with questions he could not answer about the 'meaning and purpose of life'.

> The intelligentsia had assimilated and accepted the teaching of contemporary science on evolution. Tolstoy could find no satisfaction in its vagueness : 'Everything evolves and I evolve with it, and why it is that I evolve with, everything will be known some day'.
>
> Leo Tolstoy, *A Confession*, p.13: Mudford, p.35.

In the authors whom Mudford discusses there is a continual struggle to find meaning, in the face of the nihilistic impact of scientific scepticism.

Yeats suffered too. As Leavis reported:

> 'I am very religious', says Mr. Yeats in his *Autobiographies*' and deprived by Huxley and Tyndall, whom I detested, of the simple-minded religion of my childhood'.
>
> *New Bearings in English Poetry*, p.30.

Yeats hated Victorian science with a 'monkish hate'.

(Tyndall was one of the greatest popularisers of science who published *Fragments of Science* in 1871). Edmund Gosse records how his father suffered:

> The key is lost by which I might unlock the perverse malady from which my father's conscience seemed to suffer during the whole of this melancholy winter (1857). But I think that a dislocation of his intellectual system had a great deal to

[4] See *The Knower and the Known*, Faber and Faber, 1966, Chapter 1, 'The Legacy of the Meno'. Plato, *Protagorus and Meno*, trs. W.K.C. Guthrie, Penguin Books, 1956.

do with it. Up to this point in his career, he had, as we have seen, nourished the delusion that science and revelation could be mutually justified, and that some sort of compromise was possible. With great and ever greater distinctiveness his investigations had shown him that in all departments of organic nature there were visible the evidences of slow modification of forms, of the type developed by the pressure and practice of aeons.

Father and Son, p.11.

Where was his place? Gosse felt that, as a sincere and accurate observer it must be with Darwin and Co. But what about Genesis? And the religious obligation to believe in creation?

That controversy continues. But our situation is different. It seems to use today that molecular theory, having triumphantly confirmed Darwinism, has made it all worse.

Evolutionary theory is presented and taught as a fact. And since it is a blind, chance process in Darwinian theory, it belongs to a universe which operates in the way A.N. Whitehead summarised thus:

Each molecule blindly runs. The human body is a collection of molecules. Therefore the human body blindly runs, and therefore there can be no individual responsibility for the actions of the body. If once you accept that the molecule is definitely determined to be what it is, independently of any determinism by reason of the total organisation of the body, and if you admit that the blind run is settled by the general mechanical laws, there can be no escape from this view.

(*Adventures of Ideas*: quoted by E.W.F. Tomlin).

Since the conclusion must be deterministic, meaning is impossible : poetry, as a search for meaning - indeed all art, is virtually no more than distraction. As a student points out at the time of writing, in consequence of such influences, it has seemed to writers like Joseph Conrad that any 'meaning' which is the motive of action, may be described as a 'veil of saving illusion that protects man from the underlying chaos of existence'. This problem is found in much twentieth-century literature and art, and behind it is the feeling which Michael Polanyi summarises thus:

It seems to us that there is no meaning in the universe – except possibly the subjective meanings that man tries to impart into it. But, in the end, the universe cancels these out, and Man thus sees himself like a little boy who continually repairs and rebuilds his sand-castles at the edge of the sea, only to see the waves continually washing them away ....

Meaning, p.162.

If people continue to build sandcastles in so hostile an environment, they will do so only as a diversion. Poetry, the Arts, the Humanities – and, actually, *science*, – cannot go on existing in a world of which this was true.

The problem has been with us for a long time now. As Mudford says, there was a profound shift in human attitudes during the second half of the nineteenth century - 'notably under the influence of Charles Darwin' - though this change was on the way during the previous two centuries. Darwin acted as the catalyst in the prolonged struggle between science and religion which originated in the seventeenth century. But Darwin himself certainly suffered from something of the malaise which affected the sensibility under the impact of abstract scientific ideas:

my mind has changed during the last twenty or thirty years. Up to the age of thirty or beyond it, poetry of many kinds... gave me great pleasure... But now for many years I *cannot endure* to read a line of poetry : I have tried lately to read Shakespeare, and found it so intolerably dull that it *nauseated* me. I have also almost lost my taste for pictures and music... (my italics)

Darwin excepts novels, so long as they do not end unhappily : 'A novel does not come into the first class unless it contains some person one can thoroughly love, and if it be a pretty woman - all the better'. But 'of late', Darwin complains in his middle age:

My mind seems to have become a kind of machine for grinding general laws out of large collections of facts, but why this should have caused the atrophy of that part of the brain alone, on which the higher tastes depend, I cannot conceive.[5]

---

[5] (1959 Edition, pp.138-9)

He believed this was a form of atrophy, and perhaps injurious to the intellect and the moral character.

What we are dealing with, then, seems to be something that has happened to the mind of man, under the impact of scientific discovery. Bernard Towers quotes Schrödinger on

> The old clutch of dreary emptiness which comes over everybody ...when... they first encounter the description as given by Kirchoff and Mach of the task of physics (and of science generally): 'a description of the facts with the maximum of completeness and the maximum of economy of thought'; a feeling of emptiness which one cannot master.
>
> *My View of the World*

But why should science create this sense of chill, and this nihilistic feeling about life, man and the universe! Surely something is seriously wrong in that it does? For science itself is surely one of the more remarkable products of that 'life', of which it offers such a bleak explanation?

Of course, one cannot lay all the lack of confidence in anything in which to believe at Darwin's door - after all, he was an agnostic theist. Perhaps belief was already declining - there was already a depletion of faith when Darwin appeared, and the extrapolation of misunderstood scientific theories has rushed to fill the gap, the *horror vacui*? Theories like that of the survival of the 'fittest' were hypotheses put forward to explain puzzling phenomena such as the geographical distribution of plants and animals at various stages in certain areas, and the extinction of certain species and the survival of others. Such theories originally had no philosophical implications, but by degrees an orthodox evolutionary theory developed, which made wider and wider claims about the nature and destiny of man, his being and becoming, and his position in the universe, and has become a philosophy of existence. William Paley (1793-1805), in his *Natural Theology* (1802) argued that many complex organs were so beautifully designed and so perfectly adapted to the animal's need and environment, that they could only be the result of conscious design - designed therefore by God. Evolutionary theory has thrown Paley's Watch-maker away. In consequence, teleology became taboo; yet an implicit teleological way of talking could not be fully excluded; so scientists had to find ways of squirming round the difficulty, in allowing for phenomena which cannot be denied - as E.W.F. Tomlin points out, they talk of 'apparent purpose' but later tend to drop the 'apparent'. So they live with some ridiculous contradictions.

One of the strangest contradictions is to be found in the adherence of scientists to a theory of development in which the main dynamic is a negative one ('Natural Selection' which can only eliminate the 'unfit') while endowing this negative process with creative power. Let me take an example from a recent book.

Discussing the evolution of man in *Origins*, Dr. Richard Leakey and Roger Lewin say: 'the exigencies of life aloft (i.e. in the trees) must have given natural selection a particularly *keen cutting edge, speeding along the process* of evolving sophisticated mechanisms, such as an opposable thumb, for coping with the challenges of everyday life' (my italics). Yet, in their own terms, 'what Darwin accomplished was to demonstrate how, through exceedingly gradual (passive) adaptions to the environment and through changes from generation to generation, species may diversify or simply become better attuned to its world, producing ultimately a creature which is different in form from its ancestor.' (p.30). Did these authors notice the contradiction between these statements? Since evolutionary processes - according to dogma - can only operate by strictly passive or negative means, chance mutations coming into being and surviving or not surviving as the case may be, so there can be no question of evolution by natural selection 'speeding along' with a 'keen cutting edge', purposefully, as these terms suggest. Darwinian theory rejects the Lamarckian theory that, as it were, by continually stretching their necks towards food, giraffes come to evolve a long neck.[6] Thus it must be heretical surely to suggest that, because primates went to live in the trees, some kind of 'environmental pressure' would actually generate the development of an opposable thumb even when there is 'fierce' competition for survival (in some cases where the struggle for life would

---

6 There have, however, been speculations recently that Lamarck may have been right. See Science Report in *The Times*, 22 June 1980, Pro. Nat. Acad. Sciences, U.S.A. 77.2871. 1980. Others have not been able, however, to repeat these findings. But see discussions below by Rupert Sheldrake and of F. Wood Jones

seem to be dense, there has in fact been no evolutionary development). Though the scientists use the term 'pressure' to speak of what happens from the environment in Natural Selection, it is surely not logical to believe that what appear to be ingenious 'leaps forward' in evolution requiring complex changes - like the development of a gripping thumb - could come about by the purely negative processes of chance mutation and the exigencies of selection by survival? A variant thumb might appear by chance in one individual, but this is unlikely to give any immediate advantage which could conceivably persist through the processes of 'survival of the fittest' for the structural limb-using changes would require coordinated changes such as could not come by chance, and would require changes in behaviour, that is, in brain and nervous system. (Grassé points to such problems).[7] To suggest that such a complex and considerable change as the opposable thumb came into being - a fundamentally ingenious achievement towards a new form - because so many primates without (chance) opposable thumbs fell down from the trees (? before they mated) is surely absurd? How about intermediate stages with a half-opposable thumb? The truth is surely that an opposable thumb appears *as if by design* - as if an intelligence were at work - and this requires some new kind of thinking. Established evolutionary theory does not offer a satisfactory explanation - and this is even more true of other complex and coordinated developments.

The Darwinian theory may explain minute particular developments but for the major developments in evolution, it provides no answers, while on the biggest issue of all, as to why there should be life and its forms *at all*, it can say nothing, as Marjorie Grene points out (see below p192).

Here, then, are two biological problems we shall look at again and again. Firstly, how did life originate? Marjorie Grene declares, and I believe that she is correct, that Darwin, Darwinism and neo-Darwinism have nothing to say on this point. Secondly, given the origin of life, how did it develop to become the multiplicity we know today : can this be explained in terms of the gradual accumulation of minute changes coming into existence by chance and surviving by means of 'natural selection'? Again, as Marjorie Grene and other philosophers, together with some biologists, think not : some more positive dynamic must be involved, and as yet we have no clue to its nature.

Many complex changes, as we shall see, could simply never have occured according to the theory; yet there is an exceptional reluctance to think again. Darwin himself said:

> If it could be demonstated that any complex organ existed, which could not possibly have been formed by numerous, successive, slight modifications, my theory would absolutely break down.
>
> (1859)

How long will it be before the doubts expressed by many biologists are enough to put evolutionary theory seriously in question? The molecular biologists and geneticists, as we shall see, believe they have found the solution to the problem (talking even of 'automatic foresight'). But as we should see, there are many doubts as to whether the belief to which Darwin refers can be sustained at all. If they cannot, then perhaps we in the Humanities may claim our release from the daunting metaphysical attitudes into which Darwinism mechanism has trapped us?

---

7 See below p48

2 Darwin on trial: Norman MacBeth's doubts and others

Something is seriously wrong somewhere, for as soon as we begin to examine Evolutionary theory, we begin to find flaws, illogicalities, non-sequiturs and also at the same time ways of defending entrenched positions which have all the characteristics of dogmatic religious fanaticism. Darwin's '*On the origin of Species by Means of Natural Selection; or the preservation of favoured races in the struggle for life*' was published in 1859. The title represents an exact summary of Darwin's particular creative insight, into one very important aspect of the evolutionary process. The origin and evolution of species had been much discussed during the previous hundred years, but never before in these specific terms, as Bernard Towers has pointed out.[1] Darwinism added one of its strongest planks to the platform of 'traditional' British empiricism. By 1925, Sir Arthur Keith had published 'The Religion of a Darwinist', celebrating chance and struggle as dominant forces, though Darwin himself did not think of biological variants as wholly random or fortuitous. Today, Bernard Towers suggests, established biologists express a curiously emotional comtempt for anyone who hesitates about adopting a neo-Darwinism philosophy which cannot but lead to a sense of a meaningless and empty universe, of the kind I have indicated by the quotations from Whitehead and Schrödinger. Indeed, one does get a strange sense from scientists at times that if one does not accept the full bleakness and nihilism of the strict mechanistic position, they believe one is sentimental, self-deceiving and probably superstitious : and yet their 'scientific' attitudes are held with a passionate conviction which itself implies an idealism, such as assumes that the pursuit of truth is itself a good and will bring enlightenment, and so is subjective in such a way as to be inexplicable in 'objective' terms. It is extraordinary how men can live with such contradictions, and where the implicit philosophy of life manifest in science is concerned, we are all involved, for every modern man and woman is willy nilly affected by what is offered as the scientific attitude to existence, even when they do not realise it. If the reader considers what is in his own mind, he will find there certain beliefs - some of which have been implanted there by school textbooks or science books he has read, or broadcasts he had heard, or whatever. He will adhere to these tenets because they are, he feels, 'true'. Yet in this very area, disturbingly, science as we shall see is not entirely scrupulous, not at all devoted to the strict truth, and not all that willing to obey its own principles. Possibly, one image that may come to the reader's mind is those evolutionary 'trees'. These are now recognised by many scientists to be misleading in serious works on

---

1 'The impact of Darwin's *Origin of Species* on Medicine and Biology'. Bernard Towers, *Concerning Teilhard*, Collins, 1969, p.172.
One aspect of the matter I do not discuss is that of the social theory which Darwin adopted, as by being influenced by Malthus. On the errors here see Mary Midgeley, *Times Higher Educational Supplement*, 16 April 1982. See below p198.

Evolution among the fraternity. There are many difficulties with these 'trees'. For example, while they imply a story of gradual 'progress', they also ought to show (but don't) certain organisms as *not changing at all* – some fossil forms indicate that certain sharks, fishes and amphibians have never altered their structures. The horseshoe crab, the opossum, the oyster, the platypus, the brachiopod *Lingula*, the ginkgo tree and the Australian lungfish have not changed for 50 - 500,000,000 years. Some bacteria have never changed. Why have mutations not occured to change them? Why have they never been spurred by 'the struggle for existence?' Some of the forms of ancient stock have split into a group with plasticity, and another group with almost total rigidity.

So, Norman MacBeth found that the 'trees' were being quietly dropped from the 'literature' (see MacBeth, *Darwin Retried*, p.14 and Olson, Everett C., 1960, 'Morphology, paleontology, and evolution 'in Tax, Sol, *Evolution After Darwin, The Evolution of Life*, pp 523-545). Moreover, some scientists argue that all specialized forms are dead ends : they can only evolve a little or die out. New types do not arise from specialised forms, which are blind alleys in the evolutionary sense. This applies to fossils as well as to living forms, since all known forms, extinct or extant, early or late, are already specialized. All these forms are thus ineligible as ancestors and so must be moved from the trunk of the tree to the branches. The effect is to have a 'tree' with well populated tips, but the trunk is shrouded in mystery:

> We have the paradox that the remains found in the earth's crust are not those of our ancestors, while the bodies of our ancestors have not been preserved at all. We see forms that purport to be our cousins, but we have no idea of who our common grandparents were.
>
> MacBeth, p.14

(See Robert Broom, 1933, *Evolution - is there Intelligence Behind it*? *South African Journal of Science*, 30 : 1-19; G.G. Simpson, *The Meaning of Evolution*, p.326 and Sir Julian Huxley, *Evolution, The Modern Synthesis*, pp.562 and 571. See also Michael Denton, *Evolution – A Theory in Crisis*, 1985, passim., discussed below p143). These perplexities may make the reader protest that he has seen evolutionary trees in books and museums which gave a clear and unequivocal picture of straight progressive descent. One of these was possibly that showing horses. It may be found in *Fossils, an Introduction to Prehistoric Life*, by William H. Matthews, and accredited to the American Museum of Natural History. Another may be found in Carl O Dunbar, 1960, *Historical Geology*, p.412. (In English school textbooks see Fig. 74.1 in *Illustrated Biology* B.S. Beckett, OUP, 1978; *A New Biology*, K.G. Brocklehurst and Helen Ward, English Universities Press, 1968, p.251. See also Appendix A). The truth is that although once the existing fossils of the horse seemed to indicate a straight-line evolution from small to large, from dog-like to horse-like, like the links of a chain. Later as more fossils were uncovered, the chain splayed out and it became apparent that evolution had not been in a straight line at all. Horses had now grown taller, now shorter, with the passage of time. (See MacBeth, p.15, and Hardin, G, 1961, *Nature and Man's Fate* p.225-226). Yet before these complexities were apparent the 'straight-line' picture of the development of the horse set up at the American Museum of Natural History was reproduced in educational text books - *and continues to be*. One would suppose that in such uncertainty all 'Evolutionary trees' would be thoroughly renounced and withdrawn in science teaching. But they are not, and on this question MacBeth quotes some extraordinary excuses from scientists, some of whom were responsible for the continual issue of these diagrams. One argued:

> 'The audience for whom this book is intended has very little scientific background and it is for this reason that the material is handled as it was...'

MacBeth declared this 'improper and unwise' - because it indicates a double standard. Such images may be offered as showing a series of *stages* : but they are taken, by scientists and laymen, as a series of *ancestors*. That is, they convey, in falsification of the facts, a dishonest assertion of the truth of an unproven hypothesis. So, first, we should be aware that the diagrams in our children's biology books are wrong. But then - is Evolutionary Theory a *fact*? Scientists cannot agree among themselves. It is true that Sir Julian Huxley declared,

> 'The first point to make about Darwin's theory is that it is no longer a theory but a fact'.
>
> (At Random, a television preview, 21 November 1959, in Sol Tax, *Evolution after Darwin*, 1960, p.41.)

But Professor Ernst Mayr writes

'The basic theory is in many instances hardly more than a postulate'

(*Animal Species and Evolution*, E. Mayr, 1963, p.8.)

A postulate is a position assumed without proof : so, is evolutionary theory a mere potulate that we might even discard? Ah, says the reader, you have not studied the latest 'synthetic' theory. But Professor Mayr admits:

> The fact that the Synthetic Theory is now so universally accepted is not in itself proof of its correctness.
>
> (*Animal Species and Evolution*, 1963, p.7-8.)

As Norman MacBeth says, anyone who makes a close study of any branch of science soon discovers the great illusion of the monolith. From outside you may get the impression of unanimity, of an Establishment with fixed and approved views. But the closer we get, we find people disagreeing, experiencing disorder, and bitterly contesting views. Yet often an impression of unanimity is offered, and if there is criticism the ranks close, to preserve the dogma – never mind about truth and principle! MacBeth says we must make up our own minds. But that is difficult when science has become the only authority left in our society, and seems to require specialisms to which we cannot aspire. So in any company one finds that scientists tend to support one another, often quite aggressively, in the established dogmas, or, to use a better word, paradigms. For although scientists agree to differ among themselves, they must have a paradigm which enables them to get on with their work. So for reason of survival they must firmly defend it against outside criticism. Another spur, besides the discomfort of having the paradigm dismantled upon which one's daily work is based, is the dread of ignorance. A certain (subjective) confidence is necessary in science, so the scientists are willing to support the 'no-other-feasible-solution' fallacy. Arguments from opponents disproving evolutionary theory will not be accepted unless they offer another explanation. For instance, Sir Julian Huxley wrote:

> If we repudiate creationism, divine or vitalistic guidance, and the extremer forms of orthogenesis, as originators of adaption, we must (unless we confess total ignorance and abandon for the time any attempts at explanation) invoke natural selection.
>
> (J. Huxley, *Evolution, the Modern Synthesis*, 1942, p.473.)

MacBeth asks: 'Why use the word *must*'? Socrates would have asked: Why should we not confess ignorance? This kind of argument, however is common in Evolutionary Theory. I noticed it recently in an article in the *London Review of Books*, 18 June - 1 July 1981, 'Did Darwin get it Right'? John Maynard Smith wrote:

> As I see it, a hopeful monster[2] could still stand or fall by the test of natural selection. There is nothing here to call for radical re-thinking. Perhaps the greatest weakness of the punctuationists 'i.e. those who believe evolutionary change is concentrated into relatively brief periods' *is their failure to suggest a plausible alternative to mechanism.*

The criticism of a theory may be unanswerable in that it displays that the theory is wrong: the critic does not then have to offer any alternative. The theory must simply be left in ruins, *and must be.* The 'no other alternative' argument by which scientists defend their positions on this subject is one of most fallacious devices. After all, it was T.H. Huxley, as MacBeth points out, who said of men of science: 'there is not a single belief that it is not a bounden duty with them to hold a light hand and to part with cheerfully, the moment it is really proved to be contrary to any fact, great or small'. (*Darwiniana*, 1893, p.468.). MacBeth suggests three qualifications for a sound scientific theory:

1. The theory must be reasonable, not merely the only one, or the best one.
2. The scientist must hold his theory 'with a light hand'
3. The scientist must part with his theory cheerfully the moment it is proved contrary to any fact, great or small. (p.7.)

---

[2] The theory of the hopeful monster was put forward by Richard B. Goldschmidt (1878-1958) in *The Material Basis of Evolution*, 1952. After observing mutations in fruit flies for many years Goldschmidt saw these as so 'micro' that if a thousand mutations were combined in one species, there would still be no new species, so he proposed the hypothesis of the 'hopeful monster' whereby a huge change might occur from time to time and then be preserved by a favourable environment.

MacBeth, whose useful book should be studied, exposes many other fallacies and lacunae in the literature of the scientific faith, of Darwinism, and shows well enough that evolutionists do not obey these principles. And yet evolutionary theory is the one belief over which scientists are most aggressive. Perhaps it should be pointed out here that some of the confusion and self-contradiction originates in Darwin himself. Darwin's own ideas of natural selection were based on study of the work of breeders. They looked round and saw potentialities in variations : so they selected. Nature must have done the same. But, of course, the dying out of ineffective variations, leaving the successful ones, can only be a negative process. Darwin and Darwinism were dogged by the tendency to make this natural process a positive one, like the intelligent craft of the breeder. But in Nature there is no intelligent breeder, though he lurks in the background like a ghost of God. This problem can be traced through Darwin's *Letters*: here is the basic theory:

> Man has altered, and thus improved the English race-horse by *selecting* successive fleeter individuals; and I believe, owing to the struggle for existence, that similar *slight* variations in a wild horse, *if advantageous to it*, would be *selected* or *preserved*, by nature : hence Natural Selection.
>
> (*Letters*, ii, 278).

Darwin did not, however, he says, intend to make natural selection the only cause and seems to hint at other formative dynamics:

> By the way, I think, we entirely agree, except perhaps that I use the forcible language about selection.... I have never hinted that Natural Selection is 'the efficient cause *to the exclusion of the other*', i.e. variability from climate etc. *The very term selection implies something... to be selected...*
>
> (ii. p.317.) (my italics).

Discussing a criticism of Lyell's Darwin speaks of a general 'ordering' process at work. Lyell had said that:

> If we confound 'variation' or 'natural selection' with such creational laws, we deify secondary causes, or immeasurably exaggerate their influence.

Darwin said:

> One word more on the Deification of Natural Selection : attributing so much weight to it does not exclude still more general laws, i.e. *the ordering of the whole universe.*
>
> I have said that Natural Selection is to the structure of organised beings what the human architect is to a building. The very existence of the human architect shows the existence of more general laws....
>
> *More Letters*, vol.1, p.154.

A human architect, however has an intelligence, and the capacity to imagine. If 'The very term selection implies something... to be selected', this other dynamic may be something like an architect's intelligence - that is, a force groping towards development and order. The negative process Natural Selection cannot be this, and can only operate on such a force by moderating it. Where can such a force exist then? Some critics have recognised that Natural Selection could only be a negative process : a Mr. Wallace is quoted in a footnote to Darwin's *letters.*

> The term 'survival of the fittest' is the plain expression of the fact: 'natural selection' is a metaphysical expression of it, and to a certain degree indirect and incorrect, since Nature... does not so much select special varieties as exterminate the most unfavourable ones.
>
> footnote, *Letters*, vol ii, p.46.

Darwin's inclination, however, was often to make natural selection a positive principle:

> I think it can be shown that there is such an unerring power at work: *Natural Selection...*
>
> *Letters*, ii, p.113.
>
> I can see with my prejudiced eyes no limit to the perfection of the adaptations which could be affected by Natural Selection...
>
> *More Letters*, i, p.145.

He was often accused of 'making too much of a *Deus* of Natural Selection' (*Letters*, i.p.213), and J.D. Hooker wrote:

You certainly make a hobby (i.e. a hobby-horse) of natural selection, and probably ride it too hard... If the improvement of the creation-by-variation doctrine is conceivable, it will be by unburthening you theory of Natural Selection, which at first sight seems overstrained– i.e. to account for too much.

More Letters, i.p.135.

And Darwin himself confessed to temptation:

I have for years and years been fighting with myself not to attribute too much to Natural Selection – to attribute something to direct action of conditions; and perhaps I have too much conquered my tendency to lay hardly any stress on conditions of life.

More Letters, i.p.198.

As we shall see, Darwin at first believed in God, and God's purpose in the world; but gradually became more and more sceptical, and his ambivalence about whether 'natural selection' exerted a positive force or not seems to have contributed to much of the inconsistency in present-day neo-Darwinian theory. It is interesting to note the subjective impulses behind Darwin's own progress. In his *Autobiography*, for example, he discusses his invention of natural selection in the context of his loss of religious faith, around the question of the amount of suffering in the world. That is, he reveals that his scientific objectivity, if you like, had deeply subjective roots (now being explored by Dr John Bowlby). Darwin was originally taught 'special creation' of course, and he once believed in Paley's watchmaker :

The old argument of design in nature, as given by Paley, which formerly seemed to me so conclusive, fails, now that the law of natural selection has been discovered.

The Autobiography, 1958 edition, p.87.

We may note that Darwin suggests by his way of putting it that natural selection was *a fact* that was merely 'discovered'. The truth is that certain artefacts were discovered, and certain aspects of animals and birds were observed – and natural selection was a hypothesis invented, by a great act of the imagination, to explain them. But it is significant that this follows a reference to Christian belief : 'I gradually came to disbelieve in Christianity as a divine revelation'. Moreover :

Beautiful as is the morality of the New Testament, it can hardly be deemed that its perfection depends in part on the interpretation which we now put on metaphors and allegories.

p.86.

Here we have a glimpse in Darwin of the same distrust of the poetic that we see in Galileo - and indeed may find throughout the 'scientific revolution' : a distrust, that is, of symbolic explorations of inner or moral truth. The implication is that only that is real which can be found, objectively, by mathematics, Cartesianly, and so belongs to the crisis which Husserl delineated. But Darwin goes on, and reveals one fundamental reason why 'disbelief crept over me at a very slow rate':

I can indeed hardly see how anyone ought to wish Christianity to be true; for if so the plain language of the text seems to show that the men who do not believe, and this would include my father, brother and almost all my best friends, will be everlastingly punished.

And this is a damnable doctrine.

p.87.

A footnote indicates that Mrs. Darwin objected strongly to this last sentence, and it was not published. It was strong stuff for the nineteenth century, to declare Christianity a damnable doctrine (and Mrs. Darwin declared that few would call that 'Christianity' – 'though the words are there'). Perhaps the present writer should explain that he is not a Christian, and does not seek to defend Christian doctrine: it seems to him, too, that it would be damnable to condemn people to eternal torment for being free thinkers – and Darwin's courage in rejecting belief may here be seen as a valuable blow for intellectual freedom. But, it is clear, Darwin goes on to link *design* with the whole Christian cosmology, and problems of God's responsibility for good and bad in the world. Skipping a paragraph, we read:

it may be asked how can the generally beneficient arrangement of the world be accounted for? Some writers, indeed, are so much impressed with the amount of suffering in the world, that they doubt if we look to all sentient beings whether there is more of misery or of happiness; – whether the world as a whole is a good or bad one. According to my judgement happiness decidedly prevails, though this would be very difficult to prove.

p.88.

Darwin reveals himself here as a very benign individual (he talks about 'the pleasure of our daily meals, and especially in the pleasure derived from sociability and from loving our families...'). But he cannot accept the view that what suffering there is intended to 'try us', to generate moral improvement in man.:

But the number of men in the world is as nothing compared with all sentient beings, and these often suffer greatly without any moral improvement.

And Darwin clearly has difficulties over the idea of God - the God who, according to Christian scriptures, would condemn his father, his brother and almost all his best friends to everlasting torment:

A being so powerful and so full of knowledge as a God who would create the universe, is to our finite minds omnipotent and omniscient, and it revolts our understanding to suppose that his benevolence is not unbounded, for what advantage can there be in the suffering of millions of the lower animals throughout almost endless time.

p.90.

In one sense, Darwin shows here that his invention of evolutionary theory was a great act of imagination – and this sprang from an act of imaginative sympathy with all living creatures. In this, moving towards a new sense of the relationship between man and nature, he could not accept any explanation of creation in terms of a Divine plan:

This very old argument from the existence of suffering against the existence of an intelligent first cause seems to me a strong one; whereas, as just remarked, the presence of much suffering agrees well with the view that all organic beings have been developed through variation and natural selection.

p.90.

Suffering, instead of being the inexplicable effects of the attitudes and behaviour of God, becomes merely the consequence of a generally beneficient process:[3]

Such suffering is quite compatible with the belief in Natural Selection, which is not perfect in its action, but tends only to render each species as successful as possible in the battle for life with other species, in wonderfully complex and changing circumstances.

p.90.

In another sense, Dawrin can be seen to be seeking relief from the metaphysical problems. The problems of suffering, both of men and animals, and of the meaning of life and morality (which lurk behind the question of eternal damnation, etc.) are questions which belong to philosophy - in the sense of moral science. They cannot, in fact, be solved by a scientific hypothesis about how life originates and develops, though this will of course be relevant. Darwin tries to find a principle which will enable him to see the suffering of men and animals from a detached point of view: that is, simply how things are. So we may associate the origins of the theory of natural selection with the perplexities of the nineteenth century over population - a rapidly increasing population moving into the cities and suffering terribly in the process. It is significant that Darwin began to sketch his natural selection theory jut after reading Malthus :

In October, 1838, that is, fifteen months after I had begun my systematic enquiry, I happended to read for amusement Malthus on *Population*, and being well prepared to appreciate the struggle for existence which everywhere goes on from long - continued observations of the habits of animals and plants, it at once struck me that under thee circumstances favourable variations would be preserved, and unfavourable ones be destroyed.

p.120.

There were other such ideas in the air: Halevy, in his *History of the English People* refers to

---

3 Herbert Spenser, of course, considered the principle of natural selection (he invented the term) beneficial: see below.

an obscure paphlet on the Poor Laws, written in 1786. The Poor Laws are blamed in it for preserving the weak at the expense of the strong, with all the implications of the working of Natural Selection. The author discusses an animal analogy on a remote island, in which 'the weakest of both species were amongst the first to pay the debt of nature, the most active and vigorous preserved their lives. It is the quantity of food that regulates the number of the human species...' (*Autobiography*, p.155). As we shall see, this social model has now been transferred to the struggle to exist among molecules in the primordial soup. While on the one hand Darwin's natural selection theory picks up the tendency in utilitarianism to evade the question of human suffering by examining it mathematically or logistically as the inevitable consequence of impersonal laws (a tendency which Dickens bitterly satirised in *Hard Times*), on the other hand, Darwin reveals that he was seeking some kind of quantitative theory of pleasure as against pain, which has similarities with Freud's thought: as in Freudian metapsychology, a central concept is *adaptation.* And, of course, it has analogies with Benthamism. In his *Autobiography*, Darwin makes it clear that he believes that happiness makes for survival, because when creatures are happy they multiply, while suffering can make them unable or unwilling to propagate :

> If all the individuals of any species were habitually to suffer to an extreme degree they would neglect to propagate their kind; but we have no reason to believe that this has ever or at least often occured. Some other considerations, moreover, lead to the belief that all sentient beings have been formed so as to enjoy, as a general rule, happiness. Everyone who believes, as I do, that all the corporeal and mental organs (excepting those that are neither advantageous or disadvantageous to the possessor) of all beings have been developed through natural selection, or the survival of the fittest, together with use or habit, will admit that these organs have been so formed so that their possessors may compete successfully with other beings, and thus increase in number.
>
> pp.88-89

The basic principle is *competition*, and the survival of population : and this is linked with pleasure (by contrast with *depression*) :

> Now an animal may be led to pursue a course of action which is most beneficial to the species by suffering, such as pain, hunger, thirst, and fear– or by pleasure, as in eating and drinking, and in the propagation of the species, etc., or by both means combined, as in the search for food. But pain or suffering of any kind, if long continued, causes depression and *lessens the power of action*; yet it is well adapted to make a creature guard itself against any great or sudden evil. Pleasureable sensations, on the other hand, may be long combined without any depressing effect; on the contrary, they stimulate the whole system to increased action. Hence it has come to pass that most of all sentient beings have been developed in such a manner through natural selection, that pleasurable sensations serve as their habitual guides.
>
> p.89.

In this we have a kind of mechanistic pleasure-principle theory which moves away from the metaphysical problem posed by all the suffering in the world, towards a scientific theory that the mainspring of the evolutionary process is pleasure, because that 'stimulates the whole system to increased action'. Pain, by contrast, is depressing and 'lessens the power of action', presumably also lowering the reproduction rate: it isn't that creatures fall down from the trees or get eaten before copulating, but *pre coitu contristus est animal.* It is interesting to note that this very benign, if quantitative theory, has - as in Dawkins - been turned by the neo-Darwinists into a deeply pessimistic theory in which going-on-surviving has become a ruthless 'selfishness' as a principle of life, against which any attempt at altruism has to contest. Two other views of Darwin in the *Autobiography* are worth mentioning here. One is Darwin's perplexity in the face of scientific predictions of the ultimate death of the universe:

> Believing as I do that men in the distant future will be a far more perfect creature than he is now, it is an intolerable thought that he and all other sentient creatures are doomed to complete annihilation after much long - continued slow progress.
>
> p.92.

As Darwin points out, to those who believe in immortality, it is not so bad. But he shows himself here believing in the improvement of man : and yet daunted by the metaphysical problems seemingly imposed by scientific theories. And yet he confesses himself a Theist. He speaks of :

> The extreme difficulty or rather impossibility of conceiving this immense and wonderful universe, including man with his capacity for looking backwards and far into futurity, as the result of blind chance and necessity. When thus reflecting, I feel compelled to look to a First Cause having an intelligent mind in some degree analogous to that of man.
>
> p.93.

Later, Darwin added that his conclusion was strong in his mind when he wrote *The Origin of Species*, but 'since that time it has gradually with many fluctuations become weaker'. And though his universe does contain something more than 'matter in motion'– it contains 'knowing mind' - he has a problem as to whether man's mind can be trusted :

> Can the mind of man, which has, as I fully believe, been developed from a mind as low as that possessed by the lowest animal, be trusted when it draws such grand conclusions? May not these be the result of the connection between cause and effect which strikes us as a necessary one, but probably depends merely on inherited experience?
>
> p.93.

He means, of course, when man's mind generates the idea of God. May be it would be as difficult for children, brought up under the inculcation of belief, to throw off their belief in God, 'as for a monkey to throw off its instinctive fear and hatred of a snake'– a phrase to which again Emma Darwin took great exception.

> I cannot pretend to throw the least light on such abstruse problems. The mystery of the beginning of all things is insoluble by us; and I for one must be content to remain an Agnostic.
>
> p.94.

Darwin would be tremendously surprised to learn from Richard Dawkins that he had solved the answer to the problem of human existence. Marjorie Grene quotes an interesting passage from David Lack on Darwin's doubts about man's mind:

> Darwin's 'horrid doubt' as to whether the convictions of man's evolved mind could be trusted applies as much to abstract truth as to ethics; and 'evolutionary truth' is at least as suspect as evolutionary ethics. At this point, therefore, it would seem that the armies of science are in danger of destroying their own base.
>
> *Evolutionary Theory and Christian Belief*, 1957,p.104.

As Marjorie Grene says, there is something wrong with a theory which, at root, invalidates itself. The scientist, says Lack, must be able to trust the conclusions of his reasonings. Hence he cannot accept the theory that man's mind was evolved wholly by natural selection if this means, as it would appear to do, that the conclusions of the mind depend ultimately on their survival value, and not their truth, thus making all scientific theories untrustworthy. Clearly, thus, our solution of the problem of evolutionary theory requires a reconsideration of the mind and of knowledge, as well as of life. Darwinism, Marjorie Grene concludes, represents a last and self-destroying model of the Cartesian world-machine. We need to overcome our Cartesian bias and 'open the way for a richer metaphysical vision'– admitting that while scientists try to avoid teleological problems, biology cannot but acknowledge the existence in nature of *ordered processes*, and so *ends*, towards which there is a *directiveness*. But first let us look at what Evolutionary Theory is.

3 Evolutionary theory — a layman's summary

Evolutionary theory is a question of the interpretation of the characteristics, diversity and distribution of the various forms of animal and plant life. The nature of the fossils, their location, the structure of animals, and their variations may be called the facts of the case. Evolutionary theory consists of hypotheses and postulates - by which scientists seek to account for these facts, and the theory as it is upheld at the moment concerns itself with a historical process involving descent with modification.

As soon as we begin to approach evolutionary theory, we are met with a number of hypotheses. Some theorists, for example, believe that life has originated more than once. Others believe that there was one, one-celled, ancestral group from which all forms of life have derived. Others believe that life came down to earth from outer space, the most recent proponents of which are Professors Francis Crick, Fred Hoyle and C. Wickramsinghe. Others believe life was created by the action of lightning and other forces on sun-warmed mud (or 'primordial soup') - a theory now losing ground for mathematical reasons.[1]

The main attempt is to explain distinct species or other groups from an ancestral group or groups, and to account for the natural modification of a line of ancestors and their descendants. Life in the later forms is shown to be markedly different from the earlier. The problem is how this happened, and to account for the dynamics in such processes. Homology is an important factor - the basic similarity of structure underlying many 'strands' of creatures.

The 'creationist' explanation is to invoke the theories of *special creation*, *vitalism*, and *finalism*. These have been discarded very largely by scientists because they believe they don't really explain anything, and tend to hamper research.

Special creation holds that each separate kind of species of living thing was divinely created much as it now exists, and has not altered markedly in the course of the earth's history. The fundamentalists believe this implicitly. Modern liberal theology believes that organic evolution is a method of creation employed by God as his purpose. There are some good scientists in this field who are Christians and even creationists, but most scientists take organic evolution to be an established fact, and seek to explain its processes in terms of the material aspect of nature. (*Typology* sees forms as immutable and rejects all sequential theories).

Vitalism is a theory that all living things contain a special agent, a life force, which enables them to carry out the functions characteristics of organisms. It is opposed to *mechanism* which believes that living things are merely complex machines. The vitalist believes that the behaviour of living things cannot be explained from the laws of physics and chemistry and that a unique factor must be postulated to account for this. The chief proponent of vitalism,

---

1 It seems to be as unlikely as having a continuous series of sixes on a dice for 5,000,000 throws, or a factor of 1 to 10^{46}. M. Pitman refers us to *Mathematical Challenges to the neo-Darwinian Interpretation of Evolution* by P. Moorhead and M. Kaplan

Hans Driesch (1867-1941) believed that the living cell, by contrast with inert material, is possessed of an entelechy, a kind of premonition of what it may become and of the way in which it must function in order to fulfill its potentialities. A similar conception was held by Bergson (1859-1941) whose term *élan vital* signified a life-force unconscious of its aim. This life force must be postulated to explain not only the growth and development of individual organisms, but the entire evolutionary development from the simplest living cell to man.

Finalism is a related idea, in which evolution represents, as a whole or in the main, progress towards a predetermined goal or an abstractly determined perfection of living things. In both vitalism and finalism there is the idea of a spiritual, or, at least, non-material force. Objections to the finalistic form of vitalism are found in the frequent occurence of degeneration in evolution, in the fact of the tremendous diversity of organisms and the apparently random distribution of many of their characteristics.

By contrast, *mechanism*, which is the fundamental belief of modern science, assumes that the behaviour of living things may be accounted for in terms of the physico-chemical laws which account for the behaviour of non-living things. Living things are not distinguishable from non-living things by the chemical substances that compose them. Living things are not exceptions to the physical law of the conservation of energy or the law of entropy. But on the other hand, living things are distinguished from the non-living by certain types of behaviour such as reproduction, the regeneration of destroyed tissue, and the growth of special parts to perform special functions in the maintenance of the whole organism. Scientists have not been able to deduce this behaviour from the laws of physics and chemistry, but most are unwilling to see anything 'else' in the constituents of living things. Such ideas they feel to be derived from the sense of human purpose, and they see the 'special agent' of vitalism as merely a way to avoid the difficulties merely by inventing a name : it explains nothing. To attribute purpose to living things they believe is but an animistic fallacy. The mechanist believes that our present inability to explain the special properties of living things in the terms of physics and chemistry merely represents a stage in science, and that eventually life will be explained in these terms. So, they often draw a large cheque on some future balance which they hope will turn up.

However, we may note here in passing that there is a problem first in biology over *what is a fact*. While facts in physics and chemistry are on the whole susceptible to empirical analysis and positivist principles (except perhaps when one enters the sphere of quantum theory, and the deeper levels at which notions like charm, quarks and anti-matter are put forward), the 'facts' of biological life become intractable to such approaches. For example, in a study of animal behaviour a biologist may discuss the *meaning* of a gesture or the interrelational or 'social' effect of an animal cry or a bird's song (see the work of Tinbergen, Lack and Thorpe); he must thus encounter 'centricity', the subjective, autonomous existence of a living thing, and so to some extent find the creature *as a being*. Moreover, questions may arise of purposiveness or at least ingenious intention, as with the appearance of stick insects or other forms of animal, snake venom, or the bombadier beetle.[2] Over such matters, since the understanding of living beings is involved, we cannot simply talk of 'facts' as in the strict physical sciences, while other aspects like subjectivity and autonomy of living beings (which make *choices*) cannot simply be seen as simple physical realities, but must be approached as complex manifestations of a higher organism or a subject. The facts of life are therefore beyond 'objective' physical science. When we come to human behaviour, culture and consciousness, the problem becomes limitlessly complex. Here the 'facts' cannot be approached as mere realities in natural science, and profound philosophical questions arise when attempts are made to reduce these to mere accumulations of minimals, or as products of their physical 'components', as we shall see. In man there is an additional *cultural* dimension. This interjection is inserted here to remind the reader that the kind of 'facts' of life which are encountered (say) by a biologist observing animals or even by micro-biologist and certainly by anyone studying human life, belong to a different dimension from such facts as (say) the 'fossil record'. Yet, of course, the two kinds of fact are often confused, as when

---

2 The bombadier beetle has a chamber in its rear end in which it mixes two chemicals which explode and throw hot fluid at its enemies. Can one imagine such an ingenious device coming into being by chance? See p217.

sociobiologists begin to look for a 'genetic' origin for 'compassion', or talk of how 'culture' is the 'product' of 'culturgens' or 'memes'; or, for that matter, when medical men faced with a psychological problem try to reduce it to a 'hormone profile'. Whole areas of thinking add epistemology need opening up here, if we are to avoid being inadequate to the task.[3]

Though Darwin is credited with the invention of evolutionary theory, its origins may be traced to earlier thought. In the seventeenth and eighteenth centuries the ideas of Aristotle tended to suggest a basic idea of organic evolution. Aristotle observed that organisms can be classified on a scale from the most simple to the most complex, and inferred that creation involves a perfecting principle responsible for this graduation. The English naturalist, John Ray (1628-1705), and the French naturalists, Benoit de Maillet (1656-1738), Moreau de Maupertius (1698-1759) and Leclerc de Buffon (1707-1788) all pondered evolution. More outspoken and consistent was Erasmus Darwin (1731-1802), the botanist grandfather of Charles.[4]

The problem henceforward was to produce facts and to make sober scientific assessment of what had mostly been, up to now, wild speculation. One of the first to attempt the proper work was Lamarck (1744-1829), Jean Baptiste Pierre Antoine de Monet, Chevalier de Lamarck. He published a book, *Philosophie Zoölogique*, in the year of Charles Darwin's birth, in which he argued that organic evolution was a universal principle of nature, and he tried to explain how it works. His theories, as are now commonly known, included the belief that characteristics which individuals aquire during their lifetime can be passed on to their descendents and thus become features of the species. One of the main features of life which would seem to lend endorsement to Lamarck is the thick calloused skin on the soles of our feet: this seems like an acquired characteristic which is passed on to the next generation.

Charles Darwin (1809-1882) was, of course, taught special creation in his youth, when he was trained as a minister and in *The Origins of Species* he attributes the Creation to the hand of God. But the achievement of Darwinism has been to get rid of special creation (though Darwin himself was a Theist, albeit agnostic, and did not want to give all over to brute force).[5] On the voyage of the *Beagle* he made observations which suggested evolution rather than creation, and he began to agree with his illustrious grandfather.

He wrote his notes on *the Voyage of the Beagle* in 1837 and then laboured for twenty years to produce *The Origins of Species* (1859). Today, we may note, even from the title with its reference to 'favoured species', that his thinking belongs to a particular historical period, of middle class or bourgeois attitudes in a country in which there were deep divisions between classes, in a world in which Britain was engaged in building an Empire, and in which the Industrial Revolution was in full swing, with all its social consequences. Moreover, we may see Darwinism emerging in an intellectual tradition in which utilitarianism was strong, and in which Adam Smith, Herbert Spencer and Malthus were significant figures. There is in these theories as we shall see a certain kind of 'detached' impulse to look at human life and society in quantitative, mathematical terms, which ignore individual being (a tendency against which we have noted Charles Dickens reacted in *Hard Times*). The onward march of optimistic industrialism seemed to demand such impersonality, with its inconsistent assumption that it was natural and proper for the weaker to go to the wall.[6]

Darwin's importance was not so much in showing that evolution has occurred, but in trying to give a credible scientific explanation of how it happened. Most modern biologists would agree with Darwin that evolution *has* occured. Where they differ from him is on *how* it occured. By 'Darwinism' is usually meant Darwin's particular attempt to *explain* evolution, though there have been numerous revisions - and the process is still going on.

---

3 For a discussion of philosophical inadequacy, see the chapter on 'The Selfish Gene' below, p. 96 . In a wider sense the work of Kierkegaard and Husserl is relevant. See also: Poole, R., *Towards Deep Subjectivity* and the present author's *Education and Philosophical Anthropology*, 1986.

4 D.R. Oldroyd, in *Darwinism Impacts*, has an interesting chapter on the work of Alfred Russell Wallace (1823-1913) who more or less invented the theory of evolution by natural selection independently, contemporaneously with Darwin, thereby inclining one to speculate that the invention of Evolutionary Theory was very much a product of the minds of its age and its social milieu of scientific enquiry.

5 See quotations from his *Autobiography* above p6ff.

6 See F.R. Leavis in his Introduction to *Mill on Bentham and Coleridge*, a useful volume in the critique of Utilitarianism.

Evolution is a broad general principle which serves to establish a relationship between many different items of information and brings them together into a broad picture of life as a whole. It is thus a great imaginative achievement, and its convincingness as a paradigm obviously lies in the satisfaction it accords scientists. Breeders do employ the processes examined by evolution to produce new variations. Mutations may be observed and even created in the laboratory. Some sequences among fossils do seem to be explained by evolution. So, since the various items do fit into the evolutionary picture, and into no other yet proposed, scientists take this to be evidence of the truth of evolution, at least in Britain and America.

Darwin as we have seen was much influenced by the work of breeders. Although this work was not then grounded scientifically, it did produce hereditary variations which were recombined to yield new kinds of plants and animals. Evolution as it were, took place before the eyes of the experimenter. In some cases, if left to themselves, the new forms would breed with their antecedents, and the new variety would even eventually disappear. By keeping new varieties apart, however, they could be made to continue indefinitely. Some of the best known work in this field has been with fruit flies (*Drosophila*), and the mutation process has been accelerated by the use of X-rays and mustard gas.

It should be noted, however, that despite the enthusiasm of some experimenters, there does seem to be a limit to what men can achieve. There does seem to be a limit in Nature even to changes brought about by the planning of breeders : and so the question arises, in *what does this preservation of norms reside*? As we shall see, Darwin became doubtful later about some of his early claims about the possibilities of evolutionary development. There is no infinite departure from norms and when a pure eyeless strain of fruit flies was produced, it reverted the normal after eight or ten generations.[7]

One of the problems of Darwinism is that it hardened into dogma before being thoroughly examined. Darwin was so enthusiastic in his imaginative excitement over evolution that he went far beyond sober science into the realm of doubtful extrapolations. One may be noted here, in relation to this question of breeding. Darwin could not explain how the culmination of small variations (such as breeders could achieve) was able to bring about macro-changes and close the gaps between species and common ancestors. So, he gestured at geological time: a hundred million years could accomplish anything:

> Slow though the process of selection may be, if feeble man can do so much by his powers of artificial selection, I can see no limit to the amount of change, to the beauty and infinite complexity of the coadaptations between all organic beings, one with another and with their physical conditions of life, which may be effected in the long course of time by nature's power of selection.
>
> 1859, p.109

This virtually says that while we have no explanation for how evolutionary changes come about, *given long enough time* anything is possible, which is a fallacy, though it crops up again and again in evolutionary theory. Besides the evidence from breeding scientists have had to pay painstaking attention to material from the study of plants and animals as they occur in nature, and have had to study their reactions one with another and with their environment (ecology). It is still true to say that evolutionary theory has never been 'proved' in the laboratory.

Darwin was convinced about evolution by his work in the Galapagos Islands off the coast of South America. There he noticed twenty-one species, though these were yet closely similar to those on adjacent islands. He concluded that South American birds had somehow reached the islands and had there become differentiated by evolution in isolation. There are many such examples of evolutionary development having taken place in correlation with geographic distribution and historical circumstances.

In such contexts the question of *adaptation* has been studied. All animals and plants have structures correlated with their needs : horses have graminivorous teeth, cats carnivorous ones and so on. Animals otherwise similar have different colourings - small rodents living in dark larva areas will be darker than the same living on white sand. Studies of many such phenomena seem to provide evidence that differences in adaptation arise by evolutionary

---

7 Julian Huxley, 1957, *Evolution in Action*, p.40

processes. But how these occured still defies satisfactory explanation, and the causes of evolution remain unsolved. For example, how do we come by that callous-like thick skin on the bottom of our feet - such thickness being found in foetuses who have never set foot to ground?[8] This alone has led some scientists to keep an open mind on Lamarckism : certainly nothing in Darwinian theory seems to suggest an explanation.

Everyone, is aware, of course, of *the fundamental resemblances between the structure of different organisms.* The arm of a man, the foreleg of a horse, the wing of a bird or bat, all have different forms adapted to different functions, but yet can be seen to correspond in some way - as by comparison of the bones. Some (homologous) structures seem certainly to owe their inheritance to a common ancestor, and their differences seem due to adaptation to different ways of life. These similarities and differences were, of course, the basis of classification, by pre-evolutionary scientists like Linnaeus (1707-1778) - who rightly classified the whale as a mammal because it was warm-blooded, had lungs, gave birth and suckled its young. Such classification enable evolutionists to discuss whether, for example, whales or their ancestors were once land mammals that later took up their abode in the sea. (Some resemblances, however, are analagous, not homologous, like a Whale's flipper).

Fossil Sequences, of course, are one of the main sources of evidence for evolution : the preserved remains of ancient organisms, often petrified - that is, a calcareous deposit has replaced the original bone. Of course, many organisms were not preserved in this way, so the record can never be complete.

Fossil deposits do show, however, that life on earth has changed constantly in some species (while others have remained largely unchanged, like the shark and the oyster). Each period has its own distinctive organisms, related to, but different from those of other periods. Groups which can only be described as 'higher' are seen arising in sequence from their ancestors - jawless pre-fishes appeared in the Ordovician period and true fishes arose from them later in the Devonian period (350,000,000 years ago). Amphibians developed from fishes towards the end of the Devonian period (300,000,000 years ago) and reptiles from amphibians during the Pennsylvannian period (225,000,000 to 250,000,000 years ago). Mammals developed from reptiles later in the Triassic or early in the Jurassic period, some 150,000,000 years ago. Of course, such calculations change according to new dating techniques. But, as we shall see, there are serious problems even in the fossil record: why, instance, are there no fossils in the Precambrian rocks (over 600,000,000 years ago)?

Within the large groups there are evolutionary sequences in geologic time sharing progressive changes in structure and adaptions, some preserved in great detail. However, there are still unsolved problems - like that of the forms which have never altered. Why should the same ancient stock split into one group which shows astonishing plasticity, and another which is totally conservative? Moreover, it now seems clear that the notion of plain, progressive gradual evolution overall is wrong. The picture of *Eohippus* gradually growing into *Equus* is no longer true.

> There was a time when the existing fossils of the horses seemed to indicate a straight-line evolution from small to large, from dog-like to horse-like, from animals with simple grinding teeth to animals with the complicated cusps of the modern horse. It looked straight-line - like the links of a chain. But not for long. As more fossils were uncovered, the chain splayed out into the usual phylogenetic net, and it was all too apparent that evolution has not been in a straight line at all, but that (to consider size only) horses had now grown taller, now shorter, with the passage of time.
>
> Garrett Hardin, *Nature and Man's Fate*, 1961, Mentor USA, p.225-226.

So, many diagrams in scientific books are in fact ideograms of what evolutionist biologists want to believe, rather than depictions of any truth.

---

8 In his *Habit and Heritage* (1942) F. Ward Jones raised the question posed by the thick skin on the soles of our feet in relation to the question of whether acquired characteristics can be inherited: he pointed out that the squatting facets of the Indian Punjabis are also present in the Punjabi foetus: 'The squatting facets on tibia and astragalus are present in all Asiatic peoples who adopt the hunkered position of rest, but they are absent in all people who know the habitual use of chairs'.

p.22.

As in so many branches of science, all these areas of exploration pose more and more problems. They do not solve them. Yet, at the same time, those who seek to uphold the faith of Darwinism try to suggest otherwise. The real enigma is that Darwinism and neo-Darwinism only offer a *negative* explanation of what is patently a *positive* process. Darwinism rejects *purpose*, and the Telic. Even if Darwin spoke respectfully of God, the effect of Darwinism was to get rid of God, and many modern scientists, like Crick, Watson and Monod, are quite militant in this. They want to exclude all suggestion of God, of Paley's watch-maker, vitalism and finality, that is any recognition of movement towards goals. But this all leaves a problem. They may be right to exclude *directedness* : but the *directiveness* cannot be denied, and demands a teleological explanation. That is, there needs to be some recognition of a law or principle of movement towards more complex and better organised forms - some 'pull' towards possibilities. Since teleology is unacceptable in modern science those who adhere dogmatically to the accepted paradigms are now prepared to distrust the evidence rather than meet the unsolved questions. Scientists would prefer to cling to a purely negative theory that explains nothing rather than recognise directiveness, and the Telic theme.[9]

Yet it ought to be clear that no simple theory is going to solve the problem of evolution. There are many different themes. Predominant attempts include neo-Lamarckism, neo-Darwininism, the mutationists and the synthetic school.

Behind all these theories there will be found a tacit recognition of the dynamics in life which science would prefer not to be there, since they all defy explanation. Darwin warned against speaking of the 'higher' and 'lower' in the scale of evolution, for example. But how is it possible to deny that there is a *gradient*, from 'lower' to 'higher' – that is towards more complex and effective organisms, since the oyster and amoeba are not conscious of me as I am of them? There is order in the living world because clearly that is why it interests us. And there is development towards greater and greater autonomy and self-consciousness - towards mind. These all defy explanations so far, and remain a mystery, as does the fact that we are here at all: that is, why life came into the world at all. The self-confirming confidence of evolutionary theory should not be allowed to mask these persistent problems. There is even a sense that life came into the world against all odds, against all the (physical) 'inclinations' of matter, and the laws of entropy.

We have glanced already at Lamarckism. There are neo-Lamarckians who believe that the evolution results from the direct interaction of the organism and its environment. The habits of an animal modify its behaviour - use or misuse may make a plant larger or smaller. But the question which remains is as to whether characteristics which are acquired either way can be passed on. Some recent work in immunology with mice as we have pointed out did seem to suggest that there was a genetic path by which acquired characteristics may be passed on:[10] but later experimenters have been unable to repeat the results, and the controversy still rages in the pages of *Nature*.

The 'established' schools of thought about evolution are the Neo-Darwinians, the Mutationists and the Synthetic School.

The Neo-Darwinians have placed an almost exclusive emphasis on natural selection, which, though Darwin stressed it, he did not consider the sole or total explanation of evolution, as we have seen. Beginning from the observation that no two animals are alike, and that some of the variations are hereditary, the Neo-Darwinians believe that some of the variations make creatures better able to cope with their environment and to survive in it. If, then, these individuals have on an average more offspring than do those with other variations, their favourable hereditary characteristics will appear in a larger proportion in the next generation. If this trend continues, over many generations, there will be continuous, adaptive, evolutionary change in the population as a whole. (These beliefs may be associated with the Neo-

---

9 But there are scientists who declare, like Ludwig von Bertalanffy, that,

'While fully appreciating modern selection theory, we nevertheless arrive at an essentially different view of evolution. It appears not to be a series of accidents, the cause of which is determined only by the changes of environments during earth history, and the resulting struggle for existence, but is governed by definite laws, and we believe that the discovery of those laws constitutes one of the most important tasks of the future.'

The Problems of Life, 1952. p.105.

10 'Inheritance of acquired immunologogical tolerance to foreign histocompatibility antigens in mice'. R.M. Gorczynski and E.J. Steele, *Proc. Natl. Acad. Sci.* USA 77(1980) 2871.

Lamarckian element in Darwin which D.R. Oldroyd discusses in *Darwinian Impacts*, pp175 ff).

While no doubt there is much truth in this view, it cannot be taken as a complete explanation of evolution. It is improbable that all the characteristics which distinguish a species increase the fitness of those who possess them, for instance. But one of the major objections is that the Darwinian concept of natural selection is that of a negative process which can eliminate the unfit, or less effective. What it cannot account for is the *origin* of new characters or types of organism. In Neo-Darwinism the whole problem of how new hereditary variations originated and how they were transmitted was left unsolved.

The study of mutations in genetics seemed to offer a solution to this problem. The Mutationists found that hereditary characteristics, with some few exceptions, were transmitted from parents to offspring by well-defined microscopic entities called chromosomes, in the germ cells. Particular hereditary effects are localized in genes within each chromosome.

Variations occur, among other ways, by the inheritance of different sorts of genes or chromosomes from the two parents in sexual reproduction. If there is a mutation in a set of chromosomes, or in an individual chromosome or in a gene within a chromosome, new hereditary characteristics appear. Such changes can be induced artificially in the laboratory, though the effects cannot always be predicted. In Nature the mutations occur spontaneously, and at random.

Adherents of the mutationist school believe that new species arise from such random changes in hereditary.

It is on this strict 'random' theory of the appearance of abnormal forms that much in modern scientific explanations of the origin and development of life is based. The chance mutation of what are called large chemical molecules is now believed by some to be the origin of life: life in this theory is thus a 'cosmic accident', arriving by a 'mistake in coding' and developing according to the ruthless laws of 'gene survival'.[11] How natural selection operates here is not always clear while there are those who question whether genes can determine *form* (see below p169).

The Synthetic School has tried to synthesise conflicting general theories into a consistent whole. They recognise that the evolution of life is inherently complex, and seek to explain evolution as the result of the interaction of factors which are both random and directive, internal and external. The main internal factor is the genetic system, and the random spontaneous changes in it. The main external factor is natural selection which to them is directional and not random. Selection acts in such a way as to promote the spread of some mutations and to oppose the spread of others. Mutation alone is not directional, and selection alone is not creative. Evolution is often both.

Selection does not act on the genetic system:[12] it acts on populations of organisms whose characteristics are not determined directly or solely by genetics, but by growth determined jointly by the genetic factors and the environmental factors. It should be noted that this synthetic theory does not altogether adhere to the strict notion that the development of life proceeds by 'accidents' generating mutations which yield small advantages : it recognises other factors in the process which seem to suggest something more positive.

So far, even Synthetic Theory may be seen within the compass of Norman MacBeth's definition of Classical Darwinism:

> On the basis of data drawn from comparative anatomy, and the experience of breeders, classical Darwinism asserted that the progression from the early species to the later ones, as observed in the rocks, was a process of actual physical descent governed by natural selection through such agencies as the struggle for existence, survival of the fittest, sexual selection, and adaptation, all of which worked in small cumulative steps through vast periods of relatively undisturbed time.

This has two logical corollaries :

---

[11] See *The Selfish Gene*, discussed below, p.92ff.

[12] Stephen Dawkins in *The Selfish Gene* offers a theory in which natural selection does act on the genetic system, and the richness of gene pools is the primary goal. There does seem to be a confusion here in neo-neo-Darwinism, as if molecular biologists are transferring the theme of competition in the struggle for existence from the macro-level of organisms in populations to a cunning struggle among genes or gene-pools, or even among large life molecules for the available energy and source-matter. Can this extension really be valid? See p99 below.

first, in the evolution of any structure or function, every intermediate stage must be of advantage to the species; second, natural selection tends to make each being only as perfect as, or slightly more perfect than, the other inhabitants of the same area, and does not produce absolute perfection.

MacBeth says that to formulate the synthetic theory is an almost impossible task, and sends us to Everett C. Olson's essay on 'Morphology, Paleontology and Evolution' in Sol Tax, *Evolution After Darwin*, 1960, p.525-527. It should be noted that, by this definition, evolution happens by negative processes, still. The innovative origin and growth of new complex forms are not accounted for, especially where these require coordination and could not give immediate advantages.

In the last year or so we have been offered the concept of 'molecular drive'. How does this affect Evolutionary Theory?

The idea of 'molecular drive' was developed from recent discoveries of the concerted evolution of repeated elements in eukaryote genomes and these are believed to shed new light on our understanding of the nature of genetic organisation. The idea poses new challenges to current theories of natural selection and speciation. A number of papers on this idea have been written by Dr. Gabriel Dover who is at the Department of Genetics at the University of Cambridge.[13]

Before we can grasp the implications of the idea of molecular drive, we have to look a bit more closely at the concept of genetic drift, which is a concept from the Modern Synthesis in Darwinism of the 30s.

Genetic drift is a concept within the school of thought known as the 'Modern Synthesis' established by Sir Ronald A. Fisher, a leading British geneticist of the thirties. He defined natural selection as 'a mechanism for generating a high degree of improbability' and sought to delimit the nature of the interdependence of Darwin and Mendel

His insights arose from his statistical appreciation that without natural selection there could be no net change over time in the genetic composition of a generation. This is so because the two key processes of sex are random processes: involving first the production of sperm and eggs with half the usual number of chromosomes, and secondly the fusion of eggs with sperm to reconstitute the full chromosome complement. The random segregation of chromosomes during the formation of either sperm or eggs in an individual ensure that it is highly improbable that any one egg or sperm would acquire a complete set of either of the two practical sets of chromosomes inherited by the individual. It is also highly improbable that any particular sperm would preferentially fuse with any particular egg. These random sexual processes ensure that the proportions of different types of genes in a sexual population in any given generation would appear in similar proportions in the next. The relative frequencies of different genes are maintained in stable equilibrium from generation to generation. The Medelian laws of inheritance, which are based on the stochastic[14] processes of sex, are the starting points from which such predictions of long- term stability can be made.

But only in ideal populations, mathematically defined as of infinite size, can future probabilities of gene frequencies be clearly assigned. There are accidental fluctuations in the numbers of sperm, eggs and individuals, and these disturb the predictions. In extreme cases there could be a *chance drift* of a population to only a few of the many potential types of genes that are available from preceding generations.

Two further geneticists, Sewall Wright and Motoo Kimura, developed evolutionary theories which developed the idea of genetic drift. But while there was a recognition of accidental genetic drift towards the preponderance of a certain type, mathematical calculations made it clear that such improbable shifts from equilibrium could only take effect through the agency of some external force - and that force is supposed to be natural selection. There could be no essential change in evolution, unless it were a disturbance of balance of genes selected out by natural selection. And this only operates if the gene trend gives some *adaptive advantage.*

---

[13] See 'Molecular Drive : a cohesive mode of species evolution', *Nature*, vol.299, 9 September 1982; 'A Molecular Drive Through Evolution', *Bioscience*, American Institute of Biological Sciences, Vol.32, No.6, June 1982; 'Molecular drive: a Hard force in evolution', *New Scientist*, 9 December 1982: *Forces of Evolution, The Sciences*, New York Academy of Sciences, June 1983.

[14] Stochastic Process : A process which has some element of *probability* in its structure. My account here is based on an article by Dr Dover in *The Sciences*, New York Academy of Science.

Again, thus, genetic drift is a purely negative process, whereby changes like long ears in a bat-eared fox come by accident, persist if they give immediate 'survival' advantage; and of course it is assumed that such gene shift can alter morphology.

The notion of molecular drive does not challenge the influence of the random sexual processes of chromosome assortment and mating, in the relationship between the forces of selection and the vagaries of drift. But it develops from a belief that the behaviour of the chromosomes and the behaviour of the genes are no longer considered to be one and the same thing. If the genes and chromosomes are in different dynamic states and out of synchrony during the sexual cycle, then the prediction of long-term equilibrium of genes is invalidated.

Genes are of fixed location on the chromosomes and are passively carried by them to new destinations. Previously, the genes were not considered to be capable of exhibiting any behaviour or dynamics of their own, other than their ability to mutate.

But molecular drive is the concept of a new mechanism in evolution which is capable of changing the genetics of a population *independently of natural selection* in the first instance. It is a consequence of the internal dynamics of change within the genes themselves. The internal forces that are the powerhouse of molecular drive arise out of the dynamics of turnover within the DNA and are capable of spreading new mutations in a population.

One striking feature of sexual organisms is that the genes which govern any particular trait often exist in multiple copies - 'families' of genes. For instance in one species of frogs there are 20,000 genes per individual which code for an essential cellular component (the 55 RNA). Biologically important multigene families are found extensively in all forms of life above bacteria. In addition there are hundreds of families consisting of repetitive stretches of DNA whose function is unknown. Some believe that non-genic families of genes have something to do with the architecture of the chromosomes (as during the formation of sperm and eggs), and long processes of control of gene activation during development.

Members of a family of genes, whether genic or non-genic, exhibit unusually high levels of identity between themselves, no matter on how many chromosomes they might be distributed.

The homogeneity is below the level of discrimination between individuals, so it is, as it were, out of touch of natural selection, which can only operate on individuals.

The homogenisation in gene families is explained by the geneticist by a number of recognised molecular mechanisms which, within families, ensure that *one consistent gene would replace gradually all other genes : this is molecular* drive.

The process is called 'Turnover'. In one version, two initially slightly different genes can become identical, as one gene 'converts' another to its own DNA composition. The molecular basis of conversion bias is not known. Both random fluctations in the directions of conversion and a persistent bias in one direction ensure, in time, that all genes in all individuals are of the same DNA type.

A second mechanism in turnover is 'unequal exchange'. In normal circumstances two chromosomes can exchange on exactly equivalent segment of their DNA. But if the exchange is unequal, one chromosome acquires an extra copy of the segment while the other loses the particular segment. Unequal exchange within families leads to random fluctations in the numbers of alternative genes and, as in gene conversion, one variant gene may replace eventually all the others throughout a sexual population.

A third mechanism is called 'transposition'. Segments of DNA can make extra copies of themselves which may re-insert into completely different chromosomes. Random gain-and-loss of variant genes through such 'pumping' mechanisms can eventually homogenise a family in all individuals.

The activities of all three mechanisms of turnover place the DNA in a constant state of flux and ensure the non-reciprocal transfer of DNA conformation between chromosomes.

This perhaps may have for more radical implications for evolutionary theory than Dr. Gabriel Dover realises. While these 'turnover' processes are described as 'random' they seem clearly a fundamental aspect of the dynamics of life - molecules : while they are still seen as 'accidents' they are neither pathological nor caused by 'mistakes in coding': they happen all the time, out of the nature of the life-stuff itself. May be here we have a potentiality manifest in life-molecules that seeks to realise possibilities?

The three activities place the DNA in a contant state of flux and ensure the non-reciprocal transfer of DNA information between chromosomes. So, individual genes of multigene families can no longer be considered as independent and passive components of chromosomes whose appearance in future generations depends on the random sexual processes of chromosome assortment and mating. The ability of mutations to spread between chromosomes by any of the three mechanisms of exchange ensures that they escape from the constriction and stablising consequences of the Mendelian ideas of inheritance. (And presumably, many of the elaborate mathematical calculations on which synthetic theory is based become out of date or irrelevant)?

Molecular drive is a process which can change unperceptibly the genetics of a population by slowly moving the mean DNA composition of a family from one extreme to the other without generating a large genetic difference between individuals, throughout the period of change. This implies that the form and behaviour of all individuals could evolve in unison without the generation of large biological incompatibilities between them.

What influence might natural selection have in this process? Need biologists adhere to the absolute necessity of selection for survival? Professor Stephen Gould (who has discussed some of his doubts about evolutionary theory) and his colleague Elizabeth Vrba have coined the term 'exaptation' for developments which are non-adaptive : developments which have no survival function. Exaptations under single gene control might accumulate in frequency either by genetic drift or by close linkage to genes under natural selection. Molecular drive can be understood as a powerful and universal mechanism for the origins of all exaptations which are under the control of multigene families.

A mutant copy, spread throughout a gene family, may alter some aspects of the animal in form or physiology (phenotype). But the change in phenotype could be gradual and concurrent in all individuals; the genetic variation and differences in fitness between individuals in the population would be kept to a minimum. By the time the significant change in phenotype occurs, all individuals in the family would have a similar proportion of the new variant and so would be physically similar. It is this change 'in unison' that is a distinctive feature of molecular drive.

Dr. Gabriel Dover offers the concept of molecular drive as a third force operating along with genetic drift and natural selection. Of course, evolution is bound to be a highly complex process. But in this new concept we have surely something that makes adherence to strict Darwinism mechanism no longer necessary. Dr. Dover is careful to dissociate himself from those who are in this sphere pursuing explanations in terms of 'developmental fields' since he believes they explain nothing (and do not recognise the truisms of biology in its basic gene theory). He also urges us to distinguish molecular drive from the concept of the 'internal' forces of evolution developed by L.L. White in the 60's. He is anxious not to be put in the same class as those seeking some 'other' morphological force like Sheldrake.

Yet surely we must see this ferment as an attempt to find some 'other' explanation of the way in which 'life strives' and changes, positively and dynamically within itself - not by 'mistake' and 'accident' as though kicked about by absurd chance, but because it seeks to *achieve*? However, to speak in such terms is declared by scientists like Dr Dover to be 'nonscientific' or 'obscure'. Yet to me it is obscure to believe, as he does, that the eye could be the product of 1,000 linear steps in mutational events - against all the arguments over probability, possibility and the need for coordination. It seems possible, at least, that by such new approaches some radical revision of evolutionary theory may be on the way, waiting for the moment when the current theology is willing to entertain heresies.

4 Darwinism as a religion

Macbeth shows quite clearly that Darwinism has become a kind of religion (his chapter 14).

There are, of course, objectors to Darwinism on religious grounds. Some Christian scientists accept the broad aspects of evolution but reject Darwin's explanations of how and why. They assert that the whole process could be the result of design just as well as chance. Some favour 'creationism' - retaining divine creation albeit not insisting on a literal interpretation of Genesis. Many of the criticisms made are not to be dismissed, such as those in *Evolution and Christian Thought Today* (Eerdmans, 1959) : some scientists who are Christians have revealed some major gaps and flaws in evolutionary theory.

But Macbeth is also able to show that Darwinism itself, with its belief in evolution by 'chance', accidental mutation, and survival of the fittest is itself held with the passions of a faith. Sir Julian Huxley declared that he was an atheist, and that Darwin's real achievement was to remove the whole idea of God as the creator of oganisms from the sphere of rational discussion. (Huxley, 'At Random', a television preview on November 1959, in Sol Tax, *The Issues of Evolution* 1960 pp.41-65). George Gaylord Simpson has attacked 'the higher superstitions' meaning religion in general (*This View of Life*, 1964, page 4 : see also pp.viii and 12) and, as Macbeth points out, whenever he finds he can't explain something, Simpson uses the tactic of showing how much weaker would be the explanation offered by vitalists, finalists or Christian creationists, which is of course another fallacy of the 'no alternative explanation' kind. Simpson rejects all 'non-physical postulates', and insists that the progress of knowledge rigidly requires that no such thing must ever be admitted in connection with the study of physical phenomena. (See G.G. Simpson, *Tempo and Mode in Evolution*, 1944, p.76 n. and *This View of Life*, 1964, p.151).

This persisting rejection of Christianity is but a reflection of the more significant fact, says Macbeth, that Darwinism itself has become a religion (see p.126). He quotes Conklin, late Professor of Biology at Princeton:

> The concept of organic evolution is very highly prized by biologists, for many of whom it is an object of genuinely religious devotion, because they report it as a supreme integrative principle. This is probably the reason why severe methodological criticism employed in other departments of biology has not yet been brought to bear on evolutionary speculation.
>
> *Man Real and Ideal*, 1943, p.147.

As we have seen, there is a profound paradox in Darwinism, because, having rid biology of creationism or design, and opted for a totally *negative* theory of development, evolutionists worship this negative process as a 'supreme integrative principle', when it cannot by its very nature be anything of the kind. Macbeth directs our attention to other discussions of the religious fervour in this area of science (e.g. Jacques Barzun, *Darwin, Marx, Wagner*, 1958, pp.63-69 : Joseph Wood Krutch, *The Great Chain of Life*, 1956, pp.197-198). Macbeth lists five traits which seem to him to show that Darwinism is a religion.

The first trait is the air of belonging to the ranks of the converted. Simpson, for example, declares that 'man is true to himself when he presses home the question, How has this marvellous system of Animate Nature come to be as it is'? To abandon the scientific problem as insoluble, he says, is an 'impiety'. This suggests (in Biblical language too - there is a reference to a 'mess of pottage') that sceptics are not true to themselves, and that 'All who are not with me are against me'. (See *The Meaning of Evolution*, 1949, pp.123, 159, 272-3 also Huxley *Evolution, the Modern Synthesis*, p.473). This Macbeth calls 'reproof of the fainthearted'. His fervour seems to impel Simpson to press on hastily, for he admits in one book (*Tempo and Mode*) that 'for almost every topic discussed in the following pages the data are insufficient'. (p.xviii).

This suggests a *missionary zeal*. We also find, as Macbeth shows, that, as in Huxley, this zeal does not stop at extrapolating into wider areas of the Humanities.

> It is essential for evolution to become the central core in any educational system... Human history is a continuation of biological evolution in a different form.
>
> in Tax, 1960, p.42

Here we have the seeds of E.O. Wilson's attempt to use biology to usurp the Humanities. History, morality and ethics are to be solved by microbiology on this principle.[1]

Then there is the principle of the 'one thing needful', the ultimate clue to life. Simpson also asserts at times that he and his colleagues have found the ultimate solution to all biological riddles :

> We seem at last to have a unified theory... which is capable of facing all the classic problems of the history of life and of providing a casualistic solution of each... Within the realm of what is clearly knowable, the main problem seems to me and many other investigators to be solved.
>
> 1949, pp.278-9

The Cartesian impulse in this is clear : at last the mystery of life is reduced to a simple and direct idea. Yet of the synthetic theory of which Simpson is writing here, Ernst Mayr has said, as we have seen, that 'the basic theory is in many instances hardly more than a postulate'. So how can such a postulate be such a *perfect faith*? And 'clearly knowable'?

Then in evolution one finds what Macbeth calls *Millenarianism*. In this there is a hubristic impulse to *control life*. G. Leyard Stebbings said, 'The control by man of organic evolution is now an attainable goal' (*Variations and Evolution in Plants*, 1950, p.561). Simpson declares that, 'it is unquestionably possible for man to guide his own evolution (within limits) along desirable lines' (*This View of Life*, p.285).

> It is probable that the incidence of mutations can be controlled within broad limits... control over the direction of mutation... is another eventual possibility... Growing knowledge of the chemical nature and structure of genes holds the possibility that genes or in the end, even whole genetic systems, can be made to order. The guidance of evolution could then become a simple matter of following specifications.
>
> 1969, pp.284-285.

Today, one often meets scientists and people in medical research, who actually believe this kind of thing is possible.[2] Some are especially anxious to control the mind by drugs : to them consciousness is only chemicals. But since some argue that the complex systems of genes cannot be explained in terms of the reduction to the laws of chemistry, might there not be serious dangers in such manipulations while the level of our knowledge is still inadequate? Simpson has surely forgotten his own remarks:

> How evolution occurs is much more intricate, still incompletely known, debated in detail, and the subject of most active investigation...
>
> 1964, p.10.

---

1 As Mary Midgley argues, also on this principle, shall science and mathematics also be reduced to microbiology?
See *Beast and Man*, Harvester Press, 1981.

2 In *The Eighth Day of Creation* Sidney Brenner is quoted as declaring that it should soon be possible to grow a hand using a computer linked to tissue culture.

Later he admitted that the 'gene system is so intricately balanced that insertion of a foreign element, however well specified in itself, would probably have disastrous effects': *Biology and Man*, 1969, p.119, (see also pp.58-9). The hubris some scientists now display, combined with their logical confusion and despite their occasional admissions of ignorance, surely make one shiver to contemplate the possible results of their experiments in genetic engineering and such manipulations : to say nothing of their monumental overall philosophical naivety, of which they seem totally unaware.[3]

I have followed Norman Macbeth closely here, because his excellent analysis reveals the *inconsistency* in much scientific argument, and its *emotional*, subjective, nature - even at times its psychopathological nature. 'There are', says Macbeth, 'wild swings from arrogance to humility, from boasts of wisdom to confessions of ignorance'. The scientists should cultivate sobriety, he suggests. But it is not the characteristic of religious faiths, passionately held, to be sober, and their zeal blinds scientists to their worrying philosophical innocence, their inability to examine their own thoughts.

In Darwinisn faith God and the Watchmaker have been replaced, as we have seen, by a deified 'natural selection'. Here, some of the lapses are prodigious : a totally negative process, deified, becomes capable of forethought. Simpson, for example, says: 'it is certain that if we can see any advantage whatever in a small variation (and sometimes if we cannot) *selection sees more...*' (my italics). This is, of course, quite absurd : a mere process of ineffective types not surviving can 'see' nothing. but, as I have suggested, the origin of this confusion is to be found in Darwin himself, who tended to endow a purely negative process with a positive power to *improve*:

> Natural selection is daily and hourly scrutinising every variation, even the slightest; rejecting that which is bad, preserving and adding up all that is good, silently and insensibly working... at the improvement of each organic being.
>
> *The Origin of Species*, p.83, (1888 Edn,p.65).

Stebbins speaks of selection as a guiding force (see: 1950, *Variation and Evolution in Plants*, pages 103, 119, 500, 501). This is one of the most serious fallacies in evolutionary theories. The fallacy seems to spring from the fact that evolutionists are forced to be aware that there is a dynamic in nature towards new forms : but they are disqualified by mechanistic dogma from admitting it while teleological approaches are heretical. So, they have somehow to smuggle in a directive force into their universe in which no such concept ought to be allowed.

As E.W.F. Tomlin has pointed out, the directive element in nature, though denied, is continually implicitly recognised by such scientists by such words as 'purpose', or even 'prediction'.[4]

The general view taken over into the popular mind from science in the Humanities is perhaps best shown by this passage which Macbeth quotes from Robert Ardrey, the playwright whose excursions into biology elsewhere seem so doubtful :[5]

> Never to be forgotten, to be reflected, to be derided, is the inconspicuous figure in the quiet back room. He sits with head bent, silent, waiting, listening to the commotion in the streets. He is the keeper of the kinds.
>
> Who is he? We do not know, nor shall we ever. He is a presence, and that is all. But his presence is evident in the last reaches of infinite space beyond men's probing eye. His presence is guessable in the last reaches of infinite smallness beyond the magnification of selection or microscope. He is present in all living beings and in all inanimate matter. His presence is asserted in all things that ever were, and in all things that will ever be. And his command is unanswerable, his identity is unknowable. But his ancient concern is with order.
>
> *African Genesis*, p.353.

---

3 Some scientists have serious doubts. See *The Recombinant DNA Debate*, David A. Jackson and Stephen P. Stich, eds, Prentice-Hall, 1979. 'The debate has become a principal focus for many of the most vexing questions concerning the proper social role of modern science', p.xiv.

4 According to Richard Dawkins, genes 'take decisions' that 'pay off', p.59. Other geneticists speak of 'automatic foresight' as we shall see.

5 See Bernard Towers and John Lewis : *Naked Ape - or Homo Sapiens?*, Garnstone Press, 1969.

As Macbeth says, this is Paley's Watchmaker encountered yet again : Ardrey's paragraph is really old-fashioned religion.

In the recent past, the directiveness in nature has been recognised by some philosophers of science, as by R.G. Collingwood :

> Mutations in species arise not through the workings of the laws of chance but by steps which are somehow directed towards a higher form - that is a more efficiently and vividly alive form - of life.
>
> *The Idea of Nature.*

As E.W.F. Tomlin points out, scientists evade such questions by speaking of '*apparent*' purpose, and 'the *appearance* of end' (C.H. Waddington). Teleological language, as Marjorie Grene points out, must in some way be *smuggled* into the biological discourse after all. Because of these confusions, we may find in much scientific writing what can only be seen as a belief in magic. Sun-warmed mud generates life; given long enough time, anything can happen; processes in which inefficient creatures perish can have foresight and the capacity to plan. Marjorie Grene quotes Schindewolf :

> Nothing could be more mystifying than the prophetic vision of selection which the neo-Darwinians implicitly assume : the idea that selection selects what *will* be advantageous in another ten million years, and therefore sees advantages that will accrue to the remote descendants of the forms selected.
>
> *Grundfragen der Paläontologie*, p.413, pp.430-431, Quoted by Marjorie Grene, *The Understanding of Nature*, p.130,131.

So perhaps 'selection' really stands for some force which controls the whole process which scientists dare not openly recognise (see T. Dobzhansky, *Genetics and the Origin of Species*, 1951, p.75). Surely what is needed is some kind of concept like that of Schindewolf's, that '*life can originate novelty*'? Yet the established ideology of Darwinian evolutionary theory cannot really allow any recognition of what is implied by Collingwood's 'steps... somehow directed towards a higher form'.. because their whole life-system is based on simple mechanism, on concepts which belong to the machine, rather than to the 'organismic'. So, in much scientific writing, we have a muddle, of bleakly 'realistic' terms ('mis-copyings were made and propagated') with attempts to imbue 'natural selection' with creative power ('natural selection... actually assembles instructions in the DNA molecules').

Exploring Darwinism, Norman Macbeth found the following surprising aspects of the matter:[6]

1. Biologists in the German-speaking world, many Russians, and the French did not believe that Darwin had solved the problem of evolution.
2. Leading Darwinians showed no respect for certain tenets of classical Darwinism, viz. the biogenetic law, survival of the fittest and the 'struggle for existence'.
3. The more recent books showed no 'trees of descent' (though later he found these still in school books).
4. The stress was no longer on the accomplishments of breeders or the marvellousness of adaptions, but on mathematics and genetics.

Throughout Macbeth's book he indicates how many scientists, in this sphere, no longer adhere to reason, and how they contradict one another. Thus, Sir Julian Huxley speaks at one point of: 'when we were still ignorant of the mechanism of hereditary....' having just told his readers: 'unfortunately, the precise way genes act is still very imperfectly understood'. There are many such examples.

Macbeth even found that some scientists actually regard it as bad for a scientist to be open-minded, avoiding pre-conceived theories. (Michael G. Chiselin : *The Triumph of Darwinian Method*, 1969, p.212. 'A more pernicious fallacy could scarcely be enunciated'). So Darwinism presents an extraordinary case - for while it is passionately adhered to as a religion, the conclusion seems to be that: '*classical Darwinism is no longer considered valid by qualified biologists*', as Macbeth shows.

---

[6] The four authors he read were Sir Julian Huxley, John Maynard Smith, Garrett Mardin and Loren Eiseley. See: Bibliography.

To say this, say Macbeth, is not to deny Evolution in the larger sense, nor to defend fundamentalism, creationism or vitalism or what not.

It is simply to say that Darwinism, in any of its forms, does not offer any satisfactory exploration of the dynamics of evolution. For one thing, Darwin's notion of potential changes is too facile. Animal populations seem to have a certain persistence or inertia : they resist sudden or drastic change. 'Genetic homeostasis' is the name given to this by Ernst Mayr (1963, *Animal Species and Evolution*, pp.285-65), See Norman Macbeth, p.34 ff). Darwin himself had noted this : he quotes Goethe : 'in order to spend on one side, nature is forced to economise on the other side'.

The question then arises : if Darwin invoked the success of breeders, to show that small modifications happen, and can be exploited, what of the other aspect of the matter which has been observed- the tendency to revert, and the resistance to change? Evolutionary theory depends upon small (micro-) changes accumulating into large (macro-) changes, though there is no conclusive evidence that micro changes culminate in macro changes; scientists go on believing this because they must, to preserve the dogma. (See Macbeth, Chapter 15). But perhaps what is learnt from breeding is not the clue anyway?

Loren Eiseley, 1958, in *The Immense Journey*, says:

> It would appear that careful domestic breeding, whatever it may do to improve the quality of race-horses or cabbages, is not actually in itself the road to the endless biological deviation which is evolution. There is great irony in this situation, because, more than any other factor, domestic breeding has been used as an argument for the reality of evolution.
>
> p.223.

Some with experience of breeding declare that there is a 'pull towards a mean which keeps all living things within more or less fixed limitations'. (Luther Bubank, a breeder, Wilbur Hall, 1939, *Partner of Nature*).[7] 'There are limits to the development possible, and these limits follow a law. But what law, and why'?

The existence of these laws should put a large questionmark beside the confidence with which simple mechanism upheld by Darwinians. If there are laws of the 'pull towards the mean', *what do they exist in*? And if there can be a 'pull towards the mean' can there not also be a 'pull towards higher forms'? If life processes are merely the product of the laws of physics and chemistry how can such a 'law' or dynamic exist, which seems to preserve life against undue interference?[8] Or, to put it the other way round, if there is a directive impulse towards increased complexity or even greater effectiveness, in what might it reside? Obviously, no such 'pull' could simply be the product of the laws of physics and chemistry, or the outcome of the 'codes' of DNA molecules. In this debate there needs to be a recognition that, whether biologists notice it or not, their work rests on an implicit assumption in 'biology' is that life is 'different' and follows other laws, even if not made of different material.[9] Must we not begin to look at 'other' dimensions (by say, some such discipline as a biological 'systems engineering' and a phenomenonological study of 'primary consciousness') by which to understand the nature of life? And may there not be a continuity of dynamic form, from inorganic to organic? Some biologists believe so, as we shall see.

Here it will be sufficient to note that Darwin's imaginative speculations included a conviction that unlimited change was possible, against all the evidence (He also believed in the perfectibility of man: See above, p.15). Moreover, his whole theory depends on an assumption that micro-changes would culminate in big macro-changes. He said, in the deleted

---

7 Norman Macbeth also quotes Edward S. Deevey Jnr., 1967, *Yale Review*, p.637.

8 Similarly, throughout his book: *The Selfish Gene*, Richard Dawkins speaks of how genes 'control', 'manipulate', 'make', 'colloborate', 'interest' and act as 'agents of evolution', in such a way as to suggest an intelligence and purpose in them. But in what can this reside if the genes simply obey the laws of physics and chemistry? The problem also remains over 'DNA coding' and the complex patterns of development in life-forms: See below.

9 See Polanyi, discussed below, on how the only product of DNA 'codes' could be 'noise'.

A recent example may be given of the problems here. In work on immunology with animals it was found that the animals' system did not develop immunity to the chemical content of dead Foot and Mouth viruses but to the way in which the molecules were folded - that is, to a feature to be understood by systems engineering rather than physics.

passage already referred to:

> I can see no difficulty in a race of bears being rendered, by natural selection, more and more acquatic in their habits, with larger and larger mouths, till a creature was produced as monstrous as a whale.
>
> 1859 edition, p.184.

This was to extrapolate disastrously, as he must have realised when he cut it. Comparative anatomy and embryology showed resemblances between types or species: there were macro differences. As Macbeth says: there were horses, cows, sheep - with backbones, hearts, four legs and a skull. They were similar but not identical. What Darwin needed was a neatly converging fossil genealogy. He had to find processes by which the gaps could be bridged.

These he extrapolated from the work of breeders. Yet Malthus, who influenced Darwin, pointed out that, while sheep had been bred for short legs and head, they could hardly be bred without legs or head at all, or to be the size of a rat. A carnation could never be produced as big as a cabbage. No one has ever produced a black rose. Darwin declared that he could find no single fact on which a belief could be grounded, that there was a limit to variation in nature. But, says Macbeth, he neglected to add that he could discover no single fact on which an opposite belief could be based, either!

Here is one instance where imagination and passionate conviction got the better of Darwin the scientist - and this set the pattern for Darwinism, which has become a faith rather than a body of scientific conclusions.

In the chapters that follow I try to do no more than suggest that there is, and can be, no monolithic scientific truth in evolutionary theory to which we must all defer. There are important heretics in biology: there are profound philosophical and scientific difficulties; in fundamental ways the question is wide open. What we are up against is not science so much as convictions, and so it is quite fallacious for us to bow down to any metaphysical conclusions apparently based on the facts of science.

My survey of doubts and heresies will not be exhaustive or complete : but I hope to leave the structure of orthodox Darwinism and related dogmas severely riddled by the time I have finished, so that we in the Humanities may be free to investigate the world without feeling daunted by mechanistic nihilism.

The subjects I propose to tackle are as follows:

Firstly, Marjorie Grene's objections in *The Knower and the Known* to Darwinism as a means of extending the machine image to life itself, and especially the impulse to develop a purely negative process of small change and selection into a explanation of the creative processes of the growth and development of life. (In this Professor Grene is supported by Norman Macbeth).

Secondly, Marjorie Grene's objection, in *The Understanding of Nature*, to the whole reductionist impulse to explain life by breaking down into causal sequences of minimals : this is a rejection of a mode of thinking central to British philosophy, science and empiricism, of course, from Hume to Hobbes to modern analytical philosophy. A different mode of philosophy is to be found in Europe, which is the basis of a quite different form of philosophical biology.

Thirdly, I shall glance at the problem of the origin of life and discuss the findings of Frey-Wissling, that there is no occasion to find a discontinuity between 'inanimate' and 'animate' matter. I shall look later at some of the work of E.W.F. Tomlin on 'the category of life'. This work indicates that life cannot be reduced to explanation in terms of minimals, without serious harm to our understanding of it - and it is this that offers the most telling charge against Darwinism orthodoxy: moreover, our attention is directed to perplexities of *form* and *vitality*.

Fourthly, I shall indicate some of the dissentient views of an important heretic, Professor Pierre-Paul Grassé, who challenges the basic notion of 'chance' mutations, and raises other questions.

Fifthly, I shall discuss Rupert Sheldrake's *A New Science of Life*, a heretical work by a Cambridge biologist, demanding a new approach.

Sixthly, I shall summarise the arguments of Michael Polanyi, that life cannot be reduced to physics and chemistry - an assumption on which neo-Darwinism orthodoxy rests.

I shall examine in detail the arguments of a book which was among the influences prompting the present work, *The Selfish Gene*, which seems to me to represent the worst kind of combination of dogma and poor argument, extrapolating from science into the Humanities. I shall discuss some teaching manuals and the way evolution is offered in education, referring to the doubts expresed by a science teacher (whose book *Adam and Evolution*, Hutchinson, 1984, came too late to be absorbed here, unfortunately). And I shall summarise the devasting analysis of evolutionary theory made by a molecular biologist, Michael Denton's *Evolution: a Theory in Crisis*, Hutchinson, 1985.

I have tried to summarise and collate all the arguments I can find, to show that the bleak view of life and existence accepted as obligatory in the Humanities today is based on something of a misunderstanding of what 'science says', but also on contradictions and confusions in science itself - all contributing to a false myth. It is time to think again, about man's place in Nature, even as our technology outstrips our moral sense and our capacity to explore a possible philosophy of being.

5 Marjorie Grene and the Darwin machine

Professor Grene is a follower of Michael Polanyi, and much of her writing is devoted to resisting the dream of biologists to reduce all biology to physics and chemistry, in the wake of Polanyi's *Personal Knowledge*. Her grasp of the life-sciences is formidable.

Her critique of Darwinism, in Chapters 7 ('The Faith of Darwinism') and 8 ('The Multiplicity of Forms') is profound.

Her work in *The Understanding of Nature* has had some influence in America, and is fundamental. What has been so triumphantly successful in Darwin's theory she says is the reduction of life to the play of chance and necessity, and the elimination of organic categories from the interpretation of living things. Darwin's interpretation is Cartesian, and portrays nature as a mechanically interacting aggregate of machines. It is the extension of the machine image to life itself and it is out of that that grows the Cartesian impulse to grasp the machine clearly and to control it : to rule and manage nature and man like a machine. (The impulse may be traced back to the Renaissance: Raleigh for example wrote of 'that which... draweth out of nature's hidden bosom to human use', a theme taken up enthusiastically by Bacon, while of course the triumph of the whole empirical mode has been in its efficiency in the manipulation and exploitation of the world since).

Darwinism appeared to its first adherents, and still appears to its faithful, as a great liberating agency against bigotry and superstition, a liberator for the Spirit of Science and the heroic pursuit of truth.

Of course, one may have sympathy with the reaction against Christianity : against such horrible concepts as eternal damnation, sin, and God's jealous vengeance. For some, like Julian Huxley, there seemed to be the possibility of a new ethic, a 'morality of evolutionary direction':

> Anything which permits or promotes open development is right, anything which restricts or frustrates development is wrong.
>
> *Evolution in Action*, p.146.[1]

For others Darwinian nature, which their scientific conscience obliges them to accept, is 'the blind ongoing of fact, indifferent to value and incapable of generating value'.

The religion of Darwinism sees the 'derivation of life, of man, of man's deepest hopes and highest achievements, from the external and indirect determination of small chance errors'. *(The Knower and the Known* p.187):

> evolution is the result of selection operating on heritable variation in the form of mutation, and recombination of Mendelian genes. These processes *must have been* at work in all the evolutionary changes considered in this book.

---

1 Professor John Wisdom deals with evolutionary ethics by reviewing a book by C.H. Waddington in *Philosophy and Psycho-analysis*. Evolutionary theory cannot be the basis of an ethics because it deals with *what is* and not with *oughts*.

Sir Gavin de Beer, who wrote that in *Embryos and Ancestors*, also wrote that, like Newtonian physics, this provided

> the first general principle to be discovered applicable to the entire realm of living being
>
> 'The Darwin-Wallace Centenary' in *Endeavour* XVII, 1958, 61-76, p.76

Says Marjorie Grene :

> Biology, thanks to Darwin... has at last matured into a proper science, 'Lamarck nonsense' is disinherited, old metaphysical follies re-echo only distantly in the ears of a few foolish mystics... Matter becoming life (and mind) through natural selection of chance mutations; life spreading in ever new directions through opportunistic exploitation of the unexpected; of new riches in nature happening to fit a slightly new departure in the arrangement of established genotypes or slightly new genotypes happening to tumble into hitherto unexploited environments - this is the vision which experiment and mathematics, field observation and its statistical analysis combine to support.
>
> *The Knower and the Known* p.188

Beginning with Sir Ronald Fisher's *Genetical Theory of Natural Selection* in 1930 there has arisen a most imposing synthesis of two conceptions - that of mutations, and Darwin's view of a slow and gradual process in which slightly less fit variations were eliminated in favour of the slightly better adapted.

Modern Darwinism sees life *as* evolution : 'Evolution is progressive adaptation and nothing else' (Fisher, 'Measurement of Selective Intensity', *Proc. Roy. Soc. B.* CXXI (1936), 58-62). Selection theory has a compelling power : its view that all major trends in evolution are adaptive, and that the genesis of adaptions is explained by the gradual and external control of chance variations by selection pressure. So compelling is this argument that even when mathematical computations seem to make developments in evolution seem unlikely or impossible, the evolutionists take these objections simply to show how marvellously the process works! Marjorie Grene quotes Sir Gavin de Beer:

> Muller has estimated that on the existing knowledge of the percentage of mutations that are beneficial, and a reasoned estimate of the number of mutations that would be necessary to convert an amoeba into a horse, based on the average magnitude of the effects of mutations, the number of mutations required on the basis of chance alone, if there were no natural selection, would be of the order of one thousand raised to the power of one million.

But then he goes on:

> This impossible and meaningless figure serves to illustrate the power of natural selection in collecting favourable mutations and minimizing waste of variation, for horses do exist, and they have evolved.
>
> Sir Gavin de Beer, 'Darwin Wallace Centenary', p.63
> and *Evolution*, p.19, *The Knower and the Known*, p.189

As Marjorie Grene points out, this is an astonishing way of arguing. That 'horses do exist and they have evolved' proves nothing, since the whole purpose of the argument is to explain how they have evolved. The majority of mutations are adverse or lethal, this surely sugests (see Grassé below) that the transformation of the horse through random mutations is extremely unlikely - leaving aside the mathematical computation? The astonishing result *must* have been caused by the automatic selection in each generation, of very slightly advantageous variants, according to de Beer's argument. *But how do we know this?* In de Beer, as so often, an argument which seems to totally undermine the dogma is absorbed and turned to support it!

What kind of process, in any case, is 'selection'? By definition, it selects the better adapted alternatives. Yet as de Beer went on to point out, 'the vast majority of lines of evolution have led to extinction'. In a characteristic scathing rejection of 'providential guidance and purpose' de Beer uses his fact as evidence againt teleology. But (asks Marjorie Grene) isn't it also a grim comment on Natural Selection? If in the majority of cases mutational developments like those of the doomed Hua birds lead to extinction, how can the same principle be the source of the Evolution of all life?

Yet despite such doubts, neo-Darwinian thinking moves, in happy self-confirmation, within the tight circle of these concepts. The basic explanatory concepts are *chance* and *necessity.*

These two, Marjorie Grene points out, have been the sole permitted instruments of reductivist explanation - from Democritus in ancient times, through Hobbes, to modern physicalism. Marjorie Grene as a philosopher, points to the fact that such explanations are 'wrong enough' to accommodate all the immense achievements of modern biological research, and there is a dogma - certain modes are 'the sole permitted instruments of reductivist explanation'.

And yet, as we shall see, there are heretics. And these would agree with Sir James Gray:

> No amount of argument, or clever epigram, can disguise the inherent improbability of orthodox (Darwinian) theory...
>
> 'The Case for Natural Selection', *Nature*, CLXXIII (1954), 227

At this point perhaps we may leave Marjorie Grene in mid-chapter, to look back further at some of Norman Macbeth's discoveries, as he read his way through books on evolutionary theory. Remember that the 'trio of concepts' which Marjorie Grene says are the *sole permitted instruments* are as follows:

> *Mutations* are the chance failures of the duplicating mechanism. *Selection* is the agent of external compulsion, eliminating less well adapted variants through environmental pressure. The consequent *Evolution* is merely 'descent with modification' (no 'emergence' no 'higher or lower' allowed).

All these are *negative* processes - the resultant living creatures being those left when chance mutations, which didn't give immediate modifications, died out.

Yet (as we have seen) Darwin himself offered natural selection as a positive process; directed at 'improvement': and so do his followers everywhere.

Norman Macbeth traces the struggles of proponents of Darwinism who have striven to turn the negative processes into positive ones - as explanations of the forward movement of life.

G. Ledyard Stebbins Jr., for instance, in *Variation and Evolution in Plants* (1950) speaks of natural selection as a *guiding force* or a *directing force*, even likening it to a sculptor creating a statue by removing chips from a block of marble. (See pp.103, 108, 118, 119, 500, 501). Yet there is no principle in Darwinism to account for such creative developments.

Simpson, as Norman Macbeth points out, speaks of the 'opportunism' of evolution: but then warns his reader against 'any personal meaning or anthropomorphic implications' (1949, p.160).

Despite such warnings, however, there is a passionate impulse, as Macbeth shows, to produce a single, all-embracing principle, an absolute and final theory, which is natural selection. According to Sir Julian Huxley, 'so far as we now know, not only is natural selection inevitable, not only is it an effective agency of evolution, but it is *the* only effective agency of evolution'. (*Evolution, the Modern Synthesis*, 1957, p.35).

Surely if such a principle is so important, it is surely absolutely crucial to science to test it in the field? However, natural selection is simply not there. As Norman Macbeth shows, while natural selection is the key factor in evolutionary theory it is also impossible to see: 'The operations of natural selection, real or imagined, are not accessible to the human eye'.

> Stebbings says: 'While the demonstration that selection has occurred is not excessively difficult, the nature of action and the causes of this selective process are much harder to discover or to prove'.
>
> *Variation and Evolution in Plants*, p.107

Later the same biologist declares that it is impossible to determine what is adaptive.

> 'Obviously... a final estimate of the importance of selection in evolution must depend largely on determining what... differences are... adaptive... Unfortunately, however, the determination of the adaptive character of many types or differences between organisms is one of the most difficult problems in biology'.
>
> p.118

At the end of his book he says, 'We can, therefore, do little more than speculate'.

C.G. Simpson, of a hypothetical case, admits that 'it would be quite impossible to observe such weak selection either in the laboratory or in nature... selection may be highly effective although quite beyond our powers of observation'. (*The Major Features of Evolution* 1953, p.118). He concedes that 'it might be argued that the theory is quite unsubstantiated and has status only as a speculation'. As Macbeth says, when certain Lamarckians contended that inherited effects are so slight they cannot be detected experimentally, Sir Julian Huxley declared that 'To plead the impossibility of detection is a counsel of despair. It is also unscientific'. (*Evolution, in Modern Synthesis* 1942, p.459). But, surely, this is also true of natural selection, since there is no tangible evidence for it?

> The Darwinians contend that any given result must have been produced by natural selection working on small changes, but when asked to be exact they are helpless.
>
> Norman Macbeth, p.44

(Macbeth refers us to Goldsmith, R.B. *The Material Basis of Evolution*, 1940, pp.6-7; Ernst Mayr, 'The emergence of evolutionary novelties' in Tax, Sol, ed. *Evolution after Darwin*, vol I *The Evolution of Life*, pp.1503-1504; and Everett C. Olson also Tax, 1960, pp.540-542) Sir Gavin de Beer admits that, 'The causes of the origins of patterns, colours and of many other things are not known'. Simpson confesses to 'the sad, one might almost say the shameful fact' that he does not know what natural selection is doing, (G.G. Simpson, *Biology and Man*, p.123).

Robert Ardrey, oddly enough, is quoted as saying:

> It is fruitless to attempt to explain everything in the natural world in terms of selective value and survival necessity.

Despite all these misgivings, some scientists cling to the principle of natural selection as one which has 'enormous power' - this despite the fact that so distinguished a scientist as C.H. Waddington finds it tautologous:

> Darwin's major contribution was, of course, the suggestion that evolution can be explained by the natural selection of random variations. Natural selection, which was at first considered as though it were a hypothesis that was in need of experimental or observational confirmation, turns out on closer inspection to be a tautology, a statement of inevitable although previously unrecognised relation. It states that the fittest individuals in a population (defined as those which leave most offspring) will leave most offspring.
>
> 'Evolutionary Adaptation' in Tax, Sol, *Evolution After Darwin.* Vol.I, p.38

Waddington goes on to say this in no way detracts from the magnitude of Darwin's achievement in putting forward this principle: 'once the statement is made its truth is apparent'. How can he believe this if it is so tautologous it can tell us nothing about the nature of things?

Macbeth also indicates the way in which the theory of natural selection is diluted. In G.G. Simpson's *Tempo and Mode in Evolution*, 1944, and *The Meaning of Evolution*, 1949, he was content to define natural selection as differential reproduction. But in *The Major Features of Evolution*, 1913, he proposed

> Slightly to extend the definition used in population genetics and to define selection, a technical term in evolutionary studies, as *anything* tending to produce systematic, heritable change in one generation and the next.
>
> p.138.

Is this to include in 'natural selection' mutations, recombination of genes, processes of isolation and length of generations? Is 'natural selection' now *anything tending to produce change*?

In 1969, in *Biology and Man*, Simpson next wrote 'natural selection... is... usually and most strongly a stabilizing normalizing influence preventing or slowing and not hastening evolutionary change'. To this Macbeth adds remarks by George C. Williams in *Adaptation and Natural Selection.* 1966, p.54 : 'natural selection was first developed as an explanation for evolutionary change. It is much more important as an explanation for the maintenance of adaptation', and (p.139). '...evolution takes place, not so much because of natural selection, but to a large degree *in spite of it*'. (my italics).

Natural selection, it seems, can be anything : so it has 'enormous power' as a weapon of explanation. Simpson finds that rates of evolution cannot be explained by mutations: so there must be another factor - and 'the most reasonable probability is that that factor is selection' (*The Major Features of Evolution*, p.146). But how could this be, since selection cannot initiate anything?

According to the prevailing Darwianian view, the processes which have produced the multifarious organic scene are mutation : linkage : crossing-over and recombination which have to do with arrangement of re-arrangement of the genetic material; isolation, and natural selection. In answer to criticisms, as to how the eye could develop by such processes, the Darwinians reply that the process is long and gradual and the pool of variations available to any population is immense. Above all, *natural selection*, working on the existing gene pool would suffice. Whichever way we turn, we find the evolutionists clinging to a force or principle which can do anything - and again looks very much like God, after all.

So, despite all the confusions, contradictions, and doubts I have indicated, the scientists can still assert a broad dogma of belief. Sir Peter Medawar has said, 'There are philosophical and methodological objections to evolutionary theory...' But then elsewhere makes a declaration that seems totally at odds with such caution: '... It is too difficult to imagine or envisage an evolutionary episode which could not be explained by the formulae of neo-Darwinism'.

Yet, recognising the Popperian requirement as to the nature of scientific truth, P. Ehrlich and C.H. Birch wrote in *Nature* (1967, Vol 219, p.352). 'Our theory of evolution has become one which cannot be refuted by any possible observation'. So it is not really a tenable scientific hypothesis.

Despite this, the speculative theory which is evolutionism is universally adopted and taught as scientific fact.

So, to return to Marjorie Grene (p.191 in *The Knower and the Known*), the question arises - why does such a set of unsatisfactory concepts take such a firm hold on so many able minds?

First, though Darwinian theory is, and must be by the nature of its subject-matter, speculative, it has been held up as a model of simple Baconian induction, through the patient accumulation of facts.

Of course, Darwin did collect facts. But the species theory was a triumph of a 'leap of the imagination'. It is a question, on the lines of Polanyi has discussed, of the *subjective* appeal of a satisfactory hypothesis. The Darwinian hypothesis is so satisfying that it has *felt like fact.*

There are pieces of work which are taken to confirm the theory - such as that of H.B.D. Kettlewell on the black mutant of the common peppered moth (*Nature*, CLXXV, 1955, 943, etc). And, of course, clearly, the colour of such creatures is controlled by visibility to predators. But how did moths come into existence in the first place? And do such processes provide an explanation of the *origin* of species, and the phyla of living organisms?

At this point Marjorie Grene makes an interesting point. We take it for granted that there is a rhythm in individual development (as from the foetus to the child, to the adult and so on). Why should there not be some analogous process in the development of types? Why should not 'phylogeny be ontogeny writ large'? Why does Darwinism so strongly deny such a possibility?

Marjorie Grene's final point in her chapter concerns the implicit concept in Darwinism of what evolution 'is': it is 'progressive adaptation', to leading proponents, as we have seen. Are organic phenomena primarily explicable in terms of adaption? This has been discussed in relation in nineteenth century thought by Dr. Gertrude Himmelfarb in *Darwin and the Darwinian Revolution* (Chatto, 1959), and it reveals a concealed teleology even in Darwinism.

Adaptation is a question of means and ends : such relations are essential to the Darwinian argument. Darwinism offered to *dispense with a planner.* To some, Darwinism seems materialistic and mechanistic : yet it has a teleological content, too, for it recognises such goals as the proper habitat and the natural biological 'niche'.

> Everything in nature is defined in terms of its purpose, but an unplanned purpose in which the organism is tool, tool-user, and beneficiary all in one.
>
> Grene, p.195

Organisms are contrivances, aggregates of characters and functions good for - what?

> For survival, that is, for going on and being good for, going on and being good for - and so on *ad infinitum* ... the *summum bonum*, like the maker, is dispensed with; yet the means-end relation, the notion of 'this is useful for that', is fundamental still.
>
> Grene, p.195

For this reason, Marjorie Grene speaks of Darwinism's 'truncated teleology'.

Some biologists, Professor Grene points out, question whether adaptation is really the essential of evolution at all (e.g. A.M. Dalcq, 'Le Probleme de l'evolution est-il pres d'etre resolu?' Annales de la Societé zoologique Belgique, LXXXII (1951) p.117; O. Schindewolf,

Grundfragen der Paläontologie (Stuttgart, 1950) ; A Vandel, *L'Homme et L'Evolütion* (Paris, 1938)). Some of the central examples in Darwin are dead ends (like the Galapagos finches).

What about the great new inventions, which arise with startling suddenness? In Darwinian terms these would have no reason in the world to proliferate and persist:

> Neither the origin and persistence of great new modes of life - photo-synthesis, breathing, thinking - nor all the intricate and coordinated changes needed to support them, are explained or even made conceivable on the Darwinian view.
>
> Grene, p.197

For all the brilliance of the *Origin of Species*, if one reads with this in mind, for all its implicity of 'mechanism' and its explanations, it is 'simply not about the origin of species, let alone of the great orders and classes and phyla, at all'.

Though Darwinism is offered as explaining the origins and development of life, it does nothing of the kind : this is Marjorie Grene's devasting demolition. It deals with minute specialised adaptations, which lead, if not to extinction, nowhere. It can explain things like melanism,

> but how from single-celled (and for that matter inanimate) ancestors there came to be castor-beans and moths and snails, and how from those emerged llamas and hedgehogs and lions and apes - and men - this is a question which neo-Darwinian theory simply leaves unasked... it provides no conceptual framework in terms of which they can be admitted to exist, let alone an 'explanation' of their descent from 'lower' forms.
>
> p.197

It is impossible at such a point, not to exclaim about the extraordinary nature of our intellectual situation. And, as Marjorie Grene shows, hidden away in odd folds of the complex arguments are strange admissions - such as that she quotes from C.H. Waddington :

> the unprejudiced student is likely to derive the impression that the failure of the present theory to provide any plausible explanation for such occurences has played a not unimportant part in weighting the scales against their real existence. It would certainly seem that in this field... the adequacy of modern theory may be doubted...
>
> *The Strategy of the Genes*,
> Allen and Unwin, 1957, p.64

Macbeth uncovers some parallel admissions, as we have seen. Yet despite the admitted inadequacy, evolution is upheld in America by the Civil Rights movement as a proper subject for science teaching, against Creationism! Yet its authenticity depends often upon impossibilities : as M. Pitman argues, the odds against chance producing a specific L-form amino-acid chain of 100 units (i.e. one molecule of a simple protein) in five billion years (the supposed age of the earth) is 10^{71}.[2] Professors Fred Hoyle and C. Wickramsinghe have made parallel calculations, demonstrating the statistical impossibility of life coming into existence by chance. Even in Arkansaw, in the light of such doubts, the question should surely, at least, *be left open*?

As Marjorie Grene goes on to point out, other disciplines, like paleontology and embryology, appear to lend weight against selectionist Darwinism. One feature being discussed in paleontology is the apparent fact that there seems to be no gradual progression of adaptive relationships. In some areas a bewildering variety of new types appears - with a suddenness termed 'explosive' (this seems true of the major mamallian types).

If the bat's wing had evolved on the principles of Mendelian mutation and selective pressure, at the same rate at which it has altered since its origin, it would have had to begin developing before the origin of the earth (See George Gaylord Simpson, *Major Features of Evolution*, New York 1913, p.351).

New species which appear develop far beyond any systematic relation to adaptive needs. Others seem to have simply 'played themselves out' (See T.S. Westoll, 'Some aspects of growth in fossils', *Proc. Roy. Soc. B.* CXXXVII (1950), p.409-509).

From embryology Marjorie Grene selects views which suggest that perhaps important changes came when circumstances generated an 'inspired infantilism' : perhaps (say) the

---

[2] See the arguments of M. Pitman in an Appendix below. He discusses some of these calculations in *Adam and Evolution* 1984 p 148.

retention of a larval stage into adult life permitted the rejuvenation of the race. (A host of such arguments is quoted by Sir Gavin de Beer).

Why, despite such doubts and alternatives, is neo-Darwinisn so confidently affirmed? Because, says Professor Grene, it is not only a self-confirming scientific theory, but a theory deeply *embedded in a metaphysical faith* :

> in the faith that science can and must explain all the phenomena of nature in terms of the one hypothesis, and that a hypothesis of maximum simplicity, of maximum impersonality and objectivity.
>
> p.199

It is this philosophical assumption behind science, of course, which Polanyi has challenged. It has a symbolic, or mythological content :

> Nature is like a vast computing machine set up in binary digits : no mystery there... the machine is self-programmed : it began by chance, it continues automatically, its master plan itself creeping up on itself, so to speak, by means of its own automatism. Again, no mystery there : men seems at home in a simply rational world.
>
> p.200

It is the challenge to this (commonsense, Hobbesian-Humean, rational) philosophy that makes today's scientists want to burn books that question it. But its self-satisfaction masks the need to find more satisfactory explanations of the category of life.

Even if we could make 'life' in the laboratory, Marjorie Grene declares, it would be *we* that made it : and this would make an essential, logical, even a metaphysical difference to the import of the achievement (see A.M. Dalcq, 'Sur la notion de vie', in *Colloque Orient-Occident*, Brussels, 1958, p.81-89).

It must be emphasised that there is nothing un-scientific or anti-scientific in Marjorie Grene's critique of Darwinism. The danger is in Darwinism itself, which leads to the situation of which David Lack warned in his *Evolutionary Theory and Christian Belief* (1957) :

> the scientist... cannot accept the theory man's mind was evolved wholly by natural selection if this means, as it would appear to do, that the conclusions of the mind depend ultimately on their survival value and not their truth, thus making all scientific themes, including that of natural selction, untrustworthy.
>
> p.104

There is something wrong with a theory like Darwinism which, at its very root, invalidates itself.

What is needed are new concepts, such as those coming over from biology itself:

> concepts (such as) those of structure or *form*, and goal-directed process, or *end.* Biologists acknowledge, in the world of living things, many and diverse shapes. And they acknowledge also temporal patterns, life-histories ordered through time. In other words, however, impassioned their denial of 'teleology', they do acknowledge the existence in nature of ordered processes and therefore of ends.
>
> Grene, p.201

If we take into account such new concepts, then we 'may... overcome our Cartesian bias and... open the way for a richer metaphysical vision'. If biology proved able to achieve this, there would be a marvellous liberation, throughout the Humanities because the creative striving power of life would be seen, once again, as a central principle in the universe, a principle of which poetry, and indeed all products of the mind, are products and could not exist without.

Marjorie Grene develops her arguments against Darwinism further in *The Understanding of Nature*, a collection of essays developed at seminars at Boston University. Once more, she examines Darwinism in its place as part of the philosophy of science.

Darwin himself spoke of this kind of problem in Chapter IV of *The Origin of Species*:

> It has been said that I speak of natural selection as an active power or Deity; but who objects to an author speaking of the attraction of gravity as ruling the movements of the planets? Everyone knows what is meant and implied by such metaphorical expressions ; and they are almost necessary for brevity. So again it is difficult to avoid personifying the world nature ; but I mean by Nature, only the aggregate action and product of many natural laws, and by laws the sequence of

events as ascertained by us with a little familiarity such superficial objections will be forgotten.

(1888 edition, pp 81-89)

This is a characteristic Darwinian mode of argument - soon all objections will disappear. Yet the implied objections are not superficial at all, but primary philosophical ones. Darwin's is a form of determinism :

> Instead of leaving before us the bewildering and inexplicable diversity of biological phenomena, Darwin showed us how a set of natural laws, the same that hold good today and held good in the beginning, could account for the origin of the whole range of species populating the earth today. This is either a very general sort of determinism or nothing.
>
> *On the Nature of Natural Necessity* in *The Understanding of Nature.* Marjorie Grene, p.232

In Darwin's scheme there is a fundamental contingency : everything could have been different. But he substituted causes (heredity, slight variations, population increases, natural selection) for reasons (God's reasons) and so attempted to produce what in our way of talk is an 'accident', rather than something which came by purpose of foresight. Chance and necessity are destroyers of cosmic teleology.

If we accept this, we ought to see 'selection' as operating causally, 'favouring' one set of alternatives over another, metaphorically speaking only. Marjorie Grene declares that to call natural selection a 'truly creative' process, as Sir Ronald Fisher has done (*Creative Aspects of Natural Law*, 1950) is nonsense. Some biologists have called it simply 'a statement of certain empirical facts'. It isn't a mechanism, and it is absurd to personify it:

> natural selection explains - or... describes the necessary and sufficient conditions for systematic changes in relative gene frequency and that is all it does.
>
> Grene, ibid, p.277.

So it is important to examine the explanatory power of this concept about which there seems to be such confusion as to what it is, whether it can be seen or tested, and what kind of process selection is. Marjorie Grene examines three kinds of meanings in her essay: 'Explanation and Revolution' in *The Understanding of Nature*:

> it expresses the probability of the survival of certain genes in a population at some future time in terms of the ratio of the frequency of one or more genes to that of their alleles in a given population at the present time.
>
> p.217.

This, however, is tautological, and conveys no information about the world.

Meaning Two is a concept of efficient causality. No change, no stable or significant change, synthetic theorists hold, is *ever* produced by some 'drive' or 'tendency' of the organism itself, but only by 'environmental pressure'.

This is non-predictive, and also non-falsifiable, for anything that happens confirms it : whatever survives survives because of environmental change:

> A vast range of when-then processes governed by a single uniform set of laws mathematically expressible and fitting smoothly into our overall view of nature: what explanation could be better?
>
> *The Understanding of Nature*, p.218.

Yet it isn't really any kind of explanation at all.

Then a Third Meaning emerges : not only do the necessary conditions suffice to produce the result : the result, we are told, induces its conditions. The end point, survival, is said to be understood as the goal of what leads up to it.

Marjorie Grene asks, if this is logically possible and empirically correct? She answers: No, on both counts. Causal and teleological explanations, in answer to historical questions are reversed explanations of temporal succession. A *true* explanation explains how things are in the real world. Accounts which are both deterministic and teleological, in answer to how a system came to be, cannot both hold good at once.

Do Darwinian evolutionists really believe that the ends - the organism existent at the end - have *produced* the means to their survival? 'We can hardly hold that the speed of a future Derby winner induced his ancestors to run faster than the ones that didn't get away'.

The great glory of Darwin is supposed to be that he showed how all the marvellous adaptations of the natural world could have been produced without purpose. His alternative to teleology is natural *necessity.* Evolution is 'opportunistic'. But this is a metaphor and a meaningless one, and it leaves in the air the question of an end-directed process.

When the Darwinists invoke *survival* the question then arises : is this enough to satisfy? The alleged teleology in this explanation is as tautological as the formula that carries it : we are saying, what survives, survives. We cannot specify anything which natural selection *tends* to produce except the empty concept : whatever survives. It tends to produce whatever it does produce. Evolutionary theory here appears a kind of nonsense.

As Marjorie Grene says, it was not Paley's watch Darwin threw into the ocean, but only the Watchmaker. He retained what one may call the axiom of adaptivity, the view that everything about an organism must be a means.

This metaphysic of life, says Marjorie Grene, still haunts post-Darwinian thinking. But the principle of adaptivity, of all principles, cannot be self- sufficient. For means, or uses, must be uses to some one for some end: to the organism for the sake of its survival, or in evolutionary terms, to genes for the sake of their survival in some future gene pool. This end, however, is empty. The teleological perspective, supposed to be in this kind of evolutionary theory, may have been a jumping off point for explanation, but it enters not at all either into the description of the phenomena or into the explanation itself. It is simply that when we look back the goal seems, because of the temporal direction of our investigation, and because of our interest in the subject-matter, to be those organisms now existing. (Marjorie Grene is obviously here drawing on Polanyi's ideas of the involvement of the scientist in his knowing).

The subject matter of evolution, Marjorie Grene even suggests (*The Understanding of Nature*, p.226), does not lend itself to teleological explanation in this way. The core of the problem of biological explanation lies in finding a way to develop dual or plural explanations of form and causality, if-then and when-then factors, to deal with biological systems which are more massively organised than those of non-living nature, more strikingly improbable, and of infinitely greater interest to us as living things:

> It is organisation that is the overarching and fundamental concept for the study of living things, the framework for answering the question what does it do, what operating principles make it work.
>
> p.227

This is of course what Tomlin is saying, and what is represented by the scientists discussed in *Approaches to a Philosophical Biology.* It implies a need for a radical change of mode in the thinking of biologists, and a profound change in their education.

In the new approach, Marjorie Grene suggests, the summum bonum, like the Maker, is dispensed with: yet the means-end relation, the notion of 'this as useful for that', is fundamental. The insistence on the equation with adaptation defines the limits of Darwinism, and it is doubt of the all-inclusiveness of adaptation as a concept definitive of life that impels the most successful objections to Darwinism, as we shall see.

We can see Darwin struggling with this kind of problem in *The Origin of Species.* Analysis of the problems which arise often involve debates of the which-came-first, the-chicken-or-the-egg kind. For instance, how did the breast or mammary glands develop? They must have developed at an extremely remote period, says Darwin (*The Origin of Species*, p.189). A Mr. Mivart asked,

> 'Is it conceivable that the young of any animal was ever saved from destruction by accidentally sucking a drop of scarcely nutritious fluid from an accidentally hypertrophied cutaneous gland of the mother? And even if one was so, what chance was there of the perpetuation of such a variation'?

The case here is not put fairly, says Darwin, declaring that the mammary glands were originally developed within the marsupial sack.[3] But this is to accept that most mammals were derived from a marsupial form, a theory surely that is now discounted? And are we to believe that the breast developed by accident too?

---

[3] In the platypus its mammae differ from those of marsupials and placentals in that they consist of branched tubes without nipples: milk extrudes from the fur-ends. It has not been placed in any evolutionary sequence.

How about sucking? Darwin cannot explain this except by saying:

> the most probable solution seems to be, that the habit was at first acquired by practice at a more advanced age, and *afterwards* transmitted to the offspring at an earlier age.
>
> p.190, my italics

But how 'transmitted'? Even in the womb the foetus, if touched on the cheek, turns as if to suck: sucking is instinctual. But how did it come about that certain modifications in the larynx are developed, so that the infant is not choked? How is it that the whale's udder is so perfectly adapted to the cub's mouth? In general terms (as Mr. Mivart asked) the questions concern:

> the *first rudimentary beginnings* of such structures, and how could such incipient buddings have ever preserved the life of a single (creature) (Mr. Mivart is in this instance discussing sea-urchins)
>
> p.191 (Mr: Mivart's italics).

There are many complex organs and functions which would bestow no advantage in rudimentary form, in form not yet complete (like breasts before the larynx was elongated to prevent choking) ; half-blind; flying not yet airworthy; camouflage only half-way to looking like a leaf. The species in such a state of small half-step forward would seem more likely to be totally wiped out, *because* it was 'evolving'.

Towards the end of his book, Darwin admits, 'I am convinced that natural selection has been the main but not the exclusive means of modification'. He also admits that 'science as yet throws no light on the far higher problem of the essence or origin of life'. But Darwin's emphasis in his conclusions is on *evolution*, that is, the *development of life* by modification. But to him, 'The elaborately constructed forms, so different from each other... have all been *produced by laws acting around* us' :

> These laws, taken in the largest sense, being growth with reproduction ; inheritance which is almost implied by reproduction ; variability from the indirect and direct action of the conditions of life, and from use and disuse : a Ratio of Increase so high as to lead to a struggle for life, and as a consequence to Natural Selection, entailing Divergence of Character and the Extinction of less-improved forms.
>
> p.429

'Thus' Darwin goes on, 'from the war of nature, from famine and death, the most exalted object which we are capable of conceiving, the production of the higher animals, directly follows'. But does it? Darwin claims a 'grandeur in this view of life' : but it is still too negative. Life is continually those who survive the battlefield, survive the 'extinction of less-improved forms'. There is nothing said about why those who survive should grow towards becoming *higher* animals - the word surely indicating a progress or gradient?

6 An important heretic: Pierre-Paul Grassé

In France and Germany there is less certainty about that Darwinism which is a religion for scientists in Britain and America. Here, an important heretic is Pierre-Paul Grassé, the French biologist from the Laboratoire des Etres Organisés in Paris. In Britain we are beginning to grasp at ways by which scientists and philosophers come at the problem of the deficiencies in Darwinism and its 'truncated teleology'. Polanyi, as we shall see, argues that life transcends physics and chemistry and cannot be simply the product of these. Norman Macbeth shows how the dogma rests on insupportable bad logic - how the whole fabric is full of holes. Some scientists, admitting that the theory is full of holes, desperately try to offer some other dimension which can perhaps account for the unaccountable : Sir Alister Hardy for example, in *The Living Stream*, goes in for telepathy. While we struggle against the tide in Britain, with its powerful inheritance of reductionist, materialist, positivistic approaches to such problems, it is thus refreshing to read someone like Pierre-Paul Grassé. Reviewing his book, *L'Evolution des Etres Vivants*, T. Dobzhansky, admitting Grassé's status,[1] disagreed with such a 'full frontal attack on all kinds of Darwinism', whose purpose was to 'destroy the myth of evolution as a simple, understood and explained phenomenon'. This defence, again, is characteristic of a religion rather than an 'open' science.

Grassé argues, convincingly I believe, that the dogma, that life evolved by small chance mutations, selected by the processes of 'the survival of the fittest', is no explanation at all of the evolution of life :

> Mutations do not explain either the nature or the temporal ordering of evolutionary facts ; they do not account for innovations ; the precise arrangement of organs are beyond their capacity.

Indeed, mutation (the very basis of Darwinian Theory) 'is an accident or disease having only a remote bearing on the evolutionary process' (p.223).

And when it comes to certain complex life-forms, 'that several *adequate* mutations should occur in the same line is an amazing, almost unbelievable, stroke of fortune' (p.207). Grassé shows by a painstaking argument not only that 'the principles laid down by the founder of evolutionary theory do not entirely account for evolution' (p.210) but that the things believed by scientists are simply impossible to believe. It may be over this question 'biology is reduced to helplessness and must hand over to metaphysics'.

Grassé recognises that the science of life is a meeting point for those concerned with philosophical issues of a metaphysical kind, and naturalist scientists. He intends, he says, to try to present in his book 'only those propositions that have been well established by scientific study or directly observed' in his own experiments. He had been working in zoology and biology for half a century.

---

[1] Grassé edited 34 volumes of the *Traité de Zoologie.*

To understand evolutionary phenomena, he says, requires a thorough knowledge of zoology, paleontology, cytology, genetics, biochemistry, and even mathematics. Yet who can hope to have all this scholarship? One necessity, however, is a feeling for living animals : and with this, 'Who, having seen a seal or walrus swimming, would dare to claim that these animals survived merely because chance led them to adapt to an acquatic habitat'?

It is clearly from a deep feeling for his subject the Pierre-Paul Grassé comes forward to declare that he finds orthodox Darwinism to be a doctrine that 'falls far short of universal explanation'.

Grassé begins from the observation that while both living and non-living things are made of matter which obeys physico-chemical laws, the non-living creatures defy them, without violating them. 'Necessity does not imperatively impose its laws on the living world'. Birds fly, thereby defying gravity : creatures live in the extremes of hot and cold. Creatures display 'inventions' to cope with extraordinary conditions - a word Grassé says he realises will cause trouble in biology (though he uses the word in a factual way, without any metaphysical overtones). Life's constant victory is to oppose the laws of entropy.

Living things have laws which belong to life : oaks remain oaks and breed oaks, for instance. One other important law is the law of evolution, by which, in their 'line' of development, creatures change in the cycle.

Now we come to the problem which biologists and philosophers cannot solve. In the investigation of DNA biologists have found out about the passing of what they call 'information'. Grassé calls this *intelligence.* The molecular biologists mistakenly regard this as 'programmed' as in a computer : but it is rather 'condensed' on a molecular scale in the chromosomal DNA or in that of an organelle in each cell. But it is far more than this. It is a principle of force which is the *sine qua non* of life. 'If absent, no living being is imaginable'. Where does it come from? How does it reside where it does reside? The DNA discoveries do not answer this question : they only show what happens as this intelligence operates, not what it is or how it operates. It cannot be a product of the 'coding' in DNA structures.

Science cannot yet solve the problem of this 'intelligence' in living things : yet it is clearly a fact and a problem. 'The powers of invention in the living world are immense'. There is a huge difference in living things from non-living matter - and this resides in their capacity to invent, and seemingly to move in construction *towards a hidden goal.* 'Creatures are dependent on physico-chemical laws, but they also display certain rules which are specific to them alone'. The laws complement one another, and if discord occurs between them, the living being dies.

Evolution, Grassé agrees, is in one sense a fact not a hypothesis, and the evidence for it as a fact comes from paleontology, the fossil record. Paleontology is the true science of evolution. But what we are concerned with are the hypotheses of how evolution took place and what kind of process it was and is. Here, paleontology itself makes many such hypotheses seem questionable or illegitimate.

The evolutionary mechanism, it is implied by many, is well known in detail with a high degree of certainty. But this is not so. Darwin was much less certain. In a letter to one of his grandsons he wrote,

> But I believe in natural selection, not because I can prove in any single case that it has changed one species into another, but because it groups and explains well, it seems to me, a lot of facts in classification, embryology, morphology, rudimentary organs, geological succession and distribution.
>
> (Quoted in A. Vernet, *L'Evolution du Monde Vivant*, 1950)

Today many are misled by Darwinism, which often offers fallacious statements. Grassé quotes remarks by one Lévine which suggests that because bacteria multiply rapidly, they evolve rapidly, because of the number of mutations. But this is not true: it is to confuse *mutations* (which are DNA copying errors) and *evolution*, and to take it for granted that mutation is the basis of evolution, when this is not proven. In fact, bacteria have not transgressed the structural frame within which they have oscillated for billions of years. While bacteria are valuable for biological study, they are of little evolutionary interest, because they have remained so simple and stable for so long. It is often assumed the evolution is an on-going, universal, ruthless process - but bacteria and sharks show otherwise.

In this kind of fallacious hypothesis, presented as fact when it is really nonsense, Grassé detects a *pseudo-science*. Many biologists believe that fundamental concepts have been demonstrated as accurate, when this is really far from the case. They say things like, 'Darwinism is now so well established ...' without being critical enough : they tend to search for evidence that will be in agreement with their theories, and orient their research in this direction. This deprives observation and experiment of their objectivity, makes them biassed, and creates false problems.

Grassé suggests that it is time to sweep away all *a priori* ideas and dogmas and adhere to facts. We have seen that it is a principle of science that hypotheses must be held with a light hand, and rejected as soon as they appear unreconcilable with the facts. Facts, declares Grassé, must come first and theories must follow. Biologists must observe facts coldly, not blinded by theories :

> Today, our duty is to destroy the myth of evolution, considered as a simple, understood, and explained phenomenon which keeps rapidly unfolding before us.
>
> p.8.

Interpretations and extrapolations put forward nowadays by theoreticians as etablished truths must be shown as weak. The deceit is sometimes unconscious ; but some scientists are so sectarian that they purposely overlook reality and refuse to acknowledge the inadequacies and falsities of their beliefs. These are strong charges to be made against science, which rests confidently on its probity : but the same charges are made by Macbeth, Marjorie Grene and Polanyi.

In his first chapter Grassé makes it clear that the orderly development of living things in nature is anything but a phenomenon displaying chance (like, say, the movement of corpuscles under the influence of Brownian motion). *Evolution would not exist unless there were strict rules.* What (say) the paleontologist observes is an orderly development which seems to have a *meaning* : scientists are interested in evolutionary phenomena because of the order displayed in them:

> Paleontology exists solely because chaos is banned from the living world.
>
> p.10.

However, if we talk about 'progressive' evolution, we introduce an inappropriate moral judgement which is meaningless. As Grassé shows, Darwinism is highly moral in believing that natural selection exerts a 'pressure' which enables the 'fittest' to survive, which is 'progressive', by comparison with what happens to the 'unchanged'. This way of thinking assigns a goal to evolution, *to proceed towards the best.*

What the fossil record does show is that living beings display a dual trend towards *complexity* and *diversity*, both of which are complementary :

> A simple comparison between the uniformity of the flora of the first million years (consisting solely of bacteria and blue-green algaes), and the superabundance of the thallophytes of the Tertiary era is proof of the diversified rising of the species... biological evolution never ceased proceeding from the simple to the complex.
>
> p.11.

The original forms of simple life were formed by sedimentation 3.2 billion years ago. There are 'trails' of an undefined structure. They could be small bacteria, such as *Eobacterium isolatum*, a minute rod less than 0.7 mμ long, and with a diameter of 0.2 mμ. It is suggested that they utilized amino acids or even the proteins present in the sea. Some tiny spheres (*Archaeosphaera Barbertonensis*) might be schizophytes containing chlorophyll, the agent of photosynthesis, and this could explain how a breathable atmosphere started to take form about the earth. But here we have one of the first big questions (as posed by Marjorie Grene) - how did that great step forward, photosynthesis, come about?

We are talking about what happened between 3,200,000,000 years ago until 1,000,000,000 years ago :

> What were the evolutionary steps from the schyzophyte structure (those of bacteria and Cyanophyceae)[2] to the true cell structure (i.e. a nucleus enclosed in a membrane and containing chromosomes with a well-defined structure, in which are enclosed DNA molecules combined, in some way, with filamentous proteins)? *We*

---

[2] Chlorophyll first appeared on earth with Cyanophyceae.

have no idea. Paleontology does not reveal anything on this matter...

Here, and over the origins of metazoa and protozoa, we have only tentative and extremely hypothetical theories, and no fossil evidence. All we can say is that the evolutionary trend was towards complexity and expansion: there is a 'gradient'.

In the next few pages Grassé draws on his considerable knowledge of the paleontology of primitive life-forms. There are many problems, and it is impossible to draw an evolutionary tree without many question-marks. From evidence of the internal structure of these organisms, it is highly speculative to say they evolved from one another:

> No one knows... whether the Cnidaria and Ctenaria (both diploblastic organisms) are derived from the sponges, or whether they all stem from a common ancestor.
>
> p.15.

The fossil evidence is limited and often absent. But what is certain is the development towards complexity - and autonomy:

> There is no doubt that, as animals reached the triploblaştic stage, (i.e. the acquisition of the third cell layer, the mesoderm) they made a major step forward towards increased complexity.
>
> p.15.

They exhibit a marked body regionalization, including a 'cephalization' that aggregrates most sensory and nervous centres upon and in the anterior region of the body: external 'information' is now centralized in the brain. From the third cell layer originate muscles for rapid contraction, and invertebrates, the internal body skeleton.

How these primitive organisms developed is difficult to discover, because the 'soft' parts of their bodies did not often survive in fossil form: so there are many arguments between scientists here. What can be said is that the order of appearance of vertebrates is: agnathans, fishes, amphibians, reptiles, mammals. Fishes form a superclass consisting of several major evolutionary 'lines', and both mammals and birds derived from reptiles, but their respective ancestors belong to orders which are distantly related.

> This phylum is characterised by continuously increasing overall structural complexity as well as a steady increase in 'psychism'. This fact is supported by fossils.
>
> p.17.

Evolution, thus, may be fact : but there are many 'missing links'. And the problem is that we 'hardly know anything about the major types of organisation' and as only *suggestions* can be made:

> How can one confidently *assert* that one mechanism rather than another was at the origin of the creation of the plans of organization, if one relies entirely upon imagination to find a solution?

Our ignorance is great : 'a shadow is cast over the genesis of the fundamental structural plans and we are unable to eliminate it' (p.17). This is why it is so *unscientific* for anyone to tell us that 'the origin of life was an accident in coding'. *We simply do not know that.*

One aspect of evolution we do observe, however, is the increase of the psychic faculties. *All* mobile living beings exhibit a behaviour which is adapted to their needs and ensures their survival. During the course of evolution psychism continuously expanded and developed as a result of the elaboration and specialization of the nervous system. The most primitive organisms, like amoeba, do react to light, heat and chemical substances : they even 'choose' and reject unsuitable food.[3] Ciliate infusoria show more elaborate behaviour, acquiring conditioned reflexes. Later triploblastic organisms develop until in arthropods and vertebrates behaviour is most complex. In the more primitive organisms behaviour is largely determined by genetic mechanisms and commands. In higher organisms and especially in man we find greater plasticity - and so greater freedom.

This progress, however, has not been a linear and direct path. There have been many tortuous processes, failures, dead ends : numerous lines have been aborted or have not proved

---

[3] Grassé does not mention bacteria, but their behaviour is complex. 'There is ... good evidence that the movement of many bacteria is *directed* by tactic response of the organisms towards locations favourable to growth and away from unfavourable regions ...' Cruikshank, *Medical Microbiology*, 1968 p.24. Yet this creature is so elementary in structure, bound by a cytoplastic membrane, closely covered by a cell wall, and capsule, with flagella and fimbriae.

themselves capable of overcoming the passage of time. Some creatures would seem to have regressed. But it is difficult to talk in this way : for instance, some reptiles have become legless (apodal) snakes and have survived, while other tetrapodal reptiles, once powerful, have disappeared. Evolution of the snake seems 'suppressive' but snakes have become extremely vigorous and equipped with powerful weapons in their poisonous venoms - extraordinary chemical innovations.

How could the snake lose its legs by subtractive mutations? To crawl as they do snakes need a elongated body (with the internal organs appropriately elongated and distributed). The body would have had to become elongated before the legs disappeared, for success. How could it happen, that the legs gradually disappeared, having become no longer capable of sustaining the trunk high enough above the ground? It is when intermediate stages are considered that evolutionary theory becomes doubtful.

The movement of snakes requires a *specific* arrangement of skeltal, muscular and nervous apparatus, repeating itself along the vertebral axis. Movement by repitition, undulating locomotion, is only possible when all things work together - the muscles contracting in rhythmic sequence along the body. The body surfaces are highly reduced and cannot anchor the muscles : the great development was of intersegmentary aponeuroses on which the muscles are attached.

The arrangement of muscles and nerves is specific and suited to the mode of locomotion adapted by the animal. Snakes are not monsters or failures resulting from a series of random events. The relationship between the vertebrae and the rest is so exact that it cannot possibly have been the product even of an infinity of mutations, to enable one to produce the specific mutation that was 'tailored' to the existing structure and enabled evolution to advance.

Moreover, the loss of legs and advance to the creeping manner of motion occured as an evolutionary trend in three groups (Scincidae, Cardylidae and Anguidae).

And now the snakes are stabilized. We do not find any in the throes of changes brough about by mutations. Since the Miocene period the present saurians have undergone only very slight changes. Today snakes suffer mutations : but they retain their structural plan and way of life.

If possibilities are examined, such as a creature which could be half-plant, half-animal, it becomes clear that such an 'invention' would not 'work' because mobility would be sacrificed. To be mobile is more 'fruitful' than having a photosynthesis plant in one's head.

All of these aspects of evolution pose problems which the accepted dogma cannot answer. Moreover, such theories as we have are extremely doubtful : discussing the whole range of phyla, Grassé declares :

> From the almost absence of fossil evidence relative to the origin of the phyla, it follows that any explanation of the mechanism in the creative evolution of the fundamental structural plans is heavily burdened with hypotheses. This should appear as an epigraph to every book on evolution. The lack of direct evidence leads to the formation of pure conjectures as to the genesis of the phyla : *we do not even have a basis to determine the extent to which these opinions are correct.*
>
> p.31.

- yet it is on this lack of evidence, and on opinions about the origin of organised creatures of which there is no evidence to support, that evolutionary dogma has hardened.

To show the discrepancy between reality and theory Grassé delineates our present knowledge about the development of reptiles into mammals. The shaping of the mammalian form, which lasted from 50 to 60 million years, occurred, says Grassé, in a smooth and gradual manner. Although we do not know why, homeothermy may have been the preliminary condition to the 'mammalization' of theriodonts - that is, the development of a regulated warm-blooded state which enabled higher general metabolism, quicker and continous growth and greater activity. Here is surely one of the great leaps forward in evolution?

Other developments discussed by Grassé are the changes in teeth : these became capable in later species of cutting and grinding, while in others the jaw acquires a greater freedom which allows movements from front to back as well as traversely. The limbs of the reptiles gave way to long and slender limbs which permitted running.

In one creature there are two articulations of the jaw with the skull, a reptilian one and a mammal one.

> The coexistence of two characters, one (the new one) 'meant' to replace the other (the archaic one) is a rare enough phenomenon to merit special attention. *It is important to note that the mammalian articulation is entirely independent from the reptilian articulation.*
>
> p.42.
> (my italics)

Grassé takes this as evidence that the evolutionary process has nothing to do with *mutation* (for if there had been mutation the two forms could not co-exist).

In another creature, the tritylodants, the development of the mandible, the muscles, and the nerves, 'has been closely coordinated'.

> The transformation of the reptilian mandible into the mammalian mandible could only occur thanks to a triple co-ordination simultaneously involving bones, muscles and nerves.
>
> p.40.

The existence of this co-ordination, Grassé suggests, means that such a development, 'which is what we call evolution', could not have come about 'by a mosaic of random variations affecting just anything at any time'.

Summarising a wealth of knowledge about the evolution of theriodont reptiles Grassé asserts that the facts show that 'random mutations and subsequent selection played no part in their evolution'. Those evolutionists who argue in favour of Darwinian theory here 'only take into account the facts that fit in with the theory, strictly excluding those which indicate the weakness' (50).

In fact, observed objectively, the variations of theriodonts and early mammals did *not* occur at random : they kept accumulating and adjusting all the time and lack any pathological character. Mammalization does not either depend on selection :

> On the contrary, the diversity of sub-types... the large distances between populations, and the variety of the climates to which they are submitted do not support the theory of selection.
>
> p.50.

Criticising Simpson's account of the history of the Equidae, Grassé points out how he is with the Darwinians in denying the existence of *oriented* evolution - yet he depicts an evolutionary process that is clearly progressing (and calls it 'orthogenesis'). The fact that certain developments occured in both sides of the Atlantic rules out chance. And Grassé lists a number of progressive developments in all hippomorph lines.

Simpson does not pay much attention to the structure of the hoof : yet Grassé points out its very innovative an precise evolution. It can take an impact of over a ton.

> It could not have been formed by mere chance... it is a structure of coaptations and of organic novelties...

The hoof provides solutions to complex problems of rapid locomotion on monodactyl limbs :

> The respective lengths of the bones, their mode of articulation, the curves and shapes of the articular surfaces, the structure of bones (orientation, arrangement of the bony layers), the presence of ligaments, tendons sliding within sheaths, buffer cushions, navicular bone, synovial membranes with their serous lubricating liquid, all imply a continuity in the construction which random events necessarily chaotic and incomplete, could not have produced and maintained. This decription does not go into the detail of the ultrastructure where the adaptations are even more remarkable.
>
> p.51.

The development of the skull is also very complex, and in true mammals the rise of 'cerebralization' has been a major evolutionary theme.

Concluding, Grassé cites some pathological distortions in function caused by mutations. These bear no relationship at all to the slow and coherent attainment of new form, or of new functions, in evolution. A single gene can produce multiple effects. But what geneticist would believe that such transformation of the jaws, or other complex developments in evolution, could have been caused by a single mutated gene?

The application of genetics to the study of fossils is difficult if not impossible. No breedings are possible and the interpretation of the data must be hypothetical. Discussion of 'selection pressure' cannot be informed by any data, about population densities, ecology, weather, or environment. Evolutionary changes occured everywhere, despite various conditions. How could this be so in the face of the neo-Darwinian principle, 'each environment has its own privileged genotype which, by chance, is better preactivated to it'?

In his next chapter Grassé attacks Darwinian ideas in other ways. For instance, since 320,000,000 years ago bacteria, which reproduce at the rate of 17,520 generations a year, have produced billions of billions of mutations. During the same period hominids have produced only a few million mutations. Yet during this period the hominids have evolved new brain structures, have become bipedal and erect in stature. The bacteria, by contrast, have in no way changed their structural plan. Thus, mutations do not necessarily lead to an evolution, whatever their numbers.

Evolution has not been a continuous process. There have been changes of immense importance (e.g. the synthesis of chlorophyll, the change from schizophyte to cell). And the greatest achievements came a very long time ago (e.g. the genesis of the fundamental structural plans). *There have been four creations* : fewer than twenty phyla and eighty classes for the animal kingdom, and *they are all very ancient.*

> The last major group to date, the vertebrates, made its appearance with the Agnatha... during the Ordovician (period), and with jawed fishes during the Devonian, some 450 million years ago.
>
> p.60.

Evolution does not occur in just any fashion, since the number of its possibilities is limited. The organisms in which the new plans were embodied have partly lost their evolutionary capacities. The creative powers of evolution have gradually decreased and since the Eocene period the formation of orders has been interrupted, Eutherian mammals and birds being the last to appear. Insects were the last group to evolve : but its creative stage is now over : from the last 30-70 million years it has not gained any new orders.

Some creatures show 'evolutionary stagnation' - e.g. gastropods and bivalve molluscs. Sharks and actionopterygian fishes are thriving still. The last major evolutionary explosion occurred after the disappearance of the dinosaurs, between the Mesogoic and Oaozoic eras. It was in this period that the marsupials of Australia underwent their strange evolution, occupying that ecological niche.

In all primates evolution has been supended for a long time. But doesn't the genesis of the human brain (100,000 years ago) contradict the theory that evolution has lost its creative potential?

Grassé sees 'humanization' rather as the evolution of a branching secondary line, whose most remarkable characteristics are the development of the brain and the head. Our evolution is unusual, it doesn't equal the magnitude of the genesis of a phylum or class which creates or modifies plan. It is a counterspecialization because of the brain. This frees us, by the gift of consciousness, and thought, from automatic behaviour, into autonomy. But homo sapiens has remained physically stable for 100,000 years.

Grassé sees three stages in evolution : the period of *youth*, a very slow pioneering change : *maturity* in which there is great activity and branching : and *senescence* or relative stability:

> evolution, after its last enormous effort to form the mammalian orders and man, seems to be out of breath and drowsing off.
>
> p.71.

Evolution is the result of a series of irreversible historical phenomena. It does not repeat itself in identical sequences, nor does it reverse itself.

No new broad organisational plan has appeared for several hundred years, and for an equally long time numerous species, animal as well as plant, have ceased evolving.

> A critical analysis of zoological groups, including fossil and recent species, reveals that most of them 'froze' in their present state a very long time ago, and that the less ancient ones are going through a lull. In the last 10 to 400 million years, all of them have exhibited only slight variations.
>
> p.84.

In this sphere, biologists maintain they can not only observe evolution in the present, but describe it in action. However, the facts they describe are either nothing to do with evolution or are insignificant.

Here I must leave Grassé, and leave the reader to read his book himself. But for my purposes here, this question of evolutionary theory and the dogmas based on it are one example of the limitations of scientific thinking. For now the question may be asked - why is there, among British and American scientists, *such an aggressive metaphysical impulse over such questions despite the evident uncertainty*? Of course, in some one can see that their work consists of breaking livings things down, stripping them, putting them through filters and centrifuges, almost in a kind of play, so life itself comes to seem like a mechanism which you explain by dismantling. And if you can strip it, you ought to be able to put it together - life is surely only a structure of molecules which is 'programmed' - so it ought to be possible to programme such a system oneself? In this there grows a certain hubris: there is no mystery. But then perhaps, like Francis Crick, you find a mystery : a cell in a cell wall 'knows' where it is in the wall - whatever does that mean?[4] Or that DNA is growing more intelligent? Does that kind of observation chasten the molecular biologist who denies the validity of the phrase 'life strives'? Does he observe that there is a gradient, from the lowliest organisms to the highest, so that living matter seems to wait to rise towards greater complexity and autonomy? And towards consciousness?

If this were so, in what area of language, form or structure could this tendency exist? Where do intelligence, innovation and the co-ordinations of developments inherit? Where is ingenuity to be found?

Actually, of course, despite their aggressive materialism, such scientists have really made the problems more perplexing. Their dogma, we may declare, is actually a defence against the mysteries which perplex them: it cannot rest securely on any conclusions from their scientific work. In truth, they have no greater authority, when it comes to metaphysics, or a general philosophy of life, than we have. Moreover, in the area under discussion, one of the leading biologists of the world at least rejects the whole edifice of evolutionary theory.[5]

It may be useful, in relation to Grassé's views, to note that in the *The Material Basis of Evolution* (p.156) Richard B. Goldschmidt gives the following list of major features of life which Darwinian theory does not explain by its 'step-by-step' concepts : these features could not have come into being on the basis of small chance mutations giving an immediate advantage :

Hair in mammals
Feathers in birds
The segmentation of anthropods and vertebrates
The transformation of the gill arches in phylogeny
Teeth
The shells of molluscs
Ectoskeletons
Compound eyes
Blood circulation
The alternation of generations
Statocysts
The ambulacral system of echinoderms
The pedicellaria of echinoderms
Cnidocysts
The poison apparatus of snakes
Whalebone
Primary chemical difference like haemaglobin v. haemocyanin.

4 See Crick in *The Encylopaedia of Ignorance*, ed. Duncan R. and Weston-Smith, M., Pergamon, 1977.

5 See Norman Macbeth's chapter on the 'hopeful monster' question, chapter 17, especially p.156.

7 The work of Michael Polanyi

Behind the issues we have been discussing is a wider question : hasn't science revealed the 'mechanisms' of life, so what more is there to be said?

Michael Polanyi, himself a distinguished scientist and philosopher of science, rejects this assumption. 'Mechanisms' he declared 'are *not* wholly explicable as the resultants of the operation of chemical and physical laws'.

One important chapter in Polanyi's work here is that on 'order' in the book *Meaning*. There Polanyi summarises the recent discoveries about DNA. The pattern of initiating and developing living beings has been shown to be amazingly uniform : the chemical composition of DNA is astonishingly simple, and its sustenance and renewal of all living being is so comprehensive that it does invite the supposition that DNA theory has explained the main feature of life in chemical terms.

The most common attitude to DNA theory and to findings of molecular biology is based on a belief that the discoveries of Crick and Watson show that life can be fully explained by physics and chemistry. Crick and Watson showed that the hereditary features transmitted to an offspring are prescribed by the arrangement of four alternative chemical radicals aligned in a chain.

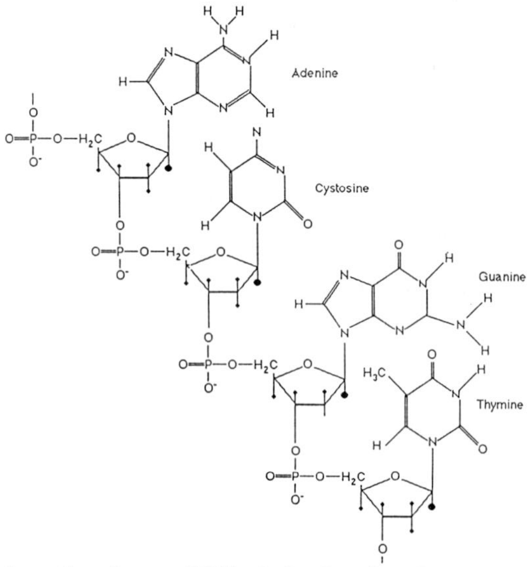

Adenine

Cystosine

Guanine

Thymine

A section from a DNA chain showing the sequence ACGT

One important aspect is that DNA remains unchanged, even though it acts as the changer : it is both continuity and transformation. It uses as its agent a similar molecule, almost the same in structure as DNA, but containing an extra OH Group of each sugar, while the methyl group of thymine is replaced by hydrogen to give urscil.

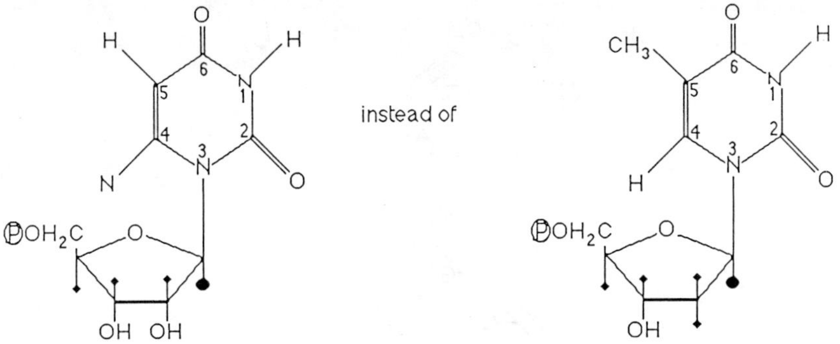

instead of

Here we come to the double helix. If the sequences of fibronucleotides (RNA) or deoxynucleotides (DNA) in the two chains are complementary, there are rules called the Watson-Crick pairing rules which disallow uncomplementary pairing. In these Watson-Crick rules, A must be lined up with T and G with C: the rules A = T and G = C govern all DNA replication.[1]

The genetic material, DNA, is always stored in cells in the form of a double helix. Then it replicates the two strands behave independently and each strand directs the synthesis of a complimentary strand as shown in the illustration. In this way two double helixes are formed: each identical with the original double helix. The details of the process are not yet understood.

DNA does not act directly in protein synthesis. Instead, one of its strands functions as a 'template' according to the Watson-Crick pairing rules - or, rather, that is how it is usually put. But leaving aside the problem of the 'template' analogy, we can see that this leads to the synthesis of strands of DNA complementary to the active DNA strand. These newly synthesized DNA molecules are known as 'messenger RNA', since they act as intermediaries carrying the genetic information stored in the DNA sequence to the protein- synthesizing apparatus. I am describing what is called 'translation'.

The whole process depends upon 'base-pair equivalence'. The enzyme, we are told, will accept a new base only in certain configurations : if mispairing occurs the enzyme 'recognises' that the orientation is wrong, and 'rejects' the incorrect base. The cell has a second line of defence : if a wrong base slips through, it is usually cut out again.

There are strange aspects of these large life molecules. A double helix in which the strands were joined by stable chemical bonds could not be separated quickly enough to fulfil a genetic function. On the other hand, the forces *between* molecules are usually so weak and non-directional that they would be unable to hold a double-helix together in a well-defined structure. DNA is held together by special inter-molecular bonds of intermediate strength, known as hydrogen bonds.

[1] Though again there do, however seem to be 'misreadings' organised by mitochondria : see *Nature*, 8 November 1979.

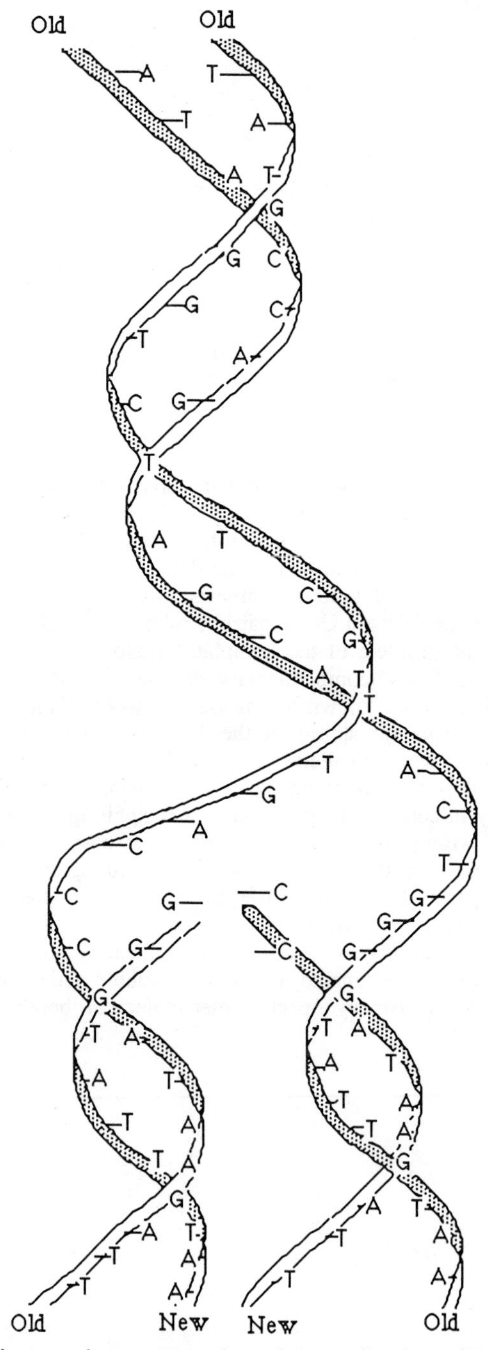

A diagramatic representation of the replication of DNA

What does this mean?

Each nucleotide in a base-pair forms two or three such special bonds, in such a way that the two members of the pair are held in exactly the correct relative configuration. The base-pairs have a remarkable property - they are geometrically equivalent. DNA is made up of four different kind of molecules, deoynucleotides, called Deoxydynodine, Deoxythymidine, Deoxyquanosine and Deoxycytidine. Whether we are dealing with adenine, thymine, quanine, or cytosine, A T G or C, these hydrogen bonds are in the same structural plane, having the same distance part and the same orientation whether we are dealing with an AT, TZ, GC or CH base pair. And, as we have seen, the Watson-Crick rules are that only these pairs are acceptable. The hydrogen bonds hold the base-pairs in their places, attached to the backbone of double-helix.

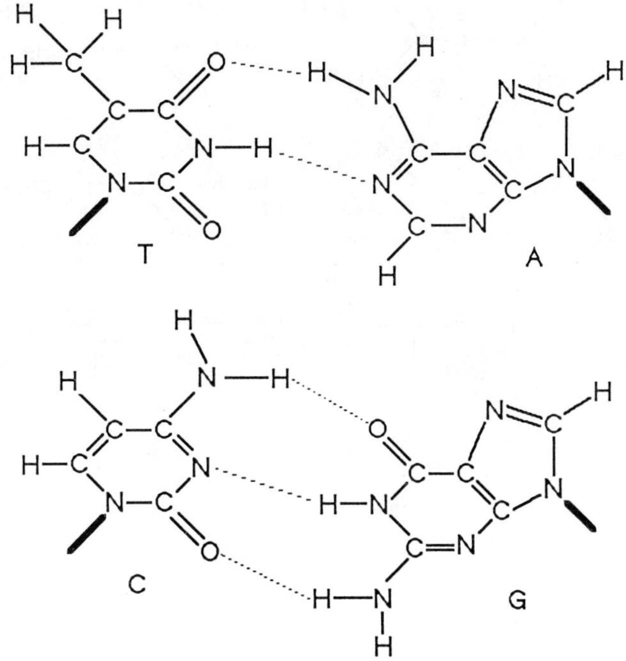

The A-T and G-C base pairs. Note the equivalent positions of the bonds joining the bases to the backbone (thick lines).

The enzymes are the catalysts : they speed up chemical reactions without getting changed in the process. They themselves are composed of amino acids of which (as I have said) there are twenty which occur naturally, polymers made of monomers. Illustrated are the twenty amino acids.

Of course, although the organic chemist uses these diagrams as a shorthand, he continually bears in mind that the letters and lines in these diagrams refer to *certain aspects only* of highly complex forms of matter, about which we know very little.

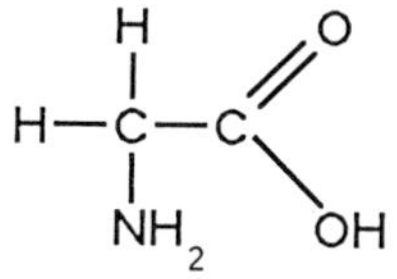

is not *all* we know about Glycine (Gly): : it is merely a shorthand way of (a) naming a substance so that we know which substance we are talking about (b) indicating what we know of its molecular structure, thus (c) showing its relationship to other substances. We may compare it with (say) Alamine (Ala) to show that the only difference is that a CH_3, configuration is substituted at

Glycine (Gly) $H-CH(NH_2)-C(=O)OH$

Aspartic acid (Asp) $HO(O=)C-CH_2-CH(NH_2)-C(=O)OH$

Alanine (Ala) $CH_3-CH(NH_2)-C(=O)OH$

Asparagine (Asn) $NH_2(O=)C-CH_2-CH(NH_2)-C(=O)OH$

Valine (Val) $(CH_3)_2CH-CH(NH_2)-C(=O)OH$

Glutamic acid (Glu) $HO(O=)C-CH_2-CH_2-CH(NH_2)-C(=O)OH$

Leucine (Leu) $(CH_3)_2CH-CH_2-CH(NH_2)-C(=O)OH$

Glutamine (Gln) $NH_2(O=)C-CH_2-CH_2-CH(NH_2)-C(=O)OH$

Isoleucine (Ilu) $CH_3-CH_2-CH(CH_3)-CH(NH_2)-C(=O)OH$

Phenylanaline (Phe) $C_6H_5-CH_2-CH(NH_2)-C(=O)OH$

Serine (Ser) $HO-CH_2-CH(NH_2)-C(=O)OH$

Tyrosine (Tyr) $HO-C_6H_4-CH_2-CH(NH_2)-C(=O)OH$

Tryptophan (Try) indole $C{=}CH$, $N-H$ $-CH_2-CH(NH_2)-C(=O)OH$

Threonine (Thr) $CH_3-CH(OH)-CH(NH_2)-C(=O)OH$

Cysteine (Cys) $HS-CH_2-CH(NH_2)-C(=O)OH$

Proline (Pro) CH_2-CH_2, CH_2, $CH-C(=O)OH$, $N-H$ (ring)

Methonine (Met) $CH_3-S-CH_2-CH_2-CH(NH_2)-C(=O)OH$

Histidine (His) $HC{=}C(-NH-C{=}N-)-CH_2-CH(NH_2)-C(=O)OH$

Lysine (Lys) $NH_2-CH_2-CH_2-CH_2-CH_2-CH(NH_2)-C(=O)OH$

Arginine (Arg) $NH_2-C(NH_2)-NH-CH_2-CH_2-CH_2-CH(NH_2)-C(=O)OH$

one point for an atom of hydrogen :

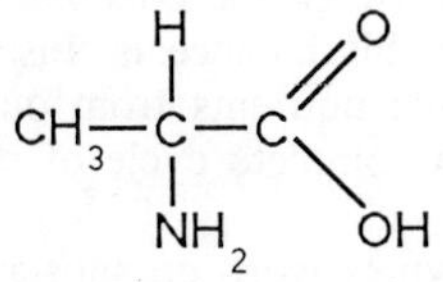

$$\begin{array}{ccccc} & H & & & O \\ & | & & \diagup\!\!\diagup & \\ CH_3 - & C & - C & & \\ & | & & \diagdown & \\ & NH_2 & & & OH \end{array}$$

And (d) such analysis serves to show that, looked at this way, the molecules of amino acids have certain striking structural similarities. For instance, they all have a basic end-structure

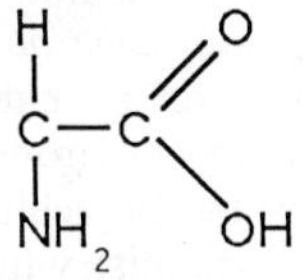

$$\begin{array}{cccc} H & & & O \\ | & & \diagup\!\!\diagup & \\ C & - C & & \\ | & & \diagdown & \\ NH_2 & & & OH \end{array}$$

However, it is also always necessary to keep in mind that our formulae simply indicate one aspect of the molecules and things themselves are little knots of whirling forces, forces which may in a sense be regarded as solid particles, but which are also 'waves'. Besides their composition, in terms of nucleus and particles, they are related to one another by complex mathematical relationships in atomic weight (descants on the figure eight) and in electrical charge. The 'bonds' between them are indicated in chemists' diagrams for the purposes of manipulating chemicals in laboratory analysis of synthesis : but these bonds also express fundamental dynamics in the universe, that hold all together.

Of course, the world of life is composed largely of proteins, and each protein is a version of a sequence of certain amino acids. The health and well-being of an organism depends upon having its 'right' sequence. Getting the sequence 'right' is bound up with the '*task*' of the nucleic acids, which are the chemicals which 'govern' heredity.

For the purposes of his analysis, the molecular biologist will write the composition of a chicken lysozyme thus:

Lys Val Phe Gly Arg Cys Glu Leu Ala Ala Ala Met Lys
Gly Leu Ser Tyr Gly Arg Tyr Asn Asp Leu Gly His Arg
Asn Trp Val Cys Ala Ala Lys Phe Glu Ser Asn Thr Gln
Gly Tyr Asp Thr Ser Gly Asp Thr Asn Arg Asn Thr Ala
Ilu Leu Gln Ilu Asn Ser Arg Trp Trp Cys Asn Asp Gly
Cys Pro Ilu Asn Cys Leu Asn Arg Ser Gly Pro Thr Arg
Ser Ala Leu Leu Ser Ser Asp Ilu Thr Ala Ser Val Asn
Asn Met Gly Asp Gly Asp Ser Val Ilu Lys Lys Ala Cys
Ala Trp Val Trp Arg Asn Arg Cys Lys Gly Thr Asp Val
Len Arg Cys Gly Arg Ilu Trp Ala Gln

But of course 'Val' is not Valine or even

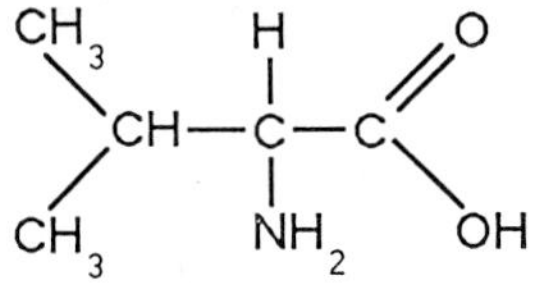

$$\begin{array}{cccccc} CH_3 & & & H & & O \\ & \diagdown & & | & \diagup\!\!\diagup & \\ & & CH - & C & - C & \\ & \diagup & & | & \diagdown & \\ CH_3 & & & NH_2 & & OH \end{array}$$

It is a molecule whose composition in one dimension is indicated by letters like that. The molecules in the whole have many other properties and systems aspects not fully indicated by such diagrammation or included in the thinking that goes with it.

Moreover, these substances only exist in complex living organisms - more often than not (in the labs) Escherichia Coli (E. Coli), a relatively simply micro-organism one which has been studied more than any other such organism in the world. It is a unicellular creature, a rod 2-3 x 10^{-4}cm long and 5 x 10^{-5}cm diameter. You could put 1,000,000,000,000 cells in a volume of one cubic centimeter.

Yet this tiny life-cell is immensely complex as a whole dynamic organism. It has a rigid cell wall and a cell membrane just within it. The membrane acts as a barrier to keep essential molecules inside the cell : and acts as a barrier to keep harmful ones out : so, it must be able to distinguish between very many molecules. The membrane also plays a more active part in growth, for it contains 'pumps' which 'select' useful molecules from outside and concentrates them inside. It maintains the correct balance between the contents of the cell and the external environment. To do this, it must 'know' what this balance is. Inside the cell hundreds of co-ordinated chemical reactions take place, to take nutrients from outside into the organism, to make more E. Coli. Under good conditions, a complete cycle of replication takes place every twenty minutes.

The work is done by enzymes and the enzymes work on substances called substrates. Each minute, a typical enzyme transforms a few thousand substrate molecules into other products. The enzymes are proteins, made up of amino-acids, polymers having many part, the parts being monomers, or single parts. Twenty kinds of monomer, amino acids, are used in the construction of proteins. Inside E.Coli there are *several hundred* enzymes carrying out 'production steps', and others can be brought into being under special circumstances - if, say, the cell finds an essential substance missing in the environment, and an alternative must be found. The activity of an enzyme is regulated by the intracellular environment.

But at this point, clearly, difficulties arise. A scientist will write, 'as in a modern assembly plant, raw materials are transformed into finished products by the co-ordination of large numbers of simple operations'. But the phrase 'assembly plant' is somewhat misleading, as an assembly plant is run by men : what takes the place of human intelligence in this? Are all the complex operations of life 'merely' the products of the operations of physics and chemistry, and are they to be explained by being reduced to chemical and physical laws? As we shall see, there are biological processes which cannot simply be the product of these laws, since the same molecules in a complex behave in quite different ways. In what, then, does the government of the processes reside? We still have no satisfactory account of the thousands of operations of molecules inside a single E.Coli cell, of how it began, how it is controlled, how it may be modified in certain circumstances.

Conventional positivism finds life to be differentiated by its capacity to reproduce : but this is only one manifestation of its autonomy. Reproduction is only the culmination of a whole complex of processes, by which an organism feeds, maintains its equilibrium, excretes, moves (E.Coli can move towards quite distant sources of necessary nutrients), defends itself and survives.

Since a cell reproduces by dividing in two, it must be able to duplicate its cellular machinery, and thus be capable of making new protein molecules with exactly the same sequences as those in the parental cell.

Proteins are not duplicated by being copied in detail. The shape that is to be duplicated is embodied in another polymer, and that polymer conducts the replication process. Now of course arises the question of *encoding*. Even if we avoid this word, we must use some such word as 'conducts' or 'directs'. Proteins duplicate only under the 'direction' or 'control' of nucleic acids : the nucleic acids are the only molecules in the cell that can, with the help of the appropriate enzymes, replicate directly. In their 'work' the nucleic acids are 'responsible' for all aspects of biological inheritance : and they ensure that the daughters of cells resemble their parents exactly.

It is at this point that we enter into the philosophical arguments which Polanyi raises. Crick and Watson and others often talk as though all these complex 'controls' and other operations — the 'knowledge' that seems to be there in minute biological entities — is *itself* merely a product of the laws of chemistry and physics. But, in his essay in the *Encyclopedia of Ignorance*, Crick says that a cell in a cell wall knows where it is in the wall. Grassé talks of the 'intelligence' in all life. How could such 'knowledge' or 'intelligence' be the product of the mere laws of physics and chemistry? There are, too, the questions of changes and developments in the growth of a living creature - as at adolescence, as at pregnancy and other stages of growth : how can these be mere straight-forward consequences of a 'coding' that is communicated at the stage of replication?

If we regard DNA as merely a chemical compound, and DNA is only acting chemically, then how does it come to change its behaviour in later stages? If we declare it is only acting in response to different chemical compounds, we know that these are only formed at the end of the embryo's previous stage. Or how can the same DNA make muscles, bones, and brain cells? Time here comes into the picture : and Polanyi declares that we have no theory which exists to show how complex on-going developments of this kind in growth can happen in a strictly 'chemical' way. Other perplexities are brought in by discoveries that (as Driesch showed in his work with sea-urchin embryos) some tissues in living organisms can seemingly be 'pressed' into undergoing changes they do not normally undergo - just as the simplest cells can alter their processes if so suitable nutrient is directly available to produce 'extracellular enzymes'.

If we cling to the belief that all such processes can be explained by the laws of physics and chemistry, we are presented with the fact that there must be a finite number of physical and chemical mechanisms available in any given DNA molecule, for making adaptations to different circumstances. These mechanisms can, of course, be 'triggered' by physical and chemical conditions prevailing when there is a need for much adaptations. But from the structure of DNA, as so far analysed by physics and chemistry, we cannot predict their existence - though they may show up and be discovered in unusual circumstances. But supposing they did? How would we know they were the creation of a truly new situation in an embryonic development? Or that we were seeing a built-in mechanism that was there all the time, but had never come into play before? Another problem which is explained by no chemical model is the growth of DNA alternatives from 20,000,000 in bacteria to 12,000,000,000 in man. Why should these impulses in higher complexity occur? Strict mechanistic theory can offer no answer.

Chemical and physical explanations for certain aspects of living things have not yet been found, despite the unravelling of DNA. And this is not a question of our 'present state of knowledge'. Polanyi argues that *they can never be* understood or explained as the orthodox reductionists hope to explain them, in terms of a reduction to the minimals of physics and chemistry, because '*life transcends physics and chemistry*'.

The structure of living beings, says Polanyi, has the same quality. Watson and Crick have shown that the complexity of the offspring's body, the 'information content' of the morphology, is supplied by the information content of the parental DNA chain. But information content is something not reducible to mere physics. So, the morphology has the same quality as the chain, that its pattern is not 'merely' the result of forces known to physics and chemistry.

The view that life can be reduced to the laws of physics and chemistry, argues Polanyi, is based on a logical confusion : it involves the kind of simplistic mechanistic explanation, which Frey-Wissling condemns as impossible. (See a note on Frey-Wissling below p170ff).

What we have is a very subtle and complex *information* content which is capable of being flexible in many different situations. This may be the product of the evolution of wonderful systems, in biological history, and in Frey-Wissling's terms it may be that this basic system in life is a manifestation of the order and 'purposive' dynamics in all matter. But a crystal does not have an appreciable information content : nor does a combination of atoms in a chemical compound have an information content. The information content is earnest of the fact that life molecules are part of a 'delicate and harmonious system' and that in this they display features which are not to be explained in terms of being due to the mere attraction between atoms or the physical laws. Such a chain is not ordered as a chemical compound, and so it is *inexplicable* in terms of chemistry and physics. To assume that it is to make a logical confusion, a confusion between two dimensions of substance, and two dimensions of explanation: between the 'levels' referred to above.

The reason is in fact very simple. Watson and Crick showed the DNA chain conveys information of immense richness. In the case of human beings, the information content of DNA is estimated to be that of a binary code having twelve thousand million links.

But, says Polanyi, there is a condition attached to this, which he illustrates by an example which has gained some response in the theory of information. Some years before, he noticed on a railway station garden a series of white painted stones, reading 'Welcome to Wales'. This information content of the stones clearly showed that their arrangement was not due to their physical chemical interaction. The discovery of the use for which a strange object had been shaped showed that its form did not result from internal chemical forces.

And this indicates the condition which Watson and Crick have attached to the DNA chain in ascribing to it an 'information content' corresponding broadly to the length of the chain. For this condition implies that the arrangement of links in the chain is not determined by their chemical interaction.

Any interaction that would favour a particular significant arrangement of a series is a redundancy that reduces the information content of the Series. Therefore, if the configuration of the chain were fully determined by the chemical interaction of the links, the chain would have no information content at all.

Science and Man, the Nuffield Lecture, 1970,
Proc.Roy.Soc.Med.Vol.63
September 1970, p.970-971

If the 'code' were merely the product of chemical interactions and physical positions all we would have is 'noise' (in the language of communications theory): that is, a meaningless non-pattern.

Polanyi supplements his argument elsewhere by referring to a machine. If we had a machine of which we did not know the function, we could make a chemical and physical analysis of it, but we would still not arrive at an understanding of its function. This we could only arrive at by 'systems' analysis, which might help us to identify the machine as belonging to a class of machines based on certain operational principles:

We know watches and can only describe one in terms like 'telling the time', 'hand', 'face', 'marked' which are incapable of being expressed by the variables of physics, length, mass and time. The impossibility is of a logical kind, similar to that by which a poorer deductive system cannot define the terms of a richer one.

Chemical and Engineering News, Aug.21, 1967,p.59.

It is this logical confusion which lies behind assumptions about the reducibility of 'DNA' to physical and chemical explanation.

If you were defining a watch for patent purposes, it would be no use offering a chemical or physical definition, since then an imitator would only have to alter one atom, to overcome the patent. You have to define an invention in terms of *the principles of the operation.*

So, with living organisms, their anatomy and their functions : these can ony be defined in terms of how they '*serve the purposes*' of the organism or a whole.

The mere shape of a living being defeats any physical-chemical definition.

Moreover, 'no biological process ever takes place in an unstructured medium' a point made, incidentally, by Frey-Wissling:

a DNA molecule produces nothing by itself, its genetic programme being initiated within the richly structured framework of a fertilized cell, and that subsequently DNA controls morphogenesis within a steadily developing framework. Moreover, it is DNA itself that introduces within its chemical structure a pattern that acts as a controlling framework to the ensuing generative process.

DNA is both 'code' and *the stuff directed by the coding*: and it operates within a larger 'delicate' system of life. Its functional pattern must be recognised as a *boundary condition* located within the DNA molecule.

The boundary condition generating the function of a living being consists in its typical appearance, its morphology:

If DNA is regarded as bearing a pattern that forms part of an organism and as transmitting information through this pattern, then such a pattern is to be classed likewise as a morphological feature of the organism, and hence be irreducible to terms of physics and chemistry.

As E.W.F. Tomlin says in discussing 'The Concept of Life', the excitements of scientific discovery in recent years have been in the area of structure. An assumption has grown 'specify the parts and you have the whole'. But form, pattern, process, dynamic - these are indicated, as by Frey-Wissling, as major features of living systems, that 'delicate wonderful harmonious' organisation, which no simple mechanism can explain : *these*, the features, the morphology, of life are paramount - DNA and its operations serve them, and DNA itself is *an aspect of the morphology*. Moreover, while 'molecules' are seen largely as merely functioning as 'matter in motion', colliding (by chance) and so on - the overall morphology displays directiveness such as is clearly manifest in the *information* content.[2]

---

[2] As Polanyi points out (C&EN,p.57) boundaries of inanimate systems established by the history of the universe are found widely in geological, geographic and astronomic domains and their information content per unit of matter is very much less than that of a living thing.

Although DNA is a chemical compound, it displays a complex structure and transmits thereby substantial information. Because of this it is irreducible to physics and chemistry. Here Polanyi's further argument is dense, but it is an argument put forward by a distinguished chemist, and we must try to follow. All the chemical compounds consist of atoms linked in an orderly manner by the energy of chemical bonds. But the links of chemicals forming a code are peculiar. We have seen how the code is a linear series of items which are composed of groups of atoms forming a chemical substituent. Each of the series consists of one out of four alternative substituents. In an ideally functioning chemical code each alternative substituent forming a possible item of the series must have the same mathematical chance of appearing at any point of the series.

If there were a chemical law which determined that the constituents could only be aligned in one particular arrangement, this arrangement could transmit no information. In an ideal code, all alternative sequences would be equally probable : chemical laws would thus not effect it, and such a code would belong to an arithmetical or geometrical design, not explicable in chemical terms. However one looks at it, it is clear that the pattern of a substance which has 'information content' is not the mere product of chemical forces : Polanyi refers to various manipulations, such as the production of antipodes in crystals, in which also discriminations are not due to chemical forces. The pattern of information storage can no more be derived from the laws of physics and chemistry when engraved on an RNA molecule than it can be when inscribed on a tape. His conclusion is that:

> the morphological features are the boundary conditions of physical-chemical laws in living things and thus are not accountable by these laws, on which they rely for their functions.
>
> *Chemical and Engineering News*, 51, August 21, 1967, p.64.

The information content in DNA, RNA, etc. is *highly improbable* like the arrangement of the stones reading a message. The configurations in life typify a living being and serve its functions. This is observed among a large group of other configurations, which are mostly meaningless.

The living being, says Polanyi, manifests a system under the control of a non-physical-chemical principle by a *profoundly informative intervention.* The truth is quite the reverse from that implied by the reductionist dogma. In living things we see a *significant* distribution of matter, not a 'mistake' at all.

This is not to deny, of course, that there is a great deal of truth in the mechanical explanation of life. The organs of the body do work much like machines, and biology which regards living functions in terms of machines has had much success. But some biologists believe that there are processes in living creatures which are utterly *non-machinelike*, and they call these *organismic*. An example is that capacity already referred of any cell or combination of cells detached from the embryo of a sea-urchin which will develop into a normal sea-urchin. There are many such processes in life which are quite different from they way machines behave.

But now the question must arise, if we challenge the mechanistic reduction of living principles to the laws of chemistry and physics, how does this effect evolutionary theory? This Polanyi discusses in *The Tacit Dimension* in the chapter on emergence.

Just as the scientist's interest in DNA and indeed all life lies in his recognition of pattern, order and significance, so his interest in evolution arises from his recognition of the rise of higher beings from lower ones, and, principly, in the rise of man. Says Polanyi,

> A theory which recognises only evolutionary changes due to the selective advantage of random mutations cannot acknowledge this problem.

Just as it is perverse to assume that our interest in human life is really an interest in finding that it was caused by an accidental mistake in DNA coding which in turn is merely the consequence of chemical and physical 'law', so

> it is the height of intellectual perversion to renounce, in the name of scientific objectivity, our position as the highest form of life on earth, and our own advent by a process of evolution as the most important problems of evolution.
>
> *The Tacit Dimension* p.47.

It is refreshing to have a distinguished scientist say this, as there is clearly something absurd in the traditional evolutionist's evasion of this problem. The primary fact in the whole argument is that it is men trying to understand life-forms, as *higher beings*.

As we have seen, much doubt is cast on the representation of evolution as a continuous process of selective improvement. The facts show that it is not : there are gaps, some creatures have never changed : opportunities are present but 'evolution' does not take advantage of them : evolution cannot be shown to be taking place at the moment, and so on. Attention has shifted towards the question of how new populations of a new kind come into existence. But, there is a fundamental question, declares Polanyi, which is not confronted :

> How any single individual of a higher species ever came into existence.
>
> *The Tacit Dimension*, p.47.

Polanyi does not doubt the occurence of accidental mutations which may prove adaptive. But he assumes that these can be distinguished from *changes of type which achieve new levels of existence*. Polanyi believes that most paleozoologists would agree that though this distinction is often difficult, it is nonetheless valid.

If this distinction is made, an *autononomous thrust of evolutionary ascent* is as clearly manifest as the growth of an individual from a germ cell. We accept the amazing growth of an individual - the baby in the womb, the change from child to adolescent, the adolescent's lengthening back - why should we not accept growth *in the species*, and in *levels of being*?

Scientists, says Polanyi, have found a 'useful technical fiction' - evolution by chance mutation and the survival of the fittest : so, they are prepared to turn their backs on a great truth, and deny it recognition. If he is right, how incredible that, in the wake of the scientists turning a blind eye, those in the other Humanities, supposedly custodians of humanness, should turn a blind eye too!

'In terms of our present scientific knowledge' Polanyi concludes, 'the biological grounds of our existence are by no means entirely clear to us', and these grounds 'have not yet been carried back entirely to the laws of inanimate nature - *and can never be*'. Living organisms are mechanisms and mechanisms can never be made wholly explicable as the result of operations of physical and chemical laws. Here Polanyi expands further the concept of a machine : the pattern in which the key parts are put together. The key to this pattern in a man-made machine is the intention of the creator of a machine - what he sets out to *achieve*. There are many examples of machines dug up from the past of which we know the chemical and physical composition : *but we do not know what they were for. How can we study the intentional element in the machines of life?* And what is implied by the evident intentionality in them? The 'pull towards the future' 'the lure of form as yet unrealised'? Many biologists and philosophers have observed these 'telic' phenomena, but we have as yet no language or paradigm by which to discuss these adequately.

Living organisms display a meaningful organisation as 'machines' : they *achieve* and they *strive*. The essence of the difference between a living system (like a bird) and an inanimate system that seems to have some autonomy and sustains itself for a time (like a thunderstorm) is that the bird can succeed or fail. Indeed its capacity to succeed or fail is something we know to be inherent in life. 'Nothing gives us the appearance of trying to 'achieve' a thunderstorm', declares Polanyi. But the whole panoply of nature *is* a display of attempts (some breath-takingly audacious, and many astonishingly inventive) to achieve something - and that 'something' is not only 'survival' : 'even a paramecium is an individual that quite apparently strives (whether consciously or unconsciously makes no difference) to adapt itself to its conditions and to stay alive and to reproduce'. If someone, however, cannot tell a thunderstorm from an elephant, there is no point in labouring the argument, say Polanyi. Yet some scientists still deny such realities, to preserve their theories and paradigms at all costs.

Polanyi next refers again to his view that the way DNA functions as a code must be due to properties which are extraneous to its physical and chemical nature, since the essence of a message-conveying medium is that it should not interact or be changed by the process. If the DNA molecule were only the product of the laws of physics and chemistry, it can only have come into existence by chance. Polanyi declares that this is much more improbable than the possibility that a number of rocks on a hillside might roll down to the local railway station and organise themselves into letters forming the name of the place. 'The fact is... that every living organism is a meaningful organism of meaningless matter and that it is very highly improbable that these meaningful organisations should all have occurred entirely by chance'. (p.172-3).

Polanyi argues that because of the evident evolutionary history of the achievement of more and more meaningful structures, from one-celled plants able to do little more than reproduce themselves and keep alive, to man, 'able to achieve so many things that he frequently supposes himself a god able to achieve *all* things', we cannot but conclude that 'some sort of *gradient of meaning* is operative in evolution in addition to purely accidental mutation an plain natural selection and that this gradient somehow evokes ever more meaningful organizations (i.e. boundary conditions) of matter'. (p.173).[3] But what can the concept of a 'gradient' mean, and what are the implications?

There is, of course, dogged resistance among some scientists to such ideas - and an evident desire to maintain the old positions at all costs. But, as Polanyi points out, in physics there is a recognition that inanimate nature is controlled by forces that draw matter toward stabler configuration. Here there *is* a recognised 'gradient' in nature : here is recognition of an 'end'. 'Since in quantum mechanics there is a notion of the existence in inanimate nature of a gradient in the direction of minimization of potential energy, why sould there not be a recognition in the study of life forms of a gradient of potential *shapes* in a field of shapes?...'

This permits also of an escape from determinism. Polanyi makes an analogy from his study of the mind: 'A mind responds in a *striving* manner to comprehend that which it believes to be a comprehensible but which it does not yet comprehend. Its choices are therefore, hazardous, not determined'. (p.176).

But these choices are not random: they are controlled, as they are evoked, by the pursuit of their intention. Polanyi is thus restoring to the world of consciousness the element of intentionality recognised by Husserl, and restoring the hazardness of striving to nature. Polanyi discovers in the world what we need to recognise as a primary human principle : the effort to actualize certain hidden poentialities. This Abraham Maslow, in psychology, sees as a primary impulse. 'Uncaused' actions are released - by what are not physical events but 'an imaginative thrust towards such a discovery'. Polanyi, in his exploration of the mind, makes imaginative spontaneity primary, thus over-throwing the deadness of the Behaviourist cause-and-event position. Relating this to the problem of the category of life, he transforms the universe into one in which matter has become capable of generating meaning, since we must recognise in it that 'gradient of potential shapes', and a 'gradient of deepening meaning'. Sentience, which is an inexplicable fact of both higher and lower creatures, now becomes a dimension *added to* those principles which structure the lower, insentient level. What he is opening up in a new way of looking at the world and new ways of trying to explain it. Consciousness and meaning are not only restored to the world - they are its products : *they are its meaning*!

Mind is part of the dynamic in nature to which we may apply the term 'gradient'. Science itself, says Polanyi, is in its progress 'a development evoked by a gradient of meaning operative in a field of potential meaning and problems'. Nothing could be more absurd, more *perverse*, than science, as one of the highest manifestations of the development of meaning in the universe, looking out at the universe and declaring it meaningless! 'Scientific inquiry is a thrust of the human mind towards a more and more meaningful integration of clues'. Living things are also orientated toward meaning and man's whole cultural framework including symbols, language, arts, fine arts, rites, celebrations and religions contitutes a vast complex of efforts at achieving every kind of meaning. The philosophy of Polanyi and Marjorie Grene restores these cultural achievements to the bleak universe of Galileo and Newton, from which a necessary but mistaken 'objectivity' excluded them.

Obviously, Polanyi's view of life in the universe, and his concept of what *knowing* is, go together. We have suffered too much, he declares, from a 'modern myth... of what science and knowledge are... a misunderstanding spawned by positivistic left-overs in out thinking and by allegiance to the false ideal of objectivity...' George Steiner declares (in *Bluebeard's Castle*) that we must pursue knowledge, even if it leads to the extinction of the human race - and he writes as if 'truth' would continue to exist even if there were no men to know it. This seems to me an absurd view : for scientific knowledge only exists when it is brought into being and is upheld by minds. Knowledge is only men doing something, and science only exists in living scientists. If all men died, knowledge an truth would no longer exist, though the mere brute

[3] See also E.W.F. Tomlin, *Towards a New Philosophy of Science* in *Tokyo Essays*, Japan, 1967.

reality of the universe might continue.[4] It is important not to confuse the two: the things, and the body of knowledge of them which we construct. Truth is never anything more than men trying to make sense of their experience : all knowledge (as Polanyi has shown) is personal, as well as existing, often as 'objective' as maybe, in collaboration in the collocation of natural descriptions. But the truths are also forever contingent, all the same, and are only upheld in that 'criss-cross of utterance between us', to borrow that useful phrase from F.R. Leavis, discussing where 'literature' exists.

Science thus has a subjective basis, in this life of the body, and mind, in the man, and in men.

Again, it is important to emphasise that those like Polanyi who reject or question Darwinism are not arguing against science or for 'mysticism'. They are simply concerned with how we understand living things, without entertaining any supposition about them that there is 'something' in them exempt from physico-chemical laws. They assume that living systems are made of the same kind of matter as non-living entities, and obey physical laws. The question is, however, 'do those laws state the sufficient as well as necessary conditions essential to the description and explanation of biological phenomenon'?

The orthodox case rests on assumptions about the reducibility of biology to physics and it is this that Polanyi questions. (We can leave chemistry out of it, because chemistry can *a fortiori* be reduced to physics anyway). The assumption is that there must be a one lowest-level set of laws into which all others are translatable and from which they can be derived. There is also an implicit impulse towards a single scientific principle which will explain everything.

That there is a problem here may be seen if we look at the ways in which scientists talk about 'codes'. Polanyi points out that there is a great deal of confusion in the talk about codes. The word 'code', of course, suggests anything but 'accident'. It is the *improbability* of the DNA code arrangements that is the subject of our attention (and we attend to it because we human beings are excited and satisfied by what seems its meaning):

> As the arrangement of a printed page is extraneous to the chemistry of the printed page, so is the base sequence in a DNA molecule extraneous to the chemical forces at work in the DNA molecule. It is this physical in-determinancy of the sequence this produces the improbability of occurrence of any particular sequence and thereby enables it to have a meaning - a meaning that has a mathematically determinate information content equal to the numerical improbability of the arrangement.
>
> Polanyi, *Science*, 100, 1968, 1308-9.

The meaning is not reducible to physics and must be understood in other terms : ergo, biology is not reducible to physics, but demands an approach in its own terms, towards the systematic understanding of life-systems, and the intelligence, or whatever one calls it, manifest in them.

To say as much will seem shocking to the individual who believes in a strictly mechanistic approach. The advances of molecular biology in recent years have yielded such valuable results, by the reduction of many features to physico-chemical-explanations. Are those who urge a more 'whole' approach going to discourage the 'part' -analysis techniques? Anyway, isn't it quite clear, logically, that since all material systems are governed by the laws of physics, and all living systems are material, therefore all living systems must be governed by the laws of physics? Doesn't the march of science require that all laws in the end must be reduced to one fundamental plane, of a one-level physics? If we oppose reductionism, doesn't this threaten to stifle the very *Leitmotif* of science, and suggest a 'thus-far-and-no-further' approach, such as Dr. George Steiner would abhor? Are we to nail up certain doors in Bluebeard's Castle?

While there is no end to the knowledge to be gained by molecular investigation, there is also a need to recognise complex living systems and to study them at the macro-level and in their own terms. For many living systems, such as the central nervous system, cannot be made tractable : they can never be reduced to their parts, and the causes and effects thereof. They can be fruitfully studied in terms of new forms of systems theory : by exact enough methods,

[4] Though perhaps in the light of John Wheeler's philosophical investigations, this might be questionable?

maybe, but not those of physics. And when we come to man, surely we need to understand him in terms of the whole being, as philosophy tackles him?

Can we not put it rather like this (as Marjorie Grene suggests) : let us admit that living systems, since they consist of matter like everything else, obey the laws of physics and chemistry. But they also *obey other laws* as well. Yes, in a very vague sense, we are DNA making more DNA : but we are not *just* that - we are a myriad of other things as well, and we exist on other levels. Living systems are not explained by physically laws exhaustively : they must be explained in other ways, too. We do not believe, do we, that the laws of physics are the only laws there are? We need only to look at a watch or pen to see it cannot be understood in terms merely of physics and chemistry. Those who declare that 'we are *only* DNA making more DNA' are asserting that in the 'last analysis' we are *only* atoms and molecules. But we are clearly not just that, merely by the most commonsense view. Obviously, poetry and music are not produced by the laws of physics, for example : nor is science itself. They are produced by human consciousness which has its own laws and must be examined in its proper dimension. Consciousness exists in a body, which has a history and a culture.

If we insist that there are other dimensions, however, we are asking for a better metaphysics than simple, one-level physicalism. Such a metaphysic we do not have, and science defends itself from the problem by assuming an anti-metaphysical stance. Yet, clearly, those educated in the scientific world view are enlisted into a *Weltanschauung*. If we resist reductionism, we do not have such a world view to offer, not yet at least - and so we threaten scientifically-minded people with that terror of not knowing what to believe in, if they are to relinquish the simple one-level 'scientific' view.

There are comprehensive world-views of an anti-reductivist kind, as Marjorie Grene points out - A.N. Whitehead's 'philosophy of organism' for one : and there are others. She believes they are not satisfactory yet and are thus unlikely to convince the scientists.

But our urgent need is to find, in a new way, man's place in the scheme of things, his place in 'Nature'. As Michael Polanyi declares, it is not true to conclude from modern science that the world is meaningless : on the contrary, it appears more and more to be full of meaning. Scientific enquiry itself is 'a thrust of our mind toward a more and more meaningful integration of clues... living things, individually and in general, are also oriented toward meaning, and it is clear... that man's whole cultural framework, including his symbols, his language arts, his fine arts, his rites, his celebrations, and his religions, constitutes a vast complex of efforts - on the whole successful - at achieving every kind of meaning'. (*Meaning*, p.176).

From the philosophy of science and from science itself, especially biology, we now have a new kind of thinking emerging. This is prepared to include in the scope man's inner life, his subjectivity, and all that belong to it, as part of the 'facts of life' in the universe. If we restore the knower to the known, we are forced to admit certain important dynamics to reality : as Marjorie Grene points out, living things depend upon the *future* as primary. Thinking is imbued with 'the lure of form unrealised' and so are living things. We have to admit into out thinking telic considerations, not least because time, as lived time, is 'telic in structure' (Marjorie Grene, *The Knower and the Known* p.245).

There is a confusion in science, which is manifest in the use of phrases like 'apparent purpose'. In truth, modern science, especially in the area of philosophical biology, admits teleology of a kind. As Marjorie Grene herself points out, the conclusion that there are telic phenomena is inescapable, if also incredible to meet scientists and to a public nourished in the mythology of a one-level, purely 'mechanistic' science. Because of this, the recognition of the telic is suppressed as a piece of lingering suspicion or wilful obscurantism. But

> in this century alone, Bergson, Whitehead, Alexander, Collingwood, Husserl, Merleau-Ponty, Polanyi among philosophers; E.S. Russell, J.S. Haldane, Cuenot, Vandel, Spemann, R.S. Lillie, Portmann, Buytendijk, Straus, Goldstein - and a number of others - among biologists have presented irrefutable refutations of a dogmatic mechanism.
>
> *The Knower and the Known*, p.239

Some of these have suggested teleological approaches. But this new teleology must be carefully distinguished from that which seems to demand a reference to a supernatural 'Maker' of organic beings, or a new kind of Paley's matchmaker.

> It is an inner teleology, where each part of organ possessed by the subject is means to the end of - the subject itself...
>
> *Approaches*, p.94

This new approach seeks to rescue us from the twin intelligibilities of mechanism and the will of God.

The whole philosophical issue may be studied in Polanyi's and Marjorie Grene's books. But even if we do not understand the full argument, it is clear that there are ways out, even in science, from the conflict between rigorous materialism (as in Monod) and a belief which brings in 'God's purpose'. It is also clear that what the modern mind is faced with is the problem of the *nature of things*, which it is too often assumed that we have solved. We have not begun to solve it, and it is salutary to note that the rapid expansion of the exploration of space has brought this home to us rather sharply in the last decade or so.

At the end of this discussion of Polanyi's philosophy of science, I should like to interpolate an interlude, on the mystery of life. It is this mystery which the positivist scientist is to active to dispel. But the inexplicable aspects of the problem of life remain to perplex both him and us.

The is much discussion of the origins of life : and in the work of materialists there have been theories to explain it. But as some see, the leap from a macro- molecule to a living cell with all its life-support systems, fully operational within a living membrane, is 'a jump of fantastical dimensions, which lies beyond the range of testable hypothesis'. (Green and Goldberger). 'It has been calculated that, for an impossibly simple organism of only one hundred parts to be composed by chance, the odds are one chance in 10^{53}' (M. Pitman, *Evolution*, a study made by a science teacher in Cambridge). The simplest protein molecule is composed if a chain of at least four hundred linked amino acids, each of which is a specific combination of four or five basic chemical elements. Each of these, in turn, has a specific atomic construction. M. Pitman quotes E. Chargaff (*The Origin of Life on Earth*, pp.298-9). '... I believe that our science has become mechanomorphic, that we talk in metaphors to conceal our ignorance, and that there are categories in biochemistry for which we lack even a proper notation, let alone an idea of their outlines and dimensions'.

We only have to look at serious apects of macro-molecules, and simple living organisms, in the light of the problems of What is life? and How did it originate? to see the extent of our ignorance.

What is 'life"? If we define it in terms of the capacity to reproduce, are there not some crystals that satisfy that definition? There are comparatively simple forms of life, like viruses, which cannot reproduce themselves, but hook into bacteria, to manipulate their molecules, and turn them into new versions of themselves. Are these large molecules 'life'? And how did they come into being? They could not have come into being without there being in the world the bacteria upon which they prey (if that is the right word) to multiply. So they must have come into being *after* more complex life-forms had developed. How could such a subtle system have evolved by 'accident', according to the laws of strict Dawinism?

It would seem important that such questions continue to be debated openly. Several possibilities have put forward about the origin of viruses. Perhaps viruses are a degenerate kind of cell, suggests one biologist (Goodheart, 1969) - a cell that over a long period of time and by successive mutations lost one enzyme system after another so that it is now completely dependent upon the host cell's enzyme systems. In this there is an idea of a 'degenerate' cell. How can a degenerate form of life be fitted into evolutionary theory?

Did this degeneracy happen often in evolution, to give us all the different kinds of new degenerates? If so, there must be serious flaws in the processes of life - and (by analogy) the natural world could be breaking down all round us: yet this hardly squares with the facts. Or did it happen only once, and all the variety of viruses could be descended upon the one original degenerate? There seems to be an indication that one type of virus actually turns into another, though types themselves vary in character and viruses seem to be able to change.

Another suggestion is that viruses could be parts of cells that, through some odd set of circumstances, acquired autonomy in the sense of having an extracellular existence. A mitochrondrion, ribosome or other cell organelle could thus be the forerunner of modern viruses. Each different main group of viruses may have come from a different cell organelle.

It is also possible some say, that viruses represent a primitive form of life - perhaps the first aggregation of nucleic acid and protein with any resemblance of life that gave rise to all higher forms of plants and animals. But this, even those who suggest it admit, is rather a thin argument, since the virus makes such extreme demands for replication on the environment: it reproduces itself only by invading and bursting its 'hosts'. It isn't a chicken and egg argument, for evidently there would be no eggs if there had never been a chicken ever. How did this process ever start?

As we shall see, these questions are related to the investigation of the origins of life. And it is clear that 'life' has powers not yet understood.

At the level of the simplest life - molecules uncanny things are found to happen. In March, 1955, Dr. Frankel-Conrat set out to see if he could remove the protein 'lock washers' from the nucleic acid 'twisted cable' so that he might have some undamaged parts to put back together. The idea was to see if he could disassemble a virus, and then reconstitute it with some 'life' in it.[5] Dr. Frankel-Conrat used a common household detergent, sodium lauryl sulphate, to remove the protein hides off one group of viruses, dissolved in solution. Then he treated another group with an alkali, sodium carbonate; this pulled out their nucleic-acid cores. The entire process was a meticulous one, full of ticklish details and minute manoeuvres: remember, these bacteriophages are only about 100mμ, from one millionth to about 12 millionths of an inch across.

In the electron microscope Fraenkel-Conrat saw that he had what he sought: some intact hollow cylinders of protein on one hand and some long strands of nucleic acid on the other. There was no sign of a whole virus in either solution.

He took two specimens up to his greenhouse on the roof and rubbed a little of each on the tobacco plants, taking care (of course) to keep them separate. The bacteriophage will normally cause a tobacco plant to burst out in a rash overnight. Nothing happened. Neither the protein shell not the RNA core are infective by themselves.

He put the two solutions together, openly declaring nothing would come of it. A few minutes later an assistant noticed an opalescent sheen forming in the solution. Examined under the microscope, they saw perfectly formed TMV bacteriophage rods! These were 'synthetic viruses'. When applied to tobacco plant leaves these were found to be infectious - though not nearly as infectious as the natural kind.[6]

Such recombinations, of course, are the subject of complex scientific discussion. The scientific text book explains such process in chemical and physical terms. Discussing another form of recombination, one author puts it like this: the first step involves hydrogen bonding of the broken ends of the two parental DNA molecules. Phosphodiester bonds are formed. The DNA synthesis step requires the synthesis of phage-coded DNA polymerase, as the second step of recombination does not occur when the parental phages are amber mutants unable to code for this enzyme. Completion of the new molecules by permanent phosphodiester bonds is thought to be performed by the same enzyme that repairs DNA after excision of thymine photodimers. The fact that the final step can be performed in mutant *Esterischia Coli* lacking the ability to recombine suggests that phages carry the information for directing the synthesis of enzymes to reform complete molecules.[7]

But what impels the one appropriate bit of DNA to search for the other bit of protein? How does it recognise it when it sees it? How does it find the right end to join? How does it 'know' its 'partner'? We may conceive of big molecules being springy enough and patterned in such a way as to hook into one another, when (say) a bacteriophage is attached to a cell wall. But what goes on as it approaches? How does it 'feel' the right place? It doesn't (does it?) shoot its long streams of DNA wildly into the air? It *knows* better, whatever that means.

When I raise such questions with scientists I meet at table in college, I am told dogmatically that there are no problems in explaining such forms of behaviour. All that is

---

5 From an account in *Virus Hunters* by Greer Williams from 'Rebuilding a Virus', *Scientific America*, by Heine Fraenkel-Conrat, June 1956.

6 Reconstitution of Active Tobacco Mosaic Virus from its Inactive Protein and Nucleic Acid Components, PW2, *Natl. Acad. Sci.* 41;690-698.

7 From Goodheart, Clyde R., *An Introduction to Virology*, 1969. p279.'Amber' virtually means 'nonsense' here and 'nonsense' implies a highly organised form in 'correct' organisms.

required is an explanation of 'energy requirements', or 'DNA coding'. But are these satisfactory explanations?

In the impulse of a phage tube to find its head and join it on, I cannot help feeling we have something which cannot be explained by physics or chemistry in that one-level dimension. Surely some overall organising principle is in operation, in any process in living substances - and I take a bacteriophage to be alive, to be 'life'.

Kellenberger has written about these processes, after a great deal of work using sophisticated techniques in electron microscopy.[8] First, there is the astonishing way in which long DNA molecules are folded up in the head, condensed into a ball-like form, perhaps with the assistance of the internal head protein because there isn't really 'room' for them. Inside the 'head' of a bacteriophage are the DNA molecules stretched out in long filaments. In that form, DNA can replicate new DNA or transcribe its 'information' into the various kinds of RNA. What do we mean by 'information' here? We mean, I believe, that the components of certain molecules have the capacity to change other molecules into different forms, and to supervise these changes according to certain principles which are embodied in the structure of the molecule and dynamic in it, in some other dimension than the physical, chemical and mechanical: some dimension to be understood in terms of the study of forms and activities, not of substance.

When DNA is folded up, however, at prophase, in preparation or mitosis, the DNA condenses into chromosomes. Following division, the DNA takes the extended form and resumes its metabolic function. The condemned phage DNA forms a framework of scaffolding, and the head protein precursors become arranged on this scaffolding. The protein parts of phage particles are assembled randomly from available materials. Some scientists have experimented with dual infection with T2 and T4 phages. The materials responsible for the specificities that emerge, the phage tail proteins, accumulate in the infected cell, and are withdrawn randomly from the pool during assembly. So, a phage head containing T2 DNA could be assembled with T4 tails, or vice versa.

Further scientific work has confirmed the randomness. Scientists speak of 'codes', but nothing seems to regulate the development of phages. They infect a cell: there is a pause, in which there seems to be chaos, and then, after ten minutes, viral DNA develops rapidly by comparison with bacterial DNA. In certain experiments mutant T4 phages were developed, which caused the cell to make complete phage heads, but no tail parts, while others formed tails but no heads. These abnormal phage progeny were not, of course, infectious. But if extracts of cells infected with appropriate mutants were mixed together so that some parts were supplied by one extract and other parts by another, infectious phages result. It is not an enzymatic process: it simply occurs 'spontaneously'. Whatever does that mean?

Two scientists actually made some films of phage activity. The infected bacterial cells swell at an increasing rate: finally, they literally explode. Osmotic forces cause rapid uptake of water from the medium, which results in swelling and bursting. Cell lysis is a dramatic event. The cell membrane is altered in its permeability, by an enzyme synthesized within the cell by the phage DNA, and is the same lysozyme carried on the tip of the phage's tail. Positively, it may be said to represent the phage's method of finding an escape for the new progeny, so that the growth cycle can begin anew. However, the host bacterium is destroyed. The philosophical questions raised about the nature and origins of life surely make this the most dramatic of all twentieth century films?[9]

As a mere observer from the Humanities, my feeling continues to be that life-molecules tend to behave in ways *for which there is no need* and in ways which are by no means merely the operation of 'physical laws'– 'energy requirements' and so on. They seem to operate against the laws of entropy, which should oblige them to seek equilibrium: they display impulses *towards* 'something'. Yet such a concept has no place in the mechanistic explanations of reductionist science.

---

8 Vegetative Bacteriophage and the Maturation of the Virus particles, Adv. Virus Res., 8:1-61.

9 Bayne-Jones and C.A. Sandholzer (1933) Changes in the shape and size of Bacterium Coli and Bacillus Metatherium under the influence of Bacteriophage - a motion Photomicrographic Analysis of the Mechanism of Lysis. *J. Exptl. Med.* 57:279-309.

Scientists seem unable to understand the problem because it is so trivial. Living beings, as Polanyi has argued (as I point out above), cannot be wholly reduced to a concept as of a machine, since a machine does not make or start itself, does not repair itself, and so on : yet there is a sense in which living organisms *are* a kind of machine, yet machines cannot be explained in terms of physics (and here physics includes chemistry): it is as simple as that, but their paradigms inhibit many scientists from seeing these points.

Besides being a problem in evolution, how we think about life obviously affects how we try to explain its origins. The generally accepted view seems to be that life was created on this earth by a kind of spontaneous combustion. Ther is no doubt some of the chemicals found in life could have been formed by lightning and sunlight, operating on the 'primaeval soup' which materialist scientists talk about. We may note that it has been strictly materialist scientists like Haldane who have tried to argue that all life was actually built up from chance collisions between atoms and electricity. But surely this seems really another 'spontaneous generation' theory, like that of Van Kelmont (1580-1644) who demonstrated successfully that mice are created in heaps of damp bedding sprinkled with grains of wheat. Here again we have a confusion of substance and form. For even when scientists show that by heating and electrifying primaeval soup they can produce amino acids - they still have to account for how atoms have the *potentiality* to become the complex molecules of life-stuff, and to wind themselves into complex, long interwoven forms which are capable of reproducing themselves, yet remaining constant, while also allowing variations over millions of years - how life came to *strive* as substance does not. One might allow some warmed mud to become lug-worms (say) until (that is) you study lug-worms and find themselves creatures of fascinating complexity and acute wormy intelligence. But when it rained in the Precambrian era, 10,000,000,000,000 generations ago (in terms of the ancestry of modern bacteria), were the tiger, the pterodactyl, the swallow, the whale and Darwin himself inherent in the pulse that throbbed in that puddle?

Even if we accept the 'chance and necessity' paradigm, even this theory *depends* upon this directional flow (which Darwin takes for granted) of life - *things becoming, and wanting to become.*

A stone does not want to become: nor does a puddle. As soon as the puddle contains a bit of algae, or frogspawn or bacteria - it is full of becoming, of striving towards a goal.

It is easy to say there was nitrogen, carbon, hydrogen and oxygen, all in a primaeval storm, and suddenly one day there was

Lysine
(Lys)

$$NH_2 - CH_2 - CH_2 - CH_2 - CH_2 - \underset{NH_2}{\overset{H}{C}} - C \begin{matrix} \nearrow O \\ \searrow OH \end{matrix}$$

You can draw a diagram to show how there is a standard method of joining amino acids to form peptides, and before you know where you are, these letters have danced around to form the long chains of DNA: it looks as if you have solved, by a few equations, the secret of life. It is all merely a matter of the shuffling of little bits: and if you then make a diagram of the substance all you have to show is how this bit links with that bit. But this, while it diagrams the reality, does not explain why it happens at all, overall. Nothing we so far know offers anything to explain why those elemental particles we know as C,O,H,N, etc. *strive to combine* in the stuff of life, and to become inreasingly multifarious. None of this diagrammatic analysis, while it looks compelling, explains any of the fundamental question about how life originates.

Haldane and the Russian Oparin may uphold Stanley Miller's flask, and insist that the organic chemicals on which life depends were formed 'spontaneously' in an atmosphere containing hydrogen and other reducing compounds, such as methane and ammonia, transformed under the influence of lightning, volcanic heat and sunshine. But the problems remains, of the inclination in matter to behave like this.

Consider the diagram of how this life-making experiment can be done in the laboratory. It looks convincing: and any scientist can do it again, producing amino-acids from these ingredients. But the whole argument may also be seen as ikonographic: it leaves the question still with us, of how the molecules in this cookery, created by the vast intercourse of the formation of the planets, had the potential to take on autonomy and become living beings and human beings.

There has been a flurry of debate around this question of the origin of life recently. For instance there was a report from the *Nature Times* News Service, in *The Times* recently. Sir Frederick Hoyle and his collaborators had put forward a theory that life on earth had its orgins in the dust clouds that swirl around in space.[10] They had compared the organic chemicals in meteorites with those in the interstellar dust clouds. Sir Frederick and Professor E.C. Wickramsinghe had actually speculated that the dust-clouds had even been the birth-place of the 'earliest primitive gene'.

In the course of the last ten years, radio-astronomers have been increasingly successful in identifying chemicals in these clouds in the vast spaces between the stars. New stars are believed to be formed from these great clouds of dispersed matter. To their surprise, they found such chemicals up there as carbon monoxide, ammonia and formaldehyde. These molecules are only one step, chemically speaking, from the simplest amino acids, the building blocks of protein. Professor Wickramsinghe argued that the clouds also contain much more complex biochemicals, and now they have identified a class of chemicals is the cloud which could include many of the most complex substances to be found in organic matter.

A certain feature of the ultraviolet light spectrum emitted from the dust clouds is also characteristic of a wide range of chemicals that have in common a particular arrangement of chemical bonds. To obtain independent support for their theory Hoyle and Wickransinghe collaborated with Japanese chemists, and astronomers in an analysis of the Murchison meteorite. When a sample of that meteorite was extracted with a suitable organic solvent, the extract was found to exhibit much the same spectrum in the ultraviolet region as had been found by radioastronomical technique from the dust cloud.

Meteorites are among the most primitive objects in the solar system, dating back more than four thousand million years to the period when the earth has in the process of formation. Within those meteorites, such as the Murchison, which are known as chrondrites, are to be found many tiny glass-like spheres whose origin is generally thought to have been the dust grains in the interstellar cloud from which the earth was formed. These Professor Hoyle believes derive from the aggregated dust grains. Furthermore, he and Professor Wickramsinghe consider the similarity to show that complex organic molecules could have formed before the appearance of the earth.

So, the origin of life could have been before the 'primordial soup'; and instead of the 'spontaneous' emergence of complex organic chemicals in the sun-warmed mud - which I for one find it difficult to swallow - at least some of the chemicals which 'led' to life could have arrived pre-formed, in meteorites. The theory, says the letter in *Nature*, is attractive to those who find the period between to formation of the earth and the appearance of life on it too short, to suit any 'primordial soup' theory.

Of course, the theory also suggests that there is extra-terrestrial life. The letter says, that these molecules 'could have led to the start and dispersal of biological activity on earth and elsewhere in the galaxy'. The aggregate dust grains with their trapped organic molecules 'could have served as the host system for the earliest primitive gene'. The best system for protecting the organic chemicals against, for example, ultraviolet light, would be 'in the nature of a biological cell wall' - but such a thing has yet to be found 'out there'. The problem is how we would explain the survival of complex life molecules, in the very hostile environment of space.

If Hoyle and Wickramsinghe are right, life is not a product of chance and necessity. Life could be seen as an eternity of beginning and likely to go on unto eternity. Life molecules are everywhere, because energies and electricities in matter tend towards such possibilities. Life could be seen as seeking to fulfil itself, to find a place, and a chance to develop into its potential myriads of miracles, wherever conditions are propitious. Life strives everywhere, and when it finds a benign home such as earth, it can become man, who here became conscious, and a bearer of the capacity to look out at that dusty ocean from which he emerged, and to

[10] *Nature*, March 17 (266,241; 1977)

know it. If this were so, we belong to an infinite continuity, to a universal potentiality in the universe to become. This would impose upon us a terrible responsibility for trying to understand our place in the scheme of things which generated us, for, perhaps, indeed, we are its most meaningful expression. We are not trapped in a futile universe which is running down by the Laws of Entropy, towards death and nothingness. Our brief knowledge of the world and ourselves is an amazing flowering of the potentialities latent in the clouds of dust between the stars - to enjoy existence in awareness, awe, and a sense of meaning. We *are* the meaning of the stars: and there may be other meanings. We are not trapped, in the dull collisions of matter: but we are a brief flowering of a new dimension of freedom, like the bloom on a plum, or the brief flowering of a pale blue Morning Glory. And in our dimension of culture, morality, knowledge, religion, we preserve, from generation to generation, meanings which transcend the material origins of our life-stuff, in those particles of dust - even though, in the end, we shall return to them. We know, and that brings a dreadful responsibility, although what we know includes our awareness of our own nothingness.

As I report elsewhere, Francis Crick has put forward a similar theory to that of Hoyle and Wickramsinghe, in his book *Life Itself : Its Origin and Nature* (1981). According to this, life may have been sent to earth by Extra-Terrestial Intelligences: however, it is not, apparently, 'their' life but an experimental form of life. This however seems closer to space fiction than philosophy of science : and in any case leaves us with the problems of the origins of the extra-terrestial intelligences. If we think about it, it is surely really no more than a up-dated version of the *Book of Genesis*? It is significant that such strict, anti-metaphysical, mechanists should be driven - presumably on the evidence of mathematical computer calculation - to take to somewhat simplistic fantasies, about the origins of life.

What we choose to believe here has, of course, an ideological quality. Materialists must try to uphold the 'accident' theory. The Russians *must* believe in the soup. Although the possibility of life arriving on Earth from some interstellar source cannot be completely excluded, says a report in *Nature* of September 7, 1978 (No.275, p.19), a Russian study makes it an improbable hypothesis (and everyone can relax). This is a study by the members of the Space Research Institute for the USSR Academy of Science which proposes a detailed scenario for the evolution of the Earth's surface. They suggest that the Earth would have been a much more favourable substratum for the origin of organic molecules than outer space. Earth once had an original atmosphere of hydrogen and helium and this was lost: the surface was exposed to meteorites and all forms of radiation from space. Gradually a new atmosphere formed, of carbon dioxide and hydrogen vapours, which were released by volcanic activity and other forms of outgassing. A critical point was reached when the surface temperature and pressure beccame such that liquid water could exist.

The Russian scenario takes place in an aquatic environment in which water takes the place of gases absorbed into the grains of rock left by meteoric bombardment and radiation. This slow process produces a layer near the surface of oxygen, nitrogen and carbon monoxide - a new atmosphere. There are chemical reactions on the surface of the mineral layer. The porous nature of the grains makes them 'ideal sites for the formation of more complex molecules'.

But why should these complex molecules occur at all? Why should they suddenly develop the hierarchical principles which govern the complex interactions of DNA and such chains? Why should they become self-replicating? Why should they develop into the beautiful form of a herpesvirus? Or an amoeba, which, without brain or nervous system, feeds, divides, and avoids objects in its path? Or the mouse or the bat, or camel or elk or man?

The improbable hypothesis is not that life molecules came from outer space, but that life began by sheer accidental cookery, in porous holes. Yet all over the world scientists are hard are hard at work, to try to justify this hypothesis. I have a feeling there are profound subjective elements here: it all depends what you believe about *Mother Earth*.[11]

But science still goes on slaving at the metaphysics of it. In July there was another of those *Nature-Times* Reports. (*Science*, July 28, 201, 361, 1978). A chemical reaction recently discovered in America may have been an important step in the formation on the primitive Earth of the complex biological molecules necessary for the evolution of life.

---

[11] See Karl Stern, *The Flight from Woman*. Allen and Unwin, 1966, a psychoanalytical study of certain philosophical trends.

Hydrogen cyanide was a main constituent of the early atmosphere, and this is know to react with itself, in slightly alkaline conditions, giving pyrimidines, a group of compounds that constitute an essential part of nucleic acids. One of the pyridimines produced in this way is orotic acid, which is closely related to uracil, one of the building blocks for the formation of the nucleic acid RNA. If orotic acid is exposed to strong ultra-violet light, two scientists found at the Renssalear Polytechnic Institute, a reaction takes place: carbon dioxide is lost, and this could well have been the first method of the production of uracil. The reaction takes place more quickly in the presence of iron and copper compounds, and these would have been abundantly around. The scheme closely ressembles the modern pathway, in which the same transformation, however, is catalysed by an enzyme.

The report says, 'There could thus have been a smooth transition from the prebiological to the biological pathway'. But that word 'smooth' is deceptive: there is still no clue as to why the non-living should become anything so radically different as life. It simply will not do, to suppose this mystery is accounted for, even in terms of the least clue, by the analysis of *substances*. Obviously, what we are concerned with are *forms* and *activities*, dynamics which organise substance, but which cannot be explained in terms of the structure of substances, their continuity, or the way they change one into another. What we are still trying to understand is how and why one substance changes into another, and especially from an objective object like a stone to a subjective entity like a virus, or a cell, or an animal or a man. None of the many reports I have read about investigations into the 'origin of life' ever came near this problem.[12]

Traditional religious approaches have explained this intentionality in terms of 'God's mind', which conceives of such things and then brings about their realisation.

In modern thought, the questions demand new modes of thought. The eye must have been developed 'for' seeing: what can that possibly mean? The concept is as old as Aristotle: 'nature as genesis is the path to nature as goal'.[13] If the lens of the new triton is removed it regenerates from the tissue of the iris, while normally the iris originates from the mesoderm. So, when a normal service is not available, the whole organism feels a need to evoke a contribution from neighbouring tissues a contribution these would not normally make to the functioning of the whole.

How can a simple organism 'think such a thing up', or seem to manifest an idea of its functioning or wholeness, to which parts may be enlisted? It is certainly possible: so, too, then, may it be possible that there is a 'yearning' in living organisms towards potentialities: in the first flight of the Archaeopteryx there was inherent the manifold ecological niches of all the birds, toward which possibility 'life' yearned or strove.

Most biology is still based on an aggregative conception of organisms. Ultimately, our biology believes, an organism will be identifiable in terms of material particles and their spatio-temporal relationships. So Dobzhansky has said, 'the development of the organism is a by-product of the processes of the genes'. The guiding principle of biology is that organisms are aggregates of material particles moved by mechanical physico-chemical laws. All this is anti-teleological.

What we mean by the word 'telic' is that to be realistic, we must see a world in which there are levels of achievement. In this, some new concept of time is needed, as by this view, the world is full of 'protensions', of movements, forward in time, in openness to the future, towards forms as yet unrealised: everywhere in the living world, as Marjorie Grene says, the same future-drawn structure is evident.[14] Time is protensive, full of possibility, as lived time, before timekeeping and physics make it a more one-way flow, from future towards the past.

Such concepts are not so foreign to science today. In mechanics and thermodynamics scientists have accepted that inanimate nature is controlled by forces that draw matter towards stabler configuration. This is, philosophically speaking, an 'end', towards which there is a directional gradient. In certain disciplines like quantum mechanics we recognise uncaused events, controlled by probably tendencies but not necessitated or caused by these. They are

---

12 But see *The Concept of Life*, E.W.F. Tomlin, *Heythrop Journal*, Vol XVIII No.3 and also 'Towards a New Philosophy of Science', in *Tokyo Essays* by the same author, 1967. Professor Tomlin has written a book as yet unpublished entitled *The Concept of Life.*

13 *Physics II*, 1936 14-15.

14 *The Knower and the Known*, p.245

evoked, not caused.

So, the growth of an embryo may be evoked and controlled by a similar 'gradient of potential shapes in a field of shapes'. Inherent somehow in the character of the universe there is perhaps a dynamic moving towards a hierarchy of being. The tension expressed in the special shapes of the DNA molecule may be seen to be related to the tension of the human mind which seeks to know. If we could only find such principles, we would be freed from many futile impulses: the DNA molecule appears as an organism belonging to a creative impulse, in an organism responding to the gradient of deepening meaning: we need no longer seek to explain sentience in terms of the insentient - but as an overplus or super-structure of lower sentient levels of existence.

8 Robert Sheldrake's 'A New Science of Life'

There is a great deal of truth in the mechanical explanation of life, and it has its successes - not least in our own experience, of medical treatment, contraception, blood transfusion and so forth. As Michael Polanyi says,

> There is a great deal of truth in the mechanical explanation of life. The organs of the body work like machines, as they are subject to a hierarchy of mechanical principles. Biologists pursuing the aim of explaining living functions in terms of machines have achieved outstanding success. But this must not obscure the fact that these advances only add to the features of life which cannot be represented in terms of laws manifested in the realm of animate nature.
>
> *The Tacit Dimension* p.42

There is a need to recognise *organismic* processes, and still a need to find special qualities which belong to *life*.

This is what Dr Rupert Sheldrake sets out to do in his book *A New Science of Life*. He begins,

> Most biologists take it for granted that living organisms are nothing but complex machines, governed only by the known laws of physics and chemistry.
>
> p.9

He used to share this point of view, having read Natural Sciences at Clare College, Cambridge, where he was Director of Studies in biochemistry and cell biology from 1967 to 1973. (He became Consultant Plant Phyiologist at the International Crops Research Institute at Hyderabad in India). But after a period of several years he came to see that such an assumption is difficult to justify. He came to believe that 'at least some of the phenomena of life depend on laws or factors as yet unrecognised by the physical sciences'. It for this heresy that some (like John Maddox) have declared that his book should be burned.

Sheldrake held discussions with members of the Epiphany Philosophers, including Professor Dorothy Emmet, the editor of *Theoria to Theory*, the group's quarterly journal, and among his associates were Professors R.B. Braithwaite, J.B.S. Haldane and W.H. Thorpe. His booklist includes Polanyi's *Personal Knowledge* (but nothing by Marjorie Grene) and books by Bergson, Brenner, Crick, Kuhn, Medawar, Monod, Popper, Ryle, Thorpe, Tinbergen, Waddington, E.O. Wilson and A.N. Whitehead. I mention all these to suggest that we are here dealing with no eccentric or crank, though Sheldrake is clearly attracted to the religious explanation, far more, certainly, than the present author.

At present, says Dr Sheldrake, the orthodox approach to biology is given by the mechanistic theory of life. This paradigm has been predominant for over a century. It 'works', and most biologists adhere to it, because as a paradigm it provides a framework of thought within which questions about the physico-chemical mechanisms of life-processes can be asked and answered. The 'cracking' of the 'genetic code' is one of the most spectacular of its successes.

However, there are those who doubt whether all the phenomena of life can ever be explained entirely mechanistically. (Sheldrake lists E.S. Russell, *The Directiveness of Organic Activities*, 1945; W.M. Elsasser, *Physical Foundations of Biology*, 1958; Michael Polanyi, *Personal Knowledge*; J. Beloff, *The Existence of Mind*, 1962; A. Koestler, *The Ghost in the Machine*, 1867; P. Lenartowicz, *Phenotype-Genotype Dichotomy*, 1975; K.R. Popper and J.C. Eccles, *The Self and Its Brain*, 1977; and W.H. Thorpe, *Purpose in a World of Chance*, 1978).

A new theory capable of extending or going beyond the mechanistic theory would have to discover qualities or factors at present unrecognised, and show what relationship they have to known physico-chemical processes.

Sheldrake discusses vitalism, which he believes has failed to qualify because it has not satisfied the criterion of falsifiability, or refutability, or testability, laid down by Karl Popper.

Organismic or holistic philosophy seems more promising. It denies that everything in the universe can be explained in terms of the properties of atoms, or indeed of any hypothetical ultimate particles of matter. Rather, it recognises the existence of hierarchically organised systems which possess properties of their own. Sheldrake quotes A.N. Whitehead: 'Biology is the study of the larger organisms, whereas physics is the study of the smaller organisms'.

Organismic philosphy, however, has so far failed to give rise to testable predictions. However, the most important organismic concept yet put forward is that of *morphogenetic fields*. (See P. Weiss, *Principles of Development*, 1939.)

> These fields are supposed to help to account for, or describe the coming-into-being of the characteristic forms of embryos and other developing systems.
>
> p.12

So far, however, it is not clear whether the organismic philosophers of biology are thinking about *a new type of physical field* or a *new way of talking about complex physico-chemical systems*. Only if morphogenetic fields can be shown to have measurable effects can testable predictions be made, and the theories found to have scientific value.

Sheldrake seeks in his books to put forward the hypothesis that morphogenetic fields to have a measureable physical effect.

> It proposes that specific morpho-genetic fields are responsible for the characteristic form and organisation of systems at all levels of complexity.
>
> P.13

These systems are derived from the morphogenetic fields associated with previous similar systems. The structure of past systems effect subsequent similar systems by a cumulative influence which acts across both space *and time*. (p.13.)

Thus, systems are organised in the way they are because similar systems were organized with the *repetition* of forms and patterns or organisation.

In this we seem to have a recognition of the existence of overall controlling principles, such as Polanyi pointed to. The new element is the implication of an imitative or historical principle, throughout life, by which morphogenetic qualities are passed on, from living nature to living creature, by some kind of telepathic intelligence.

Sheldrake turns in his opening chapters to unsolved problems of biology. He quotes T.H. Huxley, who expressed the ambition to deduce the facts of morphology on the one hand and of ecology on the other 'from the laws of the molecular forces of matter'. The success of this approach has culminated in the marvellous discovery of the structure of DNA, the 'genetic code' and the elucidation of the mechanisms of protein synthesis. The properties of living cells can be partly explained by our knowledge of physico-chemical systems.

Sheldrake seems to concede that modern micro-biology confirms Darwinian theory:

> The way in which the parts of living organisms are adapted to the functions of the whole, and the apparent purposiveness of the structure and behaviour of living organisms, can be explained in terms of random genetic mutations followed by natural selection, such that those genes which increase the ability of an organism to survive and reproduce will be selected for; harmful mutations will be eliminated. Thus the neo-Darwinian theory of evolution can account for purposiveness; it is totally unnecessary to suppose that any mysterious 'vital factors' are involved.
>
> p.18

However, we may perhaps here recall some of Marjorie Grene's criticisms: the great leaps forward - their 'directiveness' - are not accounted for by this account. And what does 'selected for' imply, unless it is *foreknowledge*? Is Sheldrake not critical enough of Darwinian Theory?

A third problem is that of regeneration, whereby organisms repair damaged structures. A fourth problem is that the simple fact of reproduction - a part becomes a whole.

All these problems suggest something in nature which is more than machine-like: something in life which is more than the sum of the parts, and dynamics which determine or define the goals of the processes of development. How can these be investigated? Sheldrake examines the idea of genetic programmes, on the anaology with a computer. But where *is* the code? It cannot be in the chemical structure of DNA, since if all cells were programmed identically, they could not develop differently. In books and essays on the computer analogy, says Sheldrake, the problem is merely re-stated.

There is another problem, the essence of which I have already referred to. A computer is assembled and programmed by an intelligent conscious being. The computer anaology *implies* some such purposive entity, who set it going, and controls it.

If we say that the genetic programme has been built up in the process of evolution by a combination of chance mutations and natural selection, the analogy with a computer becomes meaningless, because all similarity disappears.

In fact, Sheldrake suggests, those who pursue the computer analogy have in fact smuggled in a kind of vitalist principle: only the 'vital factor' now becomes a 'genetic programme', or (at least) the (hidden) programmer's intelligence behind it. The genetic programme is the vital factor in a mechanistic guise, just as Paley's watchmaker reappears in Crick's latest book in the form of the Extra-Terrestial Intelligence that sent life here by rocket.

Sheldrake admits that no conclusive answer is yet possible to the problems he lists.

He next turns to problems of *behaviour*, which are, of course, colossal. There are the problems of how spiders know instinctively how to spin webs; how birds know how and where to fly, by instinct.

Damaged animals, like dogs who have lost a leg, still get to where they want to go (Sheldrake might have mentioned Goldstein's work with brain-damaged patients, who did not see half a world, but a whole world, albeit in a somewhat impaired way - the undamaged parts of the brain compensating). This is the problem of *behavioural regulation*.

Thirdly there is the problem of learning and intelligent behaviour: the newness in these is not a mere product of previous causes.

Microreductionist explanations cannot solve these problems. Moreover, even if the ambition of the physicalists were fulfilled - there would remain a problem of morphogenesis. That is, even if the behaviour of a simple organism like a worm could be explained in terms of the 'wiring' of its nervous system, the question would remain of how the nervous system with its characteristic patterns and 'apparent purpose' came into being.

Though Sheldrake does not say so, the whole matter is very much a question of opening up questions we either take for granted, or are blind to. There is simply a blank assumption in much science, by which the mere mention of 'DNA coding' or some such concept is taken to indicate that we have the secret of life - whereas the truth is that large parts of the secret are still missing, while many aspects of the mystery totally escape any mode of thinking we can bring to bear on it.

Moreover, there are major blanks. Sheldrake turns to these next. Many doubt, for instance, that the neo-Darwinian explanations of evolution in terms of random mutations, Mendelian genetics, and natural selection are convincing. One school believes that large-scale or macro-evolution does come as a result of long-continued processes of micro-evolution (e.g. B. Rensch, *Evolution Above the Species Level,* 1959; F. Mayr, *Animal Species and Evolution*, 1963; G.C. Stebbins, *Flowering Plants: Evolution Above the Species Level*, 1974). But others deny this, and believe that major jumps occur suddenly in the course of evolution (e.g. R. Goldschmidt, *The Material Basis of Evolution*, 1940; J.C. Willis, *The Course of Evolution*, 1940: Marjorie Grene also takes this view).

The fundamental theory that at the centre of the whole process are *chance* mutation is no more than speculative, Sheldrake declares. Opponents of the mechanistic theory argue that such mutations are due to the activity of a '*creative principle*' unrecognised by science. The 'selection pressures', too, can be considered to depend on an inner organising factor which is essentially non-mechanistic. To insist, in this area of speculation and hypothesis, that *chance* or *accident* are the only principles is dogma rather than genuine science - and now seems to fly in the face of mathematical calculations of the possibilities as to whether the 'fortuitous' theory is acceptable. Evolutionary theory is, or should be, still 'open'.

So, too, is the question of *the origins of life*. For some time the predominant hypothesis here was of life coming into existence by the operation of lightning or the sun's rays on a 'primordial soup'. This, too, has been subjected to mathematical calculation, and seems also unsatisfactory. So, we have seen that other hypotheses have been put forward. (F.H.C. Crick and L. Orgel, *Directed Panspermia, Icarus* 10, 341 - 346. : F. Hoyle and C. Wickramsinghe, *Lifecloud*, Dent, 1978.)

But (argues Sheldrake) even if it were proved that life originated in a primordial broth, this would not prove that the mechanistic argument is satisfactory, since the problem of explaining the new organismic properties would remain. (See M. Eigen and D. Schuster, *The Hypercycle*, 1979.)

Problems also arise, because of the human mind. Mechanists postulate that all the phenomena of life can be explained in terms of physics. But the question of whether mental states are mere epiphenomena of the physical states of the body is not yet scientifically proved. Moreover, it is evident that science itself depends upon mental activity. Even physics presupposes the minds of observers, and their minds and their properties cannot be explained in terms of physics. This problem has become recognised in connection with the role of the observer in science, of course. (See B. D'Espagnat, *The Conceptual Foundation of Quantum Mechanics* and, of course, Michael Polanyi and Majorie Grene, passim.)

Dr Sheldrake next discusses psychology. Behaviourism, he suggests, is effective as a methodology only by ignoring mental states. Other psychological disciplines have recognised *subjective experience* - and (though Sheldrake does not say so) what is pointed to here is the need to recognise *consciousness* as a reality.

The great mystery in psychology is memory: where are memories 'stored'? Various hypotheses have been advanced - in terms of reverberating circuits of nervous activity, or changes in synaptic connections between nerves, or specific molecules of RNA: but there is no evidence that any of these mechanistic theories account for memory. (See Boring on the history of experimental psychology, 1950, and Buchtel and Berlusshi in *The Encyclopedia of Ignorance*, edited by Ronald Duncan and Miranda Weston-Smith, 1977.)

Sheldrake also glances at parapsychology, where, he believes, despite 'frauds and deceptions', there have been found manifestations which defy explanation in terms of any known physical principles. (R.A. Ashby, *Guidebook for the Study of Psychical Research*, 1972; B.B. Nolman (Ed.). *Handbook of Parapsychology*, 1977. (It is also perhaps worth recording that Soviet Russia organised some research to disprove telepathy - which resulted in the existence of paranormal powers being confirmed: Professor L.C. Vasiliev and his team found a 'modality' beyond the grasp of modern physics).

This chapter concludes that the mechanistic theory runs into serious philosophical difficulties with the problem of the limits of physical explanation, of life and mind. The 'interactionist' theory, of the influence of 'mind' on the body seems to create more problems than it solves. But is there any fruitful opportunity in bringing to bear on all these questions, of a theory of morphogenesis, on organismic lines?

Dr Sheldrake's second chapter is on three theories of morphogenesis. First he deals with biological methodology. Living creatures and their development can be studied in various ways. These descriptive results do not themselves lead to an understanding of the causes of development - but they may suggest hypotheses. These can be tested by experiments which perturb normal development. We have looked at some of the problems which evade this kind of analysis (e.g. regulation).

Mechanism today places its faith in our knowledge of the DNA structure. Hereditary differences are known to depend on genes, which can actually be mapped and located at particular places on particular chromosomes. The classical basis of genes is known to be DNA and their specificity is known to depend on the sequences of bases in DNA. It is known how DNA acts as the basis of heredity, servicing as a template for replication and acting as the template for sequences of animo acids (through the intermediary of RNA, involving processes of 'reading off' codes). The characteristic of a cell depends on its proteins, and on these also depend on its proteins, and on these also depend its metabolism, its dealing with enzymes and the way it is 'recognised' by other cells.

So, it looks as though the mechanistic reduction of life is, as it were, fully understood. Morphogenesis is but the control of protein synthesis. Dr Sheldrake explains the processes of 'switching on' and 'switching of', as far as it is understood. These processes depend upon physico-chemical patterns within the tissue. The physical or chemical factors provide 'positional information' which the cells then interpret. (See L. Wolpert, 'Pattern formation in biological development', *Scientific American*, 239 (4), 154-164). These spheres of knowledge lead mechanistic biologists to believe that morphogenesis will be explained in purely mechanistic terms.

Problems of a basic kind do, however, remain. Polanyi argues (see below p157) that the way in which life is organised and develops cannot be merely a product of the physical and chemical laws in the structure of DNA: all this would produce would be 'noise', and what we have is that order which is described by the terms 'code' and 'information'. Sheldrake points out that in species so different as chimpanzee and man, the DNA is much the same (the overall difference between the DNA sequences of humans and champanzees is only one per cent). And then, within the same organism, different patterns take place while the DNA remains the same. The differences between arms and legs are not differences between kinds of DNA, while developments which are precise (e.g. the attachment of tendons to bones) are not clearly governed by DNA 'codes', since all the DNA they are composed of is the same. There much be another level of control or intelligence which is not simply to be explained as 'DNA coding'.

The problem also arises of how, even if the physical or chemical structure of a piece of life is shown, the patterns come to be like that in the first place. Sheldrake discusses some forms of behaviour which baffle attempts to explain them in any linear physico-chemical way (Slime-moulds; the hormone auxin in vascular differentiation). Similar problems arise over 'positional information' because different cells arranged in a suitable pattern make different protcins, and cells do not behave in a straightforward way, always producing what their molecular structure would suggest.

And so there are further questions. We know how DNA and RNA generate polypeptide chains: but how do these chains fold up into the characteristic three-dimensional structures of proteins? How do the proteins give the cells their characteristic structures? While some substances do organise themselves with forms in the test-tube (thus demonstrating that they do not require some mysterious property of the living cell within which they are normally found) there are complex processes in the development of form in life which require other physical factors - membranes influenced by the forces of surface tension, the structures of gels by colloidal properties and so on. Here processes cannot always be explained by the known laws of physics. Some life processes may require causal factors not yet recognised by physics. Mechanistic biology simply sets this question aside. It does not answer it. It thus remains open.

Dr Sheldrake next examines vitalist theories, notably those of Driesch. He believed that there is something about living organisms which remained a whole even though parts of the whole could be removed[1] : *entelechy*, a Greek word whose derivation (en-telos) indicates something which bears an end or goal in itself. Driesch called entelechy an 'intensive manifoldness', and Sheldrake discusses how there could be 'a temporary suspension of inorganic becoming': some have argued in favour of a 'will' or 'mind influence' which can mollify the spatio-temporal activity of the neuronal network, so that probability can be modified in ways which orginary physical entities would not admit of. (See Arthur Eddington, *The Natures of the Physical World*, 1935; J.C. Eccles, *The Neurophysiological Basis of Mind*, 1953).

Some theorists have postulated a 'collective mind'. The web-spinning of a spider, says one, 'may be due to the individual creature being linked up into a larger system (or common subconscious if you prefer it) in which all the web-spining experience of the species is stored up' (W. Carington, *Telepathy*, 1945). Alisteir Hardy also suggests a 'subconscious species 'blueprint'' 'to supplement the 'DNA code'.

---

1 An interesting analogy here is the hologram which, despite pieces being removed, can still give a complete three-dimensional image. However, it can only do this as part of a functioning whole, including laser and mirrors.

Entelechy, which in this kind of theory organises the cells, tissues and organs of the organism and coordinates its development, is not a type of matter or energy, and so does not belong to the physical world: it is dualistic.

> They physical world and the non-physical entelechy could never be explained or understood in terms of each other.
>
> p.49

Vitalism once seemed the alternative to mechanism, but now, says Dr Sheldrake, the organismic theory seems a better explanation, so he turns to that next.

There have been a number of influences contributing to organismic theory: some philosophers like A.N. Whitehead and J.C. Smuts; some from the field concepts in physics; some from Gestalt psychology; some from Driesch's vitalism.

Organismic themes seek to propose morphogenetic *fields*. The idea was originally put forward independently by A. Gurwitsch (Uber den Begrill des embrynalen feldes, *Archiv für Entwicklunds mechanik*, 51, 388-415 and Paul Weiss, *Principles, of Development*, 1939). Analogies were suggested between these fields and inorganic electro-magnetic systems. C.H. Waddington developed a new concept, the *chreode* (Greek for 'necessary path'). Development is canalised *towards definite end points*. Réné Thom, a mathematician has tried to develop a mathematical theory embracing morphogenesis, but this is descriptive rather than an explanation. (Waddington, too, regarded his *chreode* as a 'descriptive convenience'.)

The problem remains of how the subject of organismic theories can have a causal role, when, as a sophistication of mechanistic theory, the quality or force it seeks is not electrical, chemical or physical. How can something so ambiguously discussed be *primary* in the life processes?

Sheldrake gives in one of his footnotes a fascinating quotation from C.H. Waddington - revealing a good deal about the problems scientists have with their paradigms:

> Since I am an unaggressive character, and was living in an aggressively anti-metaphysical period, I chose not to expound publically these philosophical views. An essay I wrote around 1928 on 'The Vitalist-Mechanist Controversy and the Process of Abstraction' was never published. Instead I tried to put the Whiteheadian outlook to use in particular experimental situations. So biologists uninterested in metaphysics do not notice what lies behind - though they usually react as though they feel obscurely uneasy.
>
> *Towards a Theoretical Biology*, 2
> ed. C.H. Waddington, 1969.

Here we find a retreat from a necessary new line of thinking - which shows the conservatism inherent in an established paradigm. But it also shows that a good biologist can become aware of 'something else' - some other dimension of thought necessary for the investigation of life.

Sheldrake next turns his attention to the problem of form and its causes. This is of course very much in Tomlin's sphere.

If a bunch of flowers is thrown into a furnace and reduced to ashes, the total amount of matter and energy remains the same, but the form of the flowers disappears. Form is not a quantity, nor is it conserved. We recognise forms all around us, but tend to forget that there is a gulf between our acquaintance with forms and the quantitative factors which physics and chemistry concern themselves with.

Quantities can be measured, but forms cannot. They are *recognised*: they cannot be reduced to anything else. There are serious difficulties in representing forms mathematically, because of their complexity: and, of course, the description of change of form, of *morphogenesis*, is even more difficult.

There are other difficulties: developing organisms are not closed systems, and their development is epigenetic - i.e. the complexity of form and organization increases. Although mechanistic biologists use terms like 'genetic information' these are actually misleading because they seem to have a well-defined meaning, whereas the problem is much more complex, and many aspects of it simply do not lend themselves to Information Theory from electronics at all.

Sheldrake also refers to a theme which is latent in science - what he calls Pythagorean mathematical mysticism: the belief that there is a transcendent order in the universe which is capable of being revealed by mathematics.

An alternative approach sees the pursuit of knowledge as the mere superimposition of human concepts of order of mathematical models applicable to isolated area of the world, and cannot lead to any fundamental understanding of reality.

But there is a third alternative which is the attempt to understand forms in terms of more fundamental *forms* - an approach which seems close to the ideas of Rom Harré as discussed by Majorie Grene. (Below p172.) Here Sheldrake discusses the Platonic tendency in A.N. Whitehead, relating to Plato's notion that forms in the world were imperfect reflections of archetypal ideas. Whitehead postualted that all actual events involved what he called Eternal Objects. As Yeats put it:

Plato thought Nature but a spume that plays
Upon a ghostly paradigm of things . . .

But this would seem to have much relevance to science. The problem is to open up a new hypothesis for the causation of form through morphogenetic fields in both biological and non-biological systems.

Sheldrake is groping, evidently, towards new modes of thinking, about how forms are created.

In Newtonian physics, he points out, all causation was seen in terms of *energy*. All moving things have energy, and static things may have a potential energy due to their tendency to move.

Sheldrake then discusses modern ideas of fields: in modern physics phenomena are explained in terms of a combination of spatial fields and energy: energy is still the cause of change, but the ordering of change depends upon the spatial structure of the fields (which include the gravitational field, equated with space-time, and the electromagnetic field: there is also the quantum field theory of matter in which subatomic particles are thought of as quanta-of-excitation-of-matter-fields).

Energetic considerations determine (for instance) which is the most stable state of a chemical structure, but they do not determine its spatial characteristics. These depend upon spatial patterns given by the fields of matter and electromagnetism.

Sheldrake now turns to the concept of entropy. The second law of thermodynamics declares that spontaneous process in a closed system tend towards a state of equilibrium. As we have seen, the notion that everything is running down has has a disturbing effect on the minds of many in the Humanities, while there are those who protest that the evolution and development of living organisms appears to contradict the principle of increasing entropy. This confusion, Sheldrake suggests, arises from a misunderstanding of the limitations of the science of thermodynamics. Firstly, it applies only to closed systems, whereas living organisms are open systems, exchanging matter and energy with their environment. Secondly, it only deals with the interrelations between heat and other forms of energy - it does not account for the structure of biological entities and how these arose in the first place. Thirdly, it is not correct to link the technical definition of entropy with the non-technical conception of *disorder* as the end of entropy. The greater structural complexities of organic structures are not recognised in discussions of how entropy reduces entities.

How limited thinking is in the consideration of entropy Sheldrake shows by discussing crystals. In a process in which, according to the physics of entropy, there has been a decrease in 'order', there may be a considerable increase in form, from a morphological point of view. Similarly, when an animal embryo grows and develops, there is an increase in entropy of the thermodynamic system. The second law emphasises the dependency of living systems on external sources of energy, but it does nothing to explain these specific forms.

As temperature changes, forms appear in substances. Analysis of these in terms of energy alone cannot explain the way in which forms appear: they appear spontaneously - and obviously exploration by some other principle is needed, to explain the origin of the forms.

Sheldrake next considers the limitations of quantum mechanics. While it explains understanding of chemical bands and certain aspects of crystals, it has not enabled the forms of even simple molecules and crystals to be predicted from first principles. These questions affect our attitude to the assumption that physics and chemistry provide a firm foundation for the mechanistic understanding of life. It is often assumed today that complex chemical and biological structures can be fully explained in terms of existing physical theory. But this is not so: the properties of substances cannot be calculable by known methods.

In the structure of living forms, it would seem that some factor other than energy determines how a particular structure is realised rather than some other form. In some macromolecules the polypeptide chains twist, turn and fold into complicated three dimensional forms. In some experiments the complex forms of proteins have been made to unfold: but when they are placed again in appropriate conditions, they grope towards the same structural end-point (see C.B. Anfinsen and H.A. Scheraga 'Experimental and theoretical aspects of protein folding', *Advances in Protein Chemistry*, 29, 205-300, 1975).

The stable end-point of such a form may be a 'minimum energy structure', but it may not be the *only* structure with the same minimum energy, and the problem remains of why a macromolecule *seeks* this form. It does not seem to 'test' all possible minimum energy forms first. The synthesis and folding of a protein chain can be accomplished in about two minutes: for it to try out all the possibilities (10^{45} for a chain of 150 amino acid residues) it would take 10^{26} years (a hundred million million million years).

The folding process seems to be 'directed' along certain pathways towards one particular conformation of minimum energy. Existing theories of physics can explain the range of possible minimum energy structures, but cannot explain why one solution is realised. Might it not be that some other factor then energy 'selects' between the possibilities and determines the specific structure?

The hypothesis of formative causation proposes that morphogenetic fields play a causal role in the development and maintenance of the forms of systems. Though morphogenetic fields can only work in conjunction with energetic processes, they are not in themselves energetic.

Sheldrake uses the analogy of a plan of a house - though he does not suggest conscious planning. The plan is the 'cause' of the house: it plays a causal role in the physical arrangements of bricks and mortar. The account of physical reality has, of course, been inclined to include the way in which physical systems influence one another at a distance (e.g. gravity). These are 'non-material', but yet, of course, aspects of matter. Sheldrake suggests that morphogenetic fields are parallel spatial structures detectable only by their morphogenetic influences on material systems: 'they too can be regarded as aspects of matter if the definition of matter is widened still further to include them'. (p.72)

These morphogenetic fields may influence the forms not only of living things, but also forms such as proteins, nitrogen atoms, water molecules, sodium chlorate crystals, muscle cells of worms, and cells of plants and animals.

In the organismic theory of morphogenesis, systems or organisms are hierarchically organised at all levels of complexity. They are *morphic units*, organised in hierarchies, under the influence of an overall morphogenetic field (Sheldrake refers us here to Arthur Koestler in *Beyond Reductionism* where he suggests the term *holon* for 'self-regulating' open systems which 'display both the autonomous properties of wholes and the dependent properties of parts.' p.210-211).

Sheldrake now turns to a feature of the problem of our understanding of life which often escapes notice: the processes we are discussing take place in a whole complex of processes and fields - and there is an immense history behind them. The mechanist approach tries to understand things by breaking them down into minimals, and this is the very basis of our philosophy of science: perhaps more might be gained by putting things back into their context?

Morphogenesis can only begin from an already organised system which serves as what Sheldrake calls a *morphogenetic germ*. A morphogenetic germ becomes surrounded by a particular morphogenetic field because of its characteristic form: and it is part of the *system-to-be*.

The morphogenetic field can be thought of a structure surrounding or embedding the morphogenetic germ, and this field orders events within its range of influence in such a way that the virtual form is actualised. It is something like a magnetic field, but it is not to be associated with the magnetic field that is sometimes found to be around organisms.

Because of the virtual form inherent in the morphogenetic field, it is possible (as in Driesch's sea-urchins) for regulation to take place when organisms are damaged. And as in regeneration deviations in form can be corrected.

Dr Sheldrake then discusses morphogenesis both in inorganic and organic materials. He illustrates how molecules can have virtual forms before they are actualised. With crytals, seeding seems to indicate how the virtual forms of repetitions of the lattice structure given by the morphogenetic field extend outwards from the surfaces of the growing crystal, in the action of catalysts there seems to be morphogenetic effects taking place.

With large life molecules, like proteins, their folding seems to follow canalized pathways of change, and these, Sheldrake suggests, can be regarded as chreodes (that is 'necessary paths', *Chré*, it is necessary: *hodos*, route or path). A morphogenetic germ must be present, which already has the characteristic structure in virtual form. The idea of such morphogenetic starting points has been suggested in the literature of protein folding. (See C.B Anfinsen, 'Principles that govern the folding of protein chains', *Science*, 181, 1973, pp 228-230.))

The idea that morphogenetic fields may influence the shapes of inorganic as well as organic entities seems alien to thinking of contemporary physics. But, says Sheldrake, in resistance to such ideas we may detect relics of nineteenth century atomism. The conventional assumption is that no new physical principles or fields come into play at levels of organisation above that of the atom. This, he declares, seems arbitrary. In the area of theoretical physics at the atomic level there have been many changes in thinking. Quantum theory was primarily elaborated in connection with simple systems like hydrogen atoms. As time went on new fundamental principles were introduced to account for empirical observations on those of the fine structure of the spectra of light emitted by atoms. The original quantum number characterizing discrete electronic orbitals were supplemented by another set referring to angular momentum, and then yet more referring to 'spin'. The latter is an irreducible property of particles, as is electrical charge. Still more irreducible factors have been introduced, much as 'strangeness' and 'charm': as observations are made which cannot be fitted into already accepted quantum theory, so new concepts, as of matter fields, have been introduced to make them explicable. Atoms are no longer regarded as ultimate and indivisible, and so the original theoretical justification for nineteenth century atomism has vanished. Of course, the new concepts in quantum theory cannot be extrapolated to deal with larger and more complex entities and morphogenetic fields, but developments there suggest new modes of thinking are required, to understand morphogenesis.

Chemical and physical processes in living cells are immensely complex: they include crystalline, liquid and lipid phases: colloidal solids and gels; electrical potentials across membranes; 'compartments' containing different concentrations of inorganic ions and other substances, and so on. The number of energetically possible patterns of change is enormous and Sheldrake sees the morphogenetic fields as imposing pattern on these probabilistic processes.

Some forms in living things may be random or generated by minimum-energy configurations. But physical explanations of biological forms have had only limited success (see Sir D'Arcy Thompson, *On Growth and Form*, 1942 and Peter Medawar, *The Art of the Soluble*, 1968) and this suggests to Sheldrake that most aspects of biological form must be determined by morphogenetic fields. His conclusion here runs parallel to one of Michael Polanyi: living things obey the laws of physics and chemistry, *but they obey other laws too*, as of overall hierarchic principles.

Sheldrake then discusses some actual examples from biology, of how morphogenetic fields might be responsible for organising processes - as in the positioning of microtubules in cells, the tiny rod-like structures formed by the spontaneous aggregation of protein sub-units, as microscopic 'scaffolds' within both plant and animal cells. These guide and orientate processes such as cells division. What controls their spatial distribution? On Sheldrake's hypothesis, it is specific morphogenetic fields.

The hypothesis of formative causation holds that the forms of complex chemical and biological systems are not uniquely determined by the known laws of physics. Polanyi, as we have seen, has made it clear that they *cannot* be. The operation of the mere laws of (say) DNA replication would only produce 'noise': what we have is a significant and developing pattern and process of life.

Supposing we accept the overall control of living organisms by a hierarchical principle, such as Polanyi and Sheldrake postulate, what then determines the particular form of the morphogenetic field? Here we cannot escape metaphysical questions. We can produce

Aristotelian or Platonic explanations, in terms of eternal fixed forms or paradigms. Persumably Dr Francis Crick would believe the forms were set by Extra Terrestial Intelligences, and Professor Hoyle that they were generated by some kind of primaeval intelligences, out there in space, too. Once, even before this planet appeared, there already existed in a latent state the morphogenetic fields of all the forms which have ever existed on earth, or will ever exist.

The other possible answer, which seems to be Sheldrake's, is that there is a *causal influence from previous similar forms.* This requires the postulation of a mode of influence exerted across space and time of a kind unlike any known kind of physical action.

The form of a system, by this account, is not physically determined before it appears. If there are a number of possible forms of which one develops, future forms will take this realised form. This still leaves, of course, the question of how that first form came into being. This could be explained in terms of chance; or to a creativity inherent in matter; or to a transcendant creative agency (or God). The theory of morphogenetic formative causation does not have to solve this problem: it is concerned only with the *repetition of forms.* So, Sheldrake leaves aside for the moment the problems of choosing between a simple mechanistic faith and a metaphysical organicism.

The way in which a morphogenetic field could operate is discussed by the analogy of *resonance.* Certain things, like, say, a wine glass, respond to a note played on a violin, and vibrate when its vibrations reach a certain pitch. Morphic resonance could be a response to the oscillations in morphic units, happening in three-dimensional patterns of vibration.

Sheldrake suggests that once a new form has occured, it is as if, even in inorganic substances, these behave as if they have 'learned' something: a morphogenetic field has been set up. It is generally assumed that substances crystallise more easily, once they have begun to crystallise: because of the presence in the atmosphere of microscopic particles which act as 'seeds'. But it could be because of some kind of learning, by the setting up of a morphological field - and Sheldrake makes proposals for testing this hypothesis experimentally.

Once the final form of a morphic unit is actualised, the continued action of morphic resonance from similar past forms stabilises and maintains it.

Sheldrake goes on to discuss changes in organisms caused by mutation. The conventional view is that the genes responsible for these effects are involved in the control of the synthesis of proteins necessary for the normal processes of morphogenesis. On his hypothesis, the product of the gene or genes in question would not be regarded as something which switched on or 'switched off' a mere complicated series of chemical interactions, but as something which influenced the structure of a morphogenetic germ.

If heredity depends only upon the transfer of genes, and other material structures from one generation to another, then there are certain limitations on what can be passed on. The chief question here, of course, is whether acquired characteristics can be passed on to future generations. In the past there have been fierce disputations around this question, as between followers of Lamarck on the one hand and Weismann and Mendel on the other, who assumed that heredity depended only on the germ plasm in general or the genes in particular. For acquired characteristics to be passed on the germ plasm or the genes would have to be modified, and this was considered impossible - nor could the Lamarckians suggest any plausible mechanism, by which such changes could be brought about.

Yet there are still perplexing features of life- such as the patches of hard skin on the soles of our feet. Camels have callosities on their knees and baby camels are born with them. Does this not show Lamarck was right?

The Mendelians argue that such phenomena could only come about by random mutations, with any ensuing useful developments being fostered by natural selection. Although this looks like an explanation as we have seen, it offers no real explanation of how certain features which are both immensely advantageous but extremely ingenious and complex, could come into being (e.g. the apparatus of the bombadier beetle; the eye; the brain, etc.).

Many scientists have tried to show the inheritance of acquired characteristics (see R. Semon, *Das Problem der Vererbung Enworbener Eigenschaften*, 1912, and P. Kammerer, *The Inheritance of Acquired Characteristics*, 1924). In the past these theories have been bitterly fought over: Kammerer committed suicide: in Russia followers of Mendel were cruelly persecuted by adherents of Lysenko (see Z.A. Medvedev, *The Rise and Fall of T.D. Lysenko*,

1969).

Sheldrake discusses some work of C.H. Waddington which seems to show that acquired characteristics can be inherited. He seemed to show that 'if selection takes place for the occurence of a character acquired in a particular abnormal environment, the resulting selected strains are liable to exhibit that character even when transferred back to the normal environment'. He coined the word chreode to express the notion of directed, analysed development. Sheldrake believes that his hypothesis of formative causation could be seen as complementing Waddington's interpretation, that 'the selection did not merely lower a threshold, but determined in what direction the developing system would proceed once it got over the threshold'. The inheritance of acquired characteristics Waddington showed in fruit flies depends, suggests Sheldrake, *both* on genetic selection, *and* on a direct influence by 'morphic resonance' from the organisms whose development was modified in response to abnormal environments.

Rupert Sheldrake next to the question of evolutionary theory. Very little is known about evolution in the past, and evolution is not readily observable in the present. Even on a time scale of millions of years, the origin of a new species is rare and of genera, families and orders rarer still.

The evidence is scanty - melanism in moths is referred to again as the one example - and so evolutionary theory is virtually speculation - elaborated from assumptions about the nature of inheritance and the sources of variation.

Sheldrake summarises orthodox neo-Darwinian theory, which differs from Darwin's original theory in believing that the ultimate source of heritable variability is *random mutation* and that heredity is explicable in terms of genes and chromosomes.

Sheldrake's summary is as follows:

i. Mutations take place at random.

ii. Genes are recombined by sexual reproduction, the 'crossing over' of chromosomes, and by changes in chromosomal structure. These processes produce new permutations of genes which may bring about new effects.

iii. The spread of a favourable mutation is likely to be more rapid in small than in large interbreeding populations.

iv. Natural selection tends to eliminate mutant genes with harmful effects. The agents of selection include predators; parasites and infectious diseases; competition for space and food; climatic conditions and sexual selection.

v. New selection pressures come into play as a result of changes in environmental conditions or behavioural patterns.

vi. If populations are separated they are likely to undergo divergent evolution.

vii. New species (especially in plants) may arise fron inter-specific hybrids.

One theory in evolution is that sudden changes may be produced by monstrous animals and plants in which structures have been transformed, reduplicated or suppressed - the 'hopeful monster' theory. This is a way of trying to account for the prodigious diversity of living organisms.

But there are critics who argue that it is hardly conceivable that all the adaptive structures and instincts of living organisms could have arisen purely by chance - even granted that natural selection works. They also argue that parallel and divergent evolution, in which very similar morphological characters appear independently, indicates the operation of unknown factors.

Other critics, Sheldrake points out, object to the implicit or explicit assumption that evolution as a whole is entirely purposeless. (The most stimulating critique of the mechanistic theory, says Sheldrake, is H. Bergson's *Creative Evolution*, 1911. He also refers the reader to W.H. Thorpe's *Purpose in a World of Chance*, 1978.)

Sheldrake, pursuing his hypothesis of formative causation, does not doubt the Darwinian principle that genetic mutations are random. Nor that these are the basis of evolutionary development. But he also believes that where mutations bring changes which are favoured by natural selection, the repetition of the new pathways of morphogenesis in increasing numbers of organisms will reinforce the new chreodes: 'not only the gene pools, but also the morphogenetic fields of a species will change and evolve as a result of natural selection.'

Sheldrake seems to suggest that morphogenetic resonance can carry forward characteristics acquired by organisms. Morphic resonance and genetic inheritance together account for the repetition of characteristic patterns of morphogenesis in successive generations. Characteristics acquired in response to the environment can become hereditary through a combination of morphic resonance and genetic selection. The morphology of organisms can be changed through the suppression or repetition of chreodes: and some striking instances of parallel evolution can be attributed to the 'transfer' of chreodes from one species to another.

I find it a disappointment here that Sheldrake fails to challenge the fundamental assumptions of neo-Darwinism, that all creative development is uniquely the product of chance mutations. Sheldrake does, after all, postulate a force which imposes a hierarchical formative principle on cell life: and he seems aware that conventional Darwinism (as Majorie Grene says) has really *nothing to say about the origin of species.* The mechanical theory cannot explain the positive creative evolution that has evidently taken place.

We get hints of Sheldrake's perception of this here:

> Neither the repetition, modification, addition, subtraction nor permutation of existing morphogenetic fields can explain the origin of these fields themselves. Nevertheless, during the course of evolution, entirely new morphic units together with their morphogenetic fields must have come into being.
>
> p.149

We are still back at the root of the problem: even if 'chance mutations' are the sole agent of change, the question still remains of the 'gradient', the way nature seems to be trying possibilities - for which dynamic there is no cause, no reason for it to exist, in conventional theory. The same would apply equally, if morphogenetic fields develop alongside or in conjunction with changes brought by mutation.

What we need is thinking about the morphogenetic field or dynamic that generated life in the first place, and which makes it *strive.* Sheldrake seems to betray his Cambridge training, by (as it seems) locking his morphogenetic theory in the chains of neo-Darwinian mechanism. All he seems to say is that even if there are 'jumps' in which morphogenetic forms originate, these cannot be explained 'within the frame-work of science in terms of preceding causes'.

What then? Sheldrake simply leads us back to metaphysics, saying that the choice between the alternatives can never be made on the basis of any testable scientific hypothesis: 'from the point of view of natural science, the question of evolutionary creativity can only be left open'. The choice remains, a creative agency pervading and transcending nature; a creative impetus immanent in nature; or a blind and purposeless chance.

It seems to me that Cartesian dualism lurks behind this separation of scientific thought from metaphysics. The *fact*, the patent *reality*, which confronts us, not least in our own autonomous thinking existence, is of a *creative agency which pervades and transcends nature.* We are here, and we know: these are the facts, as is all life, which some creative agency has brought into being. Why are we here at all? How did mice, elephants and oak-trees come into being? They have plainly been created, as if by an intelligence. The morphological formative fields may be an agent of that creativity: neo-Darwinian theory simply has nothing to say about the fact of creativity, and cannot explain it.

If science is to advance, it must surely accept the challenge, of trying to answer these questions? It will hardly do for science, when confronted with such a massive fact as the natural world, to refuse to tackle it! I find Sheldrake's mood at this point a strange capitulation to conventional materialism, despite his assertion that the question is 'open'!

Sheldrake next discusses movement and internal processes in animals and plants, and how these are controlled by morphogenetic fields which here he calls *motor* fields. These are hierarchically organised. All this seems convincing and bears out the views of some of the biologists discussed by Marjorie Grene in *Approaches to a Philosophical Biology*, who take the view that creatures must be understood as wholes. To understand any complex living system we must examine it as a whole, and not insist on breaking it down, atomistically, into its components, as the sole way of explaining it. Discussion of Regulation and Regeneration here could be seen as parallel to that of Kurt Goldstein in his experiments with brain-damage patients.

At one point (p.168) Sheldrake dismisses the notion of 'intelligence' as applied to regeneration, as when a damaged organism repairs itself. I believe this is a pity, since there does seem to be some overall force at work by which the whole organism operates in a crisis, of loss of parts or damage. It is surely an aspect of what E.W.F. Tomlin calls 'primary consciousness'? As he goes on through this area of his work, Sheldrake seems surprising to become increasingly reductionist, his morphogenetic field theory simply taking the place of behaviourist approaches. While (p.172) he dismises the view that there can be simple explanations of living processes in terms of specifically localized physical or chemical traces within the nervous tissue, he seems to want to reduce behaviour to 'formative causation', to 'morphic resonance' which has no physical existence. While Sheldrake wants to escape mechanism, one can feel him being tugged back towards it, or flying out towards magic.

Summarising many experiments K.S. Lashley declared,

> it is not possible to demonstrate the isolated localization of a memory trace anywhere within the nervous system . . .

Another researcher is quoted as saying 'memory is both everywhere and nowhere in particular'. As Sheldrake says, these observations are puzzling from a mechanistic point of view. He believes the hypothesis of formative causation provides an alternative: 'the habits of learning depend on motor fields which are not stored within the brain at all, but are given directly from the past states by morphic resonance'.

This seems absurd, however, because it is an escape from mechanism into magic. Sheldrake rejects the view that memory 'traces' are stored in the brain like inference patterns in a holograph. But it seems no solution to the problem to take the matter out of the brain altogether! As Straus writes, 'man thinks not the brain': where memory is concerned, the whole body and being is involved, including the brain and nervous systems as instruments. The mechanistic approach is at fault in trying to understand the process by reduction to minimals. In the 'whole' process dynamics like 'fields' may play a part: but undoubtedly it is the flesh and blood which remembers, with its intelligence: to relegate memory to a 'morphogenetic field' alone seems too much like introducing again a ghost in the machine, to solve the problems of dualism.

Sheldrake's discussion of learning and such processes seems to me to be too much conducted still within the behaviouristic naiveties so profoundly exposed by Erwin Straus, while his concept of learning is too passive - it lacks recognition of that kind of learning, in the sense of discovering new things, discussed by Marjorie Grene in relation to Plato's *Meno*: that is, Sheldrake fails to escape from essentially mechanistic perspectives. He sees learning as a form of behaviour merely regulated by the dynamics of formative causation. This seems the invocation of some kind of magical force, to explain away the difficulties of understanding the mysteries of learning, while his examples are too exclusively taken from behaviourist sources.

The trouble here is that the scientist is often imprisoned in his paradigm, even when he struggles to break out of it: a study of Marjorie Grene's *Approaches to a Philosophical Biology* reveals that a number of biologists are thinking outside the behaviourist paradigm, and have managed to make radical criticisms of it. Polanyi and Marjorie Grene have also emphasised the relevance of gestalt psychology to learning theory and the kind of phenomenon represented by the phrase 'a leap of the imagination': *real* learning, in which a sudden completely new apprehension is developed by the mind. Sheldrake simply applies his theory of 'morphic resonance' and 'motor chreodes' to problems of animal behaviour, with apparently no knowledge of the work done by biologists such as Adolf Portmann and Erwin Straus, which points to quite different modes.

Similarly, when he comes to the inheritance and evolution of behaviour, he confines himself to accounts of experimentation within the established (quantitative) reductionist and empirical paradigms, in which attention is focussed on the nervous system or on the functions of enzymes.

He says that,

> In mechanistic biology, a sharp distinction is drawn between innate and learned behaviour: the former is assumed to be 'genetically programmed' or 'coded' in the DNA, while the latter is supposed to result from physico-chemical changes in the nervous system.
>
> p.185

Since neither can modify the DNA, it is therefore considered impossible for the learned behaviour to be passed on, except by 'cultural inheritance' (i.e. by learning).

But Sheldrake does not being to bear on this question the implications of Polanyi's observations (to which his earlier thinking relates, as we have seen) that DNA 'coding' is not enough to explain the growth and development of life: life molecules obey physical and chemical laws, and DNA structure principles - but also (like folded proteins) obey other laws, too, and speak 'another' language, otherwise all we would have is 'noise'. The same is true of 'the nervous system' which, as Marjorie Grene says in *The Understanding of Nature*, has its own laws and principles as a whole, which are not explicable in terms of their reduction to particulates and minimals.

This philosophical problem is not solved by trying to understand the problem in terms of motor fields and morphic resonance, merely superimposed on the mechanistic approach. Some kind of ('systems') analysis of the overall hierarchical principles of large life-molecule and organic forms is necessary, but it requires a totally new dimension of making, to do with *forms* rather than structures and substances.

Sheldrake reports some experiements in which it does seem that animals of later generations did learn more quickly, as if knowledge could be passed on in a Lamarckian way. (W. McDougall, An experiment for the testing of the hypothesis of Lamarck, *British Journal of Psychology*, 17, 267-304, 1927; Second Report on a Lamarckian experiemnt, *British Journal of Psychology*, 20, 201-345, 1930).

The accounts Sheldrake gives of the experiments are fascinating. Some experimenters found that the question must remain open, only. But others found that there was a marked tendency for rats to learn more quickly in subsequent generations, when the parents has been given thorough experience of learning by solving, and that there was a real change in constitution in later generations of rats, an improvements 'whose nature' (said McDougall) 'I am unable to suggest' (J.B. Rhine and W. McDougall, Third Report on a Lamarckian experiment, *British Journal of Psychology*, 24, 213, 235, 1933).

As Sheldrake says, these results (which others seem to have confirmed) make no sense in terms of current ideas. He believes they do make sense in the hypothesis of formative causation even if they do not *prove* formative causation. There could be other reasons why the subsequent generations of rats become more intelligent, for instance.

But I feel there needs to be more radical assessment of the nature of this kind of experiment. For one thing, the rats were enclosed themselves, as it were, in a human-created environment, an intellectual cage - a paradigm itself. As Erwin Straus points out, there is in the laboratory situation, a great deal of naivety and falsification - derived from Pavlov's over-simplification of the problems, and associated with a blindness to the real problems, which have to do with kinds of causation. (See *The Primary World of Senses,* 1963.)

Certainly it seems questionable whether the mechanistic model of explanation here is adequate: can existence only be dealt with by empirical generalization? As Marjorie Grene points out, the rat-and-pathway kind of experiment is essentially based on the Humean Model of constant conjunction. But in the experiments of F.J.J. Buytendijk, for example, neither the phenomena studied not the scientist's approach to them can be reasonably interpreted in this way. Buytendick was trying to construct experimental situations such that the behaviour of his individual subject could be seen in its *intrinsic* significance, phenomenolopically that is, in terms of consciousness and 'being'.

Buytendijk reports the behaviour of a rat in his *Weg zum Verständis der Tiere*, in which it seems that the rat had an 'idea' of where a block of wood was, floating on the surface of water in a tank about its head. It has an 'exact apprehension' of 'the nearest distance' to the block, even though it couldn't see it.

What Buytendijk offers is not an *explanation* of the rat's behaviour, but an attempt to *understand* the behaviour in its natural form:

> The alternative is that it is directed to form as such, conceived as apart from existence; in other words, that his method is phenomenological ...

Buytendijk emphasises that 'Back to the things themselves' (a slogan from Husserl) means ... an ever renewed participation in the phenomenal world, our world with its rich structure of significant ...' Buytendijk is looking for insights into animal experience, in the kind of consciousness in animals. He thus includes in biology the 'intrinsic interest' of living things as against dead nature - which embraces a certain subjectivity. This will enable biology to include 'levels of organisation', 'centricity' (a concept from Portmann), and such concepts as 'unaddressed display' - that it, the manifestation of life on the surface, *not* intended to signal in any functional way to any spectator. His animals are beings.

In this there is not necessarily teleological thinking, but there is a recognition of the *directiveness* (not 'directedness') of life: it entails a nonreductive, non-Galilean, ontology (see Marjorie Grene, *Approaches*, pp 140-147).

If we take these dimensions into account, who can doubt that we need to examine living creatures in relation to a gradient? That, clearly, as a fact, there are 'lower' animals and 'higher' and that there is a *direction* in the natural world towards more and more complex organisation: as Marjorie Grene puts it, 'only higher forms exhibit the full pattern of growth and maturation ... Higher animals ... show a more marked contrast of *inner* and *outer* aspects than do lower forms' - and there are other distinctions.

No mere mechanical mutation plus natural selection theory could account for the development of 'higher' forms: if we admit a gradient we must admit that *directiveness* of which conventional evolution theory can offer no explanation. So, in an experiment on learning in rats, one might well expect a real change in constitution of the whole stock which has not come by learning, but by the development in the consciousness or 'centricity' of the creatures, not passed on by the 'DNA codes', but by the other dynamics or hierarchical principles which organise the systems. Perhaps this is what Sheldrake means by his 'morphic resonance': but in proposing experiments to show this, he should surely also try to escape from paradigms and methodologies which insist on reducing the problem to the Humean idea of 'constant conjuntion', and linear explanation on the basis of animals?

Sheldrake seems to glimpse a way out of the limitations of mechanistic explanations, as by his remark that mutations or exposure to unusual environments 'could enable an animal to 'tune in' to other species' motor chreodes' (p.193). But if it is possible for animals to pass on knowledge or modes of behaviour by such 'tuning in', it is surely not necessary to try to explain everything in terms of mechanistic 'coding'?

As Sheldrake says the mechanistic, or neo-Darwinian, theory assumes that innate behaviour is 'programmed' or 'coded' in the DNA, and that new types of behaviour are caused by change mutations. Chance mutations are also assumed to give animals capacities for particular types of learning. Then animals whose survival and reproduction benefits from these capacities are favoured by natural selection.

But if we accept with Polanyi that such complex dynamics as learning capacities cannot possibly be organised by the 'DNA code', any more than complex and varied processes of physical growth, and cannot therefore be reduced to chemical laws, some other explanation is required - e.g. for such complex modes of behaviour as the way certain birds, even though they are orphaned and never meet their parents, can navigate by the stars! This cannot be coded in DNA molecules, but may be inherited in some highly complex system by which DNA molecules are organised, within a complex hierarchical structure, of foldings and interactions, tensions and fluxes - which needs to be studied holistically in its own right, as a manifestation of primary consciousness. Moreover, the dynamic to become 'higher', to 'strive' - these evident features of life are not explained by any reduction to 'codes' or 'mutations', and nor can such a complex manifestations as learning and instinct. The surprising thing is that Sheldrake does not reject mechanistic theory more forcibly, since it explains so little

For he continues to defer to mechanistic theory, apparently on the assumption that there must be a mutation before any significant development can take place, and even tries to adopt his 'motor field' theory to natural selection. Towards the end of his chapter on behaviour he seems to me to confuse the issue further, by continuing to talk about 'chreodes' even when he is discussing social patterns and cultural patterns. When we begin to take into account human society and culture we must invoke human consciousness, and very complex issues arise of perception and autonomy - the human being's whole relationship to reality.

How much of this can be profitably examined in terms of 'chreodes' which are forces which govern the forms of inorganic and organic entities at the level of dynamic form? The way in which an autonomous creature develops as though the processes of early growth in relation to the mother is far too complex to relate to such systems. In child psychiatry Winnicott discusses 'primary material proccupation' and 'creative reflection'. These are imaginative - sympathetic processes and I believe it is clear that they involve something like telepathy. Sheldrake prefers some kind of mysterious controlling force like radio waves or magnetism:

> As the process of learning begins, usually by imitation, the performance of a characteristic pattern of movement beings the individual into morphic resonance with all these who have carried out this pattern of movement in the past. Consequently learning is facilitated as the individual 'tunes in' to specific chreodes.
> p.196

Supposing one tries to apply this to the behaviour of the infant at the breast? Even there, 'instincts' are affected by dynamics of relationship and fantasy (as Winnicott has shown) so that any organic explanation will not do, even if the organic approach is organismic enough to take into account the dynamic of hierarchical organisation and system.

Morphic resonance begins to appear, actually, towards the end of Sheldrake's book as a panacea or universal principle, used to solve all problems - while Sheldrake remains at times still locked in the mechanistic paradigm - as in his deference to chance mutation theory.

He suggested, for instance, that if his morphic resonance theory is true, it should have become easier for human begins to learn to ride a bicycle, owing to the 'cumulative resonance' from the large number of people who have acquired these capacities. But this seems to fail to take into account the very complex processes of 'indwelling' such as Polayni examines: such skills are not merely functional, 'motor', capacities, but involve many tacit and subsidiary powers, efforts of consciousness and body life, potentialities of being, which we can only cooperate with: they seem irreducible to any form of experimentation which could yield results of the kind Sheldrake would like to see done. Nor will it do to say that answers to such problems as that of human creativity 'can only be given in metaphysical grounds': the question is not one of deciding between a futile mechanism and metaphysics, but of developing a more adequate science of life, which can really find the category of life and advance understanding of it: can find being, consciousness, form and time in new perspectives.

Does Sheldrake, finally, help us here? Negatively, I believe. He offers his book as a preliminary sketch, offering a testable hypothesis. He says, however, that questions posed by the origination of new forms and new patters of behaviour, and by the facts of subjective experience, can only be answered by metaphysical theories.

But this is to leave Newtonian-Galilean science untouched - and leaves the situation as Edmund Husserl diagnosed it (in *The Crisis of the European Sciences*) unaltered. There are times when Sheldrake seems still imprisoned in Cartesian paradigms, and times when he seems to be resorting to magic.

But what his work does indicate is a radical dissatisfaction with conventional science. He is aware that there is something else - some other dynamic in life - urgently requiring new forms of thinking to arrive at its understanding.

At the end of his book Sheldrake examines four possible philosophical positions, and tries to show how these are reconcilable with his theory of formative causation.

The first is modified materialism. Materialism starts from the assumption that only matter is real. Yet, as Sheldrake admits, the concept of matter has no fixed meaning in the light of modern physics. Morphogenetic fields, he suggests, can be regarded as aspects of matter. If we regard these forces as modes of hierachical organisation of matter, governing it towards the directiveness of life, then I believe modified materialism requires a radical reconsideration of all assumptions based on materialism.

In conventional materialism, says Sheldrake, brain states are considered to be determined by a combination of energetic causation and change events. In his theory, conscious states are best thought of as an aspect of epiphenomenon of the 'motor fields acting on the brain.' But this is still not to escape from the determination of materialism. The first fact of existence is consciousness, and Husserl's importance is that he restored this emphasis (which was also Descartes' before he abandoned it). 'There is something else in the universe besides matter in motion - there is knowing mind.'

If we take mind not as an epiphenomenon, but as the manifestation in us of these creative strivings which are also expressed in the morphogenetic dynamics that govern all matter and life, then we shall surely stand materialism on its head? There is 'nothing else' there, by way of entelechy of soul-stuff: but matter itself (as in modern physics) is stranger than atomism ever supposed, more dynamic, more full of order, more creative. The morphogenetic fields at every level of every hierarchy are striving to organise everything into ever more complex and higher states, and consciousness (and our minds) are the culmination of that complex in all things.

If this is so, then materialism, which is the basis of blunt and grim political theories such as dialectical materialism, needs radical reconsideration. We should not try to tackle our destiny by exerting an 'objective' science on human society in the way of the physical sciences (which is the Marxist way): we need first to try to understand life and find a more adequate place for man in nature. In this we need to look at what are man's first needs - and these are for meaning, the exercise of creativity, and order, order in the sense of being-at-one-with, in being-in-the-world. All these are violated by scientific Marxism, not least creativity: because of Marxism's essential adoption of nineteenth century atomistic science. It is for this reason that it has become dead and sterile - a monolithic and conservative tyranny which resists change and suppresses creativity, because it fails to find *being*.

The reasons why are clear from Sheldrake's final paragraph on materialism:

> according to this modified philosophy of materialism, the universe is composed of matter and energy, which are either eternal or of unknown origin, organised into an enormous variety of inorganic and organic forms which all arise by change, governed by laws which cannot themselves be explained. Conscious experience is either an aspect of or runs parallel to the motor fields acting on brain. All human creativity, like evolutionary creativity, must ultimately be ascribed to chance. Human beings adopt their beliefs ... and carry out their actions as a result of chance events and physical necessities within their brains. Human life has no purpose beyond the satisfaction of biological and social needs; nor has the evolution of life, nor the universe as a whole, any purpose or direction.
>
> p.201

This is no different from the pessimistic dogma we have looked at above: it presents a totally deterministic picture of the universe and man in it, and it implicitly denies the primary fact of the universe - that life has evolved and become mind.

Next Sheldrake examines 'the conscious self' and tries to solve the mind-body problem by seeing consciousnes as an interaction between 'motor fields' and the brain. He sees that the conscious self has properties of its own which are not reducible to those of matter, energy, morphogenetic fields and motor fields. He opts, to solve the problem of memory, for some 'extra' power of the 'direct access of the conscious self to its own past states' (referring us here to Bergson). Fundamentally, however, this account of mind and self remains within Descartes' distinction between *res mensa* and *res extensa*. Sheldrake talks about consciousness, but has not found it: he talks of 'conscious causation' simply to distinguish it from instinct. The kind of consciousness discussed by Husserl, the intentional 'I', in relationship with the world, is not here - though it is glimpsed:

> The conscious self must at some stage, in a quantitative leap, have become aware of itself as the agent of conscious causation.
>
> p.205

That 'conscious self' is an entity which can only be found phenomenologically, and is quite alien to all discussions of behaviour, instinct and appetite, such as surround it here, belonging to the reductionist paradigm. So, Sheldrake remains on the horns of a dilemma, even in his hypothesis of formative causation. The problem of evolutionary creativity remains unsolved:

> the reality of the conscious self as a casual agent is admitted, but the existence of any non-physical agency transcending individual organisms is denied.
>
> p.205

- and though this seems like the exclusion of God or some spiritual agency, it also means the exclusion of any dynamic of striving towards high states, giving rise to higher beings and men.

Under 'the creative universe' Sheldrake returns to Bergman's *élan vital*, and discusses the possibility of higher-level creative agencies, 'higher selves'. 'Under certain conditions human beings might become directly aware that they were embraced or included within them.'

(- Angels affect us oft ...?)

> ... the experience of an inner unity with life, or the earth, or the universe, has often been described, to the extent that it is indescribable.
>
> p.206

This hierarchy of conscious selves, however, could not have given rise to the universe in the first place. Nor could this immanent creativity have any goal if there were nothing beyond the universe towards which it would move.

> So the whole of nature would be evolving continuously, but blindly and without direction.
>
> p.206

This philosophy denise the existence of any ultimate creative agency transcending the universe on a whole.

Lastly Sheldrake turns to *transcendent reality.* This fourth metaphysical position affirms the causal efficacy of the conscious self, *and* the existence of a hierarchy of creative agencies immanent within nature, *and* the reality of a transcendent source of the universe.

These metaphysical positions seem to take us by stages, away from the first position of bleak meaningless materialism, through recognition of what used to be called the soul (the conscious creative self), to the embracement of angels (creative hierarchies in the universe) to God, whose transcendent consciousness would be its own goal, and would be complete in itself:

> If this transcendent conscious being were the source of the universe and of everything within it, all created things would in some sense participate in its nature. The more or less limited 'wholeness' of organisms at all levels of complexity could then be seen as a reflection of the transcendent unity on which they depended, and from which they were ultimately derived.
>
> p.207

But at the end it seems to me that Sheldrake does not solve the problems of being and mind, by returning to God.

Sheldrake's is certainly a fascinating and courageous book. It is difficult to see why orthodox scientists should, however, want to burn it, since nowhere does it threaten the scientific paradigm: like Cartesian dualism, it simply allows it to exist alongside God, and the *res mensa.* Thus, in the end, I believe, the book fails to disturb the traditional paradigm, and merely leaps back into conventional religion. This is also what Francis Crick, Fred Hoyle and Professor Wickramsinge have really also done - under various disguises - God being called 'Extra-terrestial Intelligences', the Holy Spirit, a rocket: and the spirit of God a form of 'experimental life-stuff' manufactured for testing on earth (the idea of God testing out man runs through the mythology of the Old Testament). All these, along with Creationism, and Sheldrake's wildest extrapolations of morphogenetic causation (as in his application of this principle to human behaviour), seem essentially just another form of mystical or miraculous solution.

But the best part of Sheldrake's book is his *dissatisfaction* - his recognition that another dimension is called for, both of the organisation of life, its forms and patters, and of the disciplines of thought necessary to understand it. This, from a Cambridge biologist, is worth having.

In the end, alas, he leaves the deterministic universe of materialism untouched, since his higher powers exists in a separate compartment, or like ghosts in the machine: so, too, he leaves mechanistic DNA theory untouched, too, and does not expose its radical fallaciousness as Polanyi did. So, he fails to offer a critique of evolutionary theory: it is still, in his *Weltanschauung*, operated by chance mutations - even if there is no Prime Mover. The actual evident fact of creative dynamics in the stuff of life itself escapes him - as does any real explanation, except in terms of a postulated force of 'morphic resonance', which has a shadowy existence, compared with the very real dynamics manifest (say) in Driesch's sea urchins, in regeneration: or in Tomlin's 'primary consciousness'.

But from the point of view of these of us in the Humanities and especially English, we can surely see from Sheldrake's book that the pessimistic, deterministic, materialistic philosophies we take to be the 'inescapable' lessons of 'science' are the very position from which the scientists are trying to extricate themselves - not because they are appalled at the nihilism they imply, and the way they erode meaning and values: but because they *don't work*, they don't explain the phenomena, and do not illuminate the facts, do not lead us to the truth.

9 The Selfish Gene: dangerous extrapolation from microbiology

In August 1976, I received a catalogue from Oxford University Press, and was startled to read the following advertisement :

The Selfish Gene
Richard Dawkins

'Our genes made us, body and mind. We exist for their preservation and are nothing more than their throwaway survival machines. This is how Richard Dawkins introduces us to the world of the selfish gene : a world of savage competition, ruthless exploitation and deceit - seen not only in aggression between rivals, but also in the more subtle battles between the generations and between the sexes. But what of the acts of apparent altruism found throughout nature : for example, mothers working themselves almost to death for their children, or small birds risking their lives to warn the flock of an approaching hawk? The author shows how acts like these also result from gene selfishness, and in doing so he explodes the view popularized by Konrad Lorenz and Robert Ardrey that animals behave 'for the good of the species'. The book grips like a thriller, and is alive with fascinating stories : about fish who queue up to have their teeth cleaned - and then refrain from swallowing the tiny dentist; about ants who take slaves and tend fungus gardens; about the kamikaze bees who commit certain suicide when they sting robbers of the communal honey. But the most unexpected sting in this unexpected book is the one in the tail - the last chapter, which gives us a new, even startling, way of looking at ourselves, and our unique culture. We are the only animals capable of seeing through the designs of the selfish genes, and of rebelling against them'.

There is no doubt that what is offered by the advertisement is not simply an argument in philosophical biology. It exhibits an aggressive general philosophy of life. Our minds and bodies are 'made' by our genes : we exist for the preservation of the genes, and we are '*nothing more than* their throwaway survival machines'. In this language, clearly, there is an assault on our view of ourselves, such as we are concerned with, in the Humanities, on the basis of reductivism. We are only good *against* our true nature.

In the language of the advertisement, there is a doubtful anthropomorphism. 'The world of the gene machine is one of savage competition, ruthless exploitation and deceit' - and since we are only the throwaway vehicles of genes, the whole human world is implicitly a world of savagery, too. Even altruism, (again a word with human connotations) results from the selfishness of genes : the air of 'unmasking' accords, of course, with the tendency in 'scientific' sociology and Freudian metapsychology.[1] The scientific realist is to expose the cosy view that animals behave altruistically 'for the good of the species'. The last chapter is to give us a new, even startling, way of looking at ourselves, and our unique culture.

---

1 See Paul Roubiczek's *Ethical values in the Age of Science* and the present author's *Human Hope and the Death Instinct*.

The book opens with an aggressive metaphysical tone, not from the author, but from the writer of the Foreword, Professor Robert L. Trivers, of the Museum of Comparative Zoology at Harvard. Most human thinkers, he says, see themselves as 'stepping stones to the Almighty'. But there exists no *objective* basis to elevate one species above another (the fact that no chimpanzee has yet become a Professor or even an Associate Professor of Comparative Zoology does not apparently count as an 'objective' phenomenon).

Everyone is bound together by the fact that they have all evolved 'by a process known as natural selection' :

> Within each species some individuals leave more surviving offspring than others, so that the inheritable traits (genes) of the reproductively successful become more numerous in the next generation. This is natural selection : the non-random Natural Selection has built us, and it is natural selection we must understand if we are to comprehend our own identities.
>
> p.v.

This seems to be a statement of neo-neo-Darwinism, of genetic evolutionary theory. Beginning at the end of the paragraph, it is not at all clear how the way to 'comprehend our own identities', our own nature, is to understand *natural selection.* And if we understand natural selection, since it is a negative process, it cannot have built us.

Professor Trivers goes on to say however that Darwin's theory of evolution through natural selection is 'central to the study of social behaviour' (especially when wedded to Mendel's 'genetics'), and it is clear that in this sphere of thought, as we shall see, there is a politics implied. It is assumed that natural principles, directed at the survival and enrichment of 'gene pools', *determines* social behaviour in animals, and, probably, man. It is this important body of social theory (on which R.A. Fisher, W.D. Hamilton and G.C. Williams have been writing) which is presented as a popular form by Dawkins.

Dawkins presents his case, says Trivers, with a confidence that comes from mastering the underlying theory.

> Darwinian social theory should vitalise our political understanding and provide the intellectual support for a science and medicine of psychology.

In his Preface, however, Dawkins urges us to read his book almost as if it were 'science fiction'. He promises to be 'entertaining and gripping'. Of the reader he says, 'I at least hope that the book will entertain him on a train'. Yet it is clear here too that in being so diverted we are to be swept away into a certain general philosophy of life rooted in mechanism :

> We animals are the most complicated and perfectly designed pieces of machinery in the known universe...
>
> p.x.

A scientific book, popular or serious, ought today, surely, to be very careful of not stepping beyond its limits, and falsely extrapolating into regions of thought and opinion where science has no authority?

No such doubt inhibits Stephen Dawkins. The heading of his first chapter is in fact a metaphysical question which science is not qualified to attempt : 'Why are people?'. It is in fact, as he makes plain, a child's question. As such it is a legitimate question for philosophy and poetry - but not science. It is clear from Dawkin's ebullient opening that he has no sense of the complexities of such philosophical problems – that he is naive :

> Intelligent life on a planet comes of age when it first works out the reason for its own existence. If superior creatures from space ever visit earth, the first question they will ask, in order to assess the level of our civilisation, is 'have they discovered evolution yet?' Living organisms had existed on earth, without ever knowing why, for over three thousand million years, before the truth finally dawned on one of them. His name was Charles Darwin....
>
> p1.

This seems like a re-vamping of Pope's enconium :

> Nature and Nature's laws lay hid in night :
> God said, 'let Darwin be' : and all was light!

Surely neither Newton nor Darwin – magnificent as their perceptions of *what is* may be – have provided reasons for our own existence. They may have advanced our knowledge of *how*, but not of *why* : indeed there is a sense in which they have made the 'why' problem worse. But Dawkins goes on :

> It was Darwin who first put together a coherent and tenable account of why we exist.

This is not so. As we have seen, Darwin, by a superb act of imagination, put forward hypotheses about how life evolved. He had nothing to say about how it originated : and he did not explain how species came about. Certainly, he explains nothing of *why* life exists : Darwin did not offer a philosophy of being and it is false and simplistic to argue that he did, or that a metaphysic can properly be drawn from his work, though we are dogged by such an extrapolation.

Implicitly, man and man's mind are taken for granted in Dawkin's first paragraph : 'Intelligent life on a planet comes of age when it first works out the reason for its own existence' : the proper objects of study surely must be life, *consciousness* and *existence*? But, as Marjorie Grene points out, Darwinism has a reductive effect on life :

> What was so triumphantly successful in Darwin's theory was precisely its reduction of life to the play of chance and necessity, its elimination of organic categories from the interpretation even of living things... Darwinism... is reductive, and still essentially Cartesian, in its interpretation of the organic world... nature as a mechanically interacting aggregate of machines : this is the Darwinian vision. It is the extension of the machine image to life itself.
>
> *The Knower and the Known.* p.185.

Dawkins' book is all this. In vain does one look in it for a recognition – despite the opening sentence, and the concessions to autonomy in the last chapter – of the view expressed by Marjorie Grene at the opening of her chapter on *The Faith of Darwinism* :

> Whatever I succeed in knowing, it is *I* who achieve knowledge : I in my contingent, personal existence, I-in-situation. Such an 'I' is alive. 'Minding', as Ryle calls it, is one form of living, and can be understood only as a species within that genus. But living, in turn, is one kind of natural being.

This is the new ontology, and it recognises the *Lebenswelt*, the life-world, in which alone we live. Here is the new perspective in philosophy which can save us.

Dawkins, however, quickly sets about philosophy : to him, Darwinism has solved all the problems in a new way :

> philosophy and the subjects known as 'humanities' are still taught as if Darwin had never lived....

and he quotes G.G. Simpson, 'The point I want to make is that all attempts to answer that question (What is man?) before 1859 are worthless, and that we will be better off if we ignore them completely'.

I will not list the thinkers, philosophers and poets of every age and every civilisation, which this dismisses, but again we have the expression of a blind faith : there is only one true prophet, and his name is Darwin.

Dawkins' book consists of a series of absolute or sweeping statements, followed by disavowals by which he tries to disclaim any intention to settle ethical or even philosophical issues. Yet he has an evident optimism, that ethical issues can now be settled on the basis of 'scientific fact'. He is not going to write a general advocacy of Darwinism. But he is going to explore the consequences of the evolutionary theory for a particular issue. 'My purpose is to examine the biology of selfishness and altruism'.

Moreover, this, it seems is to provide a universal psychology if not a philosophy of being :

> the human importance of this subject is obvious. It touches every aspect of our social lives, our loving and hating...

Lorenz and Ardrey, apparently, are 'totally wrong' on these matters, because they believed that natural selection worked for the good of the species. Dawkins' view is quite different :

> Unlike both of them, I think 'Nature red in tooth and claw' sums up our modern understanding of natural selection admirably.
>
> p.2.

No nonsense about that. We have only turned Dawkins' first page, but we have already been presented with the following assertions :

> Darwin was the first thinker to give a reasoned explanation of the nature and purpose of life.

All explanations before 1859, and all religious explanations are worthless.

Not only did Darwin explain how we are here : he also solves the problem of the meaning of life, and why we are here.

Philosophy and the Humanities have not attended to these theories, and their profound philosophical significance.

Evolutionary theory and the idea of natural selection are scientific truths.

Other zoologists are wrong, in believing that natural selection operates for the benefit of the species.

Natural selection and evolutionary theory show that animal life is an utterly ruthless competition, and nature is dominated by destructiveness. Man only exists in opposition to this savagery, though this is against his real nature.

There are also political implications, for it is on this kind of 'realism', of course, that many modern political brutalities are based (See : Polanyi's essay 'Beyond Nihilism' in *Knowing and Being*).

Now the *argument* is promised :

> The argument of this book is that we, and all other animals, are machines *created by our genes*.... the predominant quality to be expected in a successful gene is ruthless selfishness. This gene selfishness will usually give rise to selfishness in individual behaviour.
>
> p.2.

There is evidence of altruism in the world : this, however, can be easily explained away :

> there are special circumstances in which a gene can achieve its own selfish goals best by fostering a limited form of altruism at the level of individual animals.
>
> p.2.

As with Freud, any altruism is only apparent : only 'unmask', and beneath the surface there will be organic savagery :

> universal love and the welfare of species are concepts which do not make evolutionary sense.

Dawkins tells us 'he is not advocating a morality based on evolution. I am saying how things have evolved'. This is to avoid being misunderstood -

> by those people, all too numerous, who cannot distinguish a statement of belief in what is the case from an advocacy of what ought to be the case.
>
> p.3.

But it is not true that Dawkins is not advocating a morality based on evolution. That is exactly what he is doing, on the basis of the erroneous belief that evolutionary theory provides an adequate account of existence – so that the brutal ethic that follows must be felt to be inescapable - virtually 'fact' itself. Moreover, he here goes on :

> my own feeling is that a human society based simply on the gene's law of universal ruthless selfishness would be a very nasty place in which to live. *But, unfortunately, however much one may deplore something, it does not stop it being true.*
>
> (my italics).

Thus it must seem that to build a society based on altruism is to operate against the 'real' nature of life. His whole argument is directed at making us feel that we cannot escape from the facts (of universal selfishness) which science has established beyond all doubt.

Such assumptions are evident at every step, as in phrases like : 'genetically programmed to be altruistic', 'dominated by culture'. Some of Dawkins' statements are so absolute as to be breath-taking :

> If you look at the way natural selection works, it seems to follow that anything that has evolved by natural selection should be selfish...

Dawkins takes any animal behaviour that looks 'selfish' and uses it anecdotally to vindicate his underlying theory. He tells us that he is 'not trying to make a point by telling stories' – but that is exactly what he is trying to do. His use of these accounts do not impress with their meticulous scientific presentation : for instance, he says that praying mantis females bite off the head of their mates and, since this destroys the inhibiting nerve centres, this may actually 'improve' the male's sexual performance. (This phenomenon is now thought only to occur in captivity : see *Animal Behaviour*, vol 32, pp. 916-917 (1984)). This is absurdly anthropomorphic, in a mechanistic way : is there evidence from the animal world that quantity of physical copulation is a 'benefit' for which creatures strive? But in any case, beneath his whole argument is a quite false extrapolation into human life and culture deductions from a highly selective zoology.

Dawkins next attacks the 'good of the species' argument, which he believes is erroneous, and wrongly taught in schools. Evolutionary theory, as he points out, rests upon the 'differential survival of the 'fittest". But are we talking about the fittest individuals, the fittest races, the fittest species, or what? One theory put forward is 'group selection', and others include 'individual selection'. Though Dawkins says 'it does not greatly matter', the truth is, as we have seen, that how we decide here is *fundamental* to Darwinism, since the *only* way in evolutionary theory by which development can come is by *accidental mutations* being selected out in the *struggle for existence.* As we shall see, Dawkins applies his elastic 'selection' theory with equal verve to animals, genes and even molecules.

In the course of his argument Dawkins produces asides which confuse the human and the animal to such an extent as to be ridiculous : discussing how some individuals may be sacrificed to save the whole group from destruction he says : 'how many times must this have been said to the working people of Britain'? In this confused discussion he quotes various sources on altruism and aggression : 'In higher animals, behaviour may take the form of suicide to ensure the survival of the species' (*Nuffield Biology Teachers' Guide*); aggressive behaviour has a 'species preserving' function (Konrad Lorenz, *On Aggression*). From one source he quotes an idea based on the observation of how, where a majority of baby spiders are eaten as prey, 'perhaps this is the real purpose of their existence, as only a few need to survive in order for the species to be preserved' (BBC programme on Australian spiders). What these examples show, however, is that in order to preserve the orthodoxy, biologists are willing to twist evolutionary theory to make it applicable to any observation while at the same time conveying a kind of metaphysical implication about the human applications of biological observations. Whatever Dawkins' intentions in science, the argument is extremely doubtful by any account.

After some discussion of moral and political ideas ('this is the basis of trade unionism') in which Dawkins leaps from biology to society and back with reckless disdain for levels of argument, he insists that 'speciesism' has no proper basis in evolutionary theory.

We were under the impression that, since Darwin's work was called *The Origin of Species*, evolutionary theory was based on the survival of the species, by the effectiveness of small advances – later (in neo-Darwinism) said to be brought about by mutations. Now, however, Dawkins' kind of neo-Darwinism takes the reductionist principle further : 'the best way to look at evolution is in terms of selection occuring at the lowest level of all'. This is an argument taken over from G.C. William's *Adaptation and Natural Selection.* The fundamental role of selection, and therefore of self-interest (the parenthesis implying that these are identical) is not the species, nor the group, nor even, strictly, the individual : 'It is the gene, the unit of heredity'. (p.12).

This seems to make Dawkin's theory quite distinct from Darwinian evolutionary theory. From his observations of animals and the fossil records Darwin postulated his concepts of tiny developments adding up to macro-developments, and these being fostered by natural selection in the ruthless struggle for existence. If this theory is to be transferred wholesale and in its fundamentals to the genetic level, the arguments surely must be very sound indeed. Full allowances must be made for the differences between living beings at the species level, and the simpler gene organism?

But Darwinism, as Marjorie Grene points out, is not only a scientific theory, and a comprehensive, seemingly self-confirming theory, but a theory deeply embedded in a metaphysical faith; in the faith that science can, and must, explain all the phenomena of nature in terms of one hypothesis, and that an hypothesis of maximum simplicity, of maximum impersonality and objectivity. In genetics, as she shows, quoting G.M. Sheppard, the gene-complex can do anything, with the theory of polygenetic inheritance.

> For each character is controlled, not as used to be thought, by one gene, but by many genes, all balancing and buffetting one another; and every change in the environment is balanced against the resulting balance. Thus in stable environments natural selection is conservative, preserving the advantageous arrangements against

disruption; but let the environment begin changing ever so slightly, natural selection causes - or rather is, by definition - the slight preponderance of a genotype slightly more favourable to the new conditions.

If we try to bring conventional theory up against phenomena which seem to contradict it :

> Whatever might at first sight appear as evidence against the theory is annihilated by *redefinition* into the theory.
>
> *The Knower and The Known.* p.195.

Dawkins is a master of redefinition, and he seems to be redefining the whole of Darwinian theory, even, it seems, beyond the neo-Darwinian molecular biologists.

Evolution is axiomatically evolution by natural selection, and is at the same time progressive adaptation, since it is adaptive relationships that natural selection controls. The sketch of 1837 argues plainly, as Marjorie Grene points out, from adaptation as its most basic dictum. Adaptation is a matter of means and ends. Everything in nature is explained in terms of its purpose : evolutionary theory is confronted with the conception of utility, fitness and the like, and in this provides for what Marjorie Grene calls 'teleology decapitated'. The goal is the proper habitat, the natural niche. Organisms are conceived actually in watchmaker-made terms, as contrivances, aggregates of character and functions, for going on being good for survival, that is going on being good for...

But there is nothing in this about *origins*, despite Darwin's title.

Dawkins, however, in Chapter 2, writes his own book of Genesis. It is hard enough (he admits it) accounting for the beginning of the universe :

> I take it that as agreed that it would be even harder to explain the sudden springing up, fully armed of complex order :- life, of a being capable of creating life.
>
> p.13.

This is fundamental, and it is fundamental (as Tomlin argues) that no organic structure can derive from an uncoordinated assemblage : *omnis structura e structura.* It is simply inconceiveable that order and form can be created by random accident.

Darwin, however, according to Dawkins, 'provides a solution, the only feasible one so far suggested, to the deep problem of our existence'.

> Darwin's theory of evolution by natural selection is satisfying because it shows us the way in which simplicity could change into complexity, how unordered atoms could group themselves into ever more complex patterns until they ended up manufacturing people.
>
> p.13.

All that one can say about such an assertion is that it is simply not so : there is nothing in Darwin to cast light on the problem of the origin of life, from inorganic to organic. It is not credible that complex harmonious, self-replicating structures could arise from chaos or 'noise', from 'unordered atoms'.

Dawkins discusses collections of atoms in stable states, telling us that Darwin's 'survival of the fittest' is a special case of a general law of *survival of the stable.* 'Stable' seems now to be taking over from 'fittest'. But he himself leaps from crystals to protein molecules without a doubt, and tells us that haemoglobin molecules are 'stable' in that we are producing millions of them all the time in our bodies. But, he says, 'Haemoglobin is a *modern* molecule'. What does this mean? That in some way molecules have changed and developed, so that the molecules of the 'soup' or mud have somehow become convoluted 'thornbush' shapes containing 579 amino acids, capable of performing incredible and complex functions in a living body.

How did this happen? Has there not been 'achievement' of some kind? Dawkins may tell us that there *could have been* some 'rudimentary evolution of molecules... by the ordinary processes of physics and chemistry'. But no one has yet explained how this could happen, and it remains the *secret of life.* Dawkins merely glosses over the problem, to make out there is no mystery. We may have to agree that :

> If a group of atoms in the presence of energy falls into a stable pattern, it will tend to stay that way.

But for the next sentence to follow thus is no explanation :

> The earliest form of natural selection was simply a selection of stable forms and a rejection of unstable ones. There is no mystery about this. It had to happen by definition...

It is simply not true that the problem of life - the appearance of self- replicating life molecules - has been so simply solved. And it is not true that Darwinism explains this change.

We proceed, of course, to the next act of faith : only shake these molecules long enough and you will produce a man :

> To try to make a man, you would have to work at your biochemical cocktail-shaker so long... this is where Darwin's theory... comes to the rescue...

It is only necessary to mention Darwin for all the rough places to be made plain.

We return to the primordial soup, of course. The story takes on the air of a fairy story : 'In those days the large organic molecules could drift unmolested through the thickening broth' (p.15). Of course :

> At some point a particularly remarkable molecule was formed *by accident...*

It is not difficult to think of a molecule that makes copies of itself, says Dawkins. It only had to arise once. It is easy then to think of them joining up to from a stable chain. To the Darwinian, as Marjorie Grene says,

> Nature is like a vast computing machine set up in binary digits : no mystery there. And – what man has not yet achieved – the machine is self-programmed; it began by chance, it continues automatically, its master plan creeping up on itself, so to speak, by its own automatism. Again, no mystery there; man seems at home in a simply rational world....
>
> *The Knower and the Known.* p.200.

But what about those evident leaps forward – and the *origins*? The harmony of adaptations, the persistent structures and rhythms in phylogenetic developments – are these not mystifying matters? And the mind?

What we need here, as Marjorie Grene says, is a new metaphysic, to examine these questions.

We get nothing of the kind from Richard Dawkins. In his universe there is no gradient : nothing is *wanting to become.* Instead, we have the bleak universe of the nihilist who has absurdly extrapolated a simple-minded mechanism into a metaphysic. Here, life is that hiccup in the physico-chemical flow, and to Dawkins 'miscopying' is the first principle of creation :

> As mis-copyings were made and propagated, the primeval soup became filled by a population not of identical replicas, but of several varieties of replicating molecules, all 'descended' from the same ancestor.

There was an evolutionary trend towards greater longevity in the population of molecules.

Dawkins' view seems to be that although evolution may seem in some vague sense a 'good thing', especially since we are the product of it, nothing actually 'wants to evolve' (p.19). Evolution is a reluctant process, or a 'willy-nilly' process, happening in spite of all the efforts of the replicators (and nowadays of the genes) to *prevent* it happening. What a perverse cosmos! All the multiplicity of life is generated by 'mistakes' which operate against all the impulses of molecules and life-molecules to prevent them happening. Looking around us, can we believe this? If we murmur to ourselves 'life strives', we will be silenced by incantation : 'the mechanism is the same - natural selection' used by Dawkins to settle all argument. If we object that molecules seem to organise themselves into systems which, as it were, defy the second law of thermodynamics, well - that must prove the power of natural selection!

It is surely an important question to ask what 'life' means. Here, surely, an exact philosophical discipline is needed, at the centre of many important arguments. Here the argument seems to become extremely odd :

> Should we then call the original replicator molecules 'living'? Who cares? I might say to you, 'Darwin was the greatest man who ever lived', and you might say, 'No, Newton was', but I hope we should not prolong the argument. The point is that no conclusion of substance would be affected whichever way our argument was resolved.
>
> p.19.

The use of the 'great' in such a context is not the same as the use of the word 'live', for living things do have certain evident qualities which inorganic things do not have : they have significant forms, and a certain uniqueness. We are interested in them (living things) because of their amazing form and their intrinsic worth – we are interested in the shrimp as we are not interested in the pebbles it swims over. It is crucial to our argument about the origin and development of life, to define life – whether a DNA molecule is living, or a bacteriophage or a cell. Living things reproduce and repair themselves, and display qualities as living subjects. The question is not one of mere terminology. But having said that the achievements of Darwin or Newton remain totally unchanged whether we call them 'great' or not, Dawkins goes on :

> Similarly, the story of the replicator molecules probably happened something like the way I am telling it, regardless of whether we choose to call them 'living'.
>
> p.20.

I do not see the force of that 'similarly'. What I believe Dawkins is saying is : you hesitate to call the first mis-copied molecular replicators 'living'. I tell you that this is parallel to your denying the historical facts of Newton and Darwin's work. That I am right about them makes it clear surely (a) that I am a scientist with authority, and (b) what I say about these molecules is true. This is just a Humpty-Dumpty kind of argument.

The 'mis-copied' 'replicator' molecules in the primordial soup are all *postulates.* No evidence is offered support the view that this 'probably happened something like the way that I am telling it' : the argument remains, 'what I tell you three times is true'. Dawkins sees that terms are important :

> Human suffering has been caused because too many of us cannot grasp that words are only tools for our use, and that the mere presence in the dictionary of a word like 'living' does not mean it necessarily has to refer to something definite in the real world.

This is surely scepticism gone mad? To be a 'biologist' is to study life : the profession implies the recognition of life. It isn't the word but the concept that matters, and '*living beings*' are the proper object of our study, whether Dawkins likes it or not.

But now what was a hypothesis now begins to sound like the tablets of Moses.

> Whether we call the early replicators living or not, they were the ancestors of life; they were the founding fathers.

The primordial soup 'was not capable of supporting an infinite number of replicator molecules'. In the story that follows Dawkins seems to be adding one postulate to another, in order to merge the theory of the struggle for survival with a theory of molecular competition :

> In our picture of the replicator acting as a template or mould, we supposed it to be bathed in a soup rich in the small building block molecules necessary to make copies. But when the replicators became numerous, building blocks *must* have been used up at such a rate that they became a scarce and precious resource. Different varieties or strains of replicator *must* have competed for them.
>
> We have considered the factors which would have increased the numbers of favoured kinds of replicators. We can now see that less-favoured varieties must actually have become *less* numerous because of competition, and ultimately many of their lines must have gone extinct. *There was a struggle for existence among replicator varieties* (my italics).
>
> p.20.

But according to Dawkins' kind of biology, these complex molecules *only* obey the laws of physics and chemistry. They are therefore subservient to the second law of thermodynamics, are subject to Entropy, and operate by minimum energy requirements. Under the control of these laws how can they also operate according to Malthusian laws on the survival of the fittest principle?

This attribution of the ruthless struggle for survival among the molecules is surely quite anthropomorphic?

> Some of them may even have 'discovered' how to break up molecules of rival varieties chemically, and to use the building blocks released for making their own copies. These proto-carnivores simultaneously obtained food and removed competing rivals. Other replicators perhaps discovered how to protect themselves, either chemically, or by building a physical wall of protein around themselves. This may have been how the first living walls appeared....
>
> p.21.

Implicitly, surely, Dawkins has given his whole case away. For isn't what he is talking about now ingenuity : not only 'striving' but *cunning*? He has given away what Tomlin calls 'primary consciousness', for obviously mere physico-chemical atoms cannot 'discover' (whether you put it in inverted commas or not, like 'apparent purpose') nor 'build'. There is no doubt that living things *do* 'discover' and 'build' - besides strive to survive – and also cheat and compete. It is these directive features which make for the recognition of *life*. But neither Darwin, nor neo-Darwinism, nor Dawkinism can explain how these autonomies enter into the matter. Even if we admit (which is doubtful) that mere molecules *compete*, we cannot allow that mere molecules can discover and exercise ingenuity - even if a little thickening of the protein wall is accepted as the basis of the living cell.

If, of course, we look at the complex life of a living cell (see Tomlin below), we perceive processes, forms and developments which could never, never have come about through the mere mechanical (physical and chemical) struggle in the soup, as Dawkins delineates it.

But then in Dawkins' account comes a tremendous leap, over the most prodigious problem in the universe :

> The replicators which survived were the ones which built *survival machines* for themselves to live in.
>
> p.21.

Living organisms, to Dawkins, do not exist in their own right, but have been constructed by the minimals, for them to 'survive' in.

> Replicators began not merely to exist but to contruct for themselves containers, vehicles for their continued existence.

It seems easier to believe in angels, or souls looking for bodies. Surely this is the *reductio ad absurdum* of natural selection theory – with its teleological goal of mere survival? Darwin began from the individual organisms, and the way in which species formed : now the individual organisms are epiphenomena and the survival of their composite particles is the primary dynamic. Why then evolve at all! As Mary Midgeley might say, why not simply remain as an amoeba – which is so good at survival?

Dawkins' account proceeds like a Wellsian fantasy :

> Now they swarm in huge colonies, safe inside gigantic lumbering robots, sealed off from the outside world...

Why 'lumbering'? I discuss above the marvellous engineering feat which the hoof and leg of an antelope are : why is the whole animal world and its achievements insulted by this ridiculous picture of every living thing as a 'robot'? And while the complex living organism is reduced to a blundering survival machine, the *genes* are combined with angelic intelligence :

> they created us, body and mind...

A body is not a living entity, in all its marvellous being-in-the-world, among 'the paragon of animals' :

> A body is the genes' way of preserving the genes unaltered.

It must inevitably follow that :

> we are survival machines...

Dawkin's next chapter is on DNA. DNA molecules in their fundamental chemistry are 'rather uniform'. But the question remains : why one kind of DNA makes elephants and one man, and how DNA changes at adolescence, or when the foetus develops from its first cell-division stage to the later complex developments of blastula and emergent nervous system and limbs. Why should DNA do this when all the laws are in favour of its simply going on in its simplest form?

There follow many Wellsian conjectures. Was there once a different set of DNA, in the primordial soup days? If so, there is no trace of it now. (Perhaps, then, it never existed?) A.G. Cairns-Smith thinks that maybe everything was once but inorganic crystals or even a *bit of clay*. (p.23). But why this clay or primitive DNA should seek to become 'more interesting' is never explained. We recognise that Caesar becomes clay : but why should the clay ever become Caesar?

Dawkins manages in his tone to sustain the air of offering us convincing explanations, when in fact he dodges every question. On the 'codes' he says :

> It is as though, in every room of a gigantic building, there was a bookcase containing the architect's plans for the whole building. The 'bookcase' in a cell is

called the nucleus.

p.23.

But this analogy is unsatisfactory, for a book requires an intelligence. There is nothing in the cell to study a plan and know, unless we postulate some kind of intelligence. The evolving cell is both plan, or template, or code, and the building. The building process cannot be understood unless we postulate a boundary condition and an engineering system : and then some overall intelligence or primary consciousness, surely?

Dawkins insists, of course, 'that there is no architect' : but now our universal god re-appears :

> The DNA instructions have been assembled by natural selection....

Above, 'natural selection' was a process by which warring molecules were simply sorted out 'stable' from 'unstable'. How can it *assemble* the *instructions* in the DNA molecules? As so often, in the end, the mechanist, who denies that there is any 'purposive' force, and asserts there is only 'chance', has to allow a guiding, assembling force, 'natural selection', to do the directiveness for him - though (as Darwinism conceived it) it can do nothing of the kind.

> Proteins... also exert sensitive control over all the chemical processes inside the cell, selectively turning them on and off at precise times and in precise places... Exactly how this leads to the development of a baby...

But how does this capacity to *control* exist in the protein, if it is only a sequence of physico-chemical agents?

(When doubts arise Dawkins, simply snaps : 'It is a fact that it does...')

With genes, Dawkins admits that there is in them some form of controlling principle. They control the manufacture of bodies, though the influence is strictly one way :

> the replicators no more conscious or purposeful than they ever were... The same old processes of automatic selction between rival molecules by reason of their longevity, fecundity and copying fidelity, still go on blindly.

p.25.

To assert that such forces are blind or without purpose, while yet capable of 'control' and of producing a baby is surely untenable? Certainly, it seems no explanation. Yet he uses the word 'achieved' and the word 'triumphs', thus letting the teleological cat out of the bag.

> In recent years - the last six hundred million or so - the replicators have achieved notable triumphs of survival-machine technology, such as the muscle, the heart, and the eye...

p.25.

The word 'achieved' is a recognition of undeniable positive processes of morphogenesis and the word 'triumph' recognises a gradient, a striving. Moreover, far from being a matter of 'chance' there is coordination :

> The manufacture of a body is such a *cooperative venture....*

He goes on to say :

> Some genes act as master genes controlling the operation of a cluster of other genes

As he himself says :

> each page only makes sense in terms of cross-references to other pages...

This is a recognition of complex form and harmony in the systems by which life is controlled, a complex and organised form which cannot be understood except by reference to something like Tomlin's primary consciousness, which 'knows' its own way and speaks a controlling 'language' rather than a 'code'. Dawkins says that the 'pages' of chromosomes make 'recommendations' – a way of talking which makes far more sense in Tomlin's dimension than that of an information theory which simply fails to recognise the problems.

Dawkins now enters into complex problems of genetics where it is difficult for us to follow him. However, it is interesting to note that, on page 30, he says :

> A gene is defined as any portion of chromosomal material which potentially lasts for enough generation to serve as a *unit of natural selection.*

Surely, there cannot be any such thing. What Dawkins actually means, it turns out, is that natural selection operates by *copying fidelity*, leading to *longevity*. But Darwinism is now so far stretched that it seems inappropriate to call it Darwinism at all (Perhaps it should be called Dawkinism?) for natural selection has become virtually equated with Bergson's vital principle.

To say that a gene is a 'unit of natural selection' is to equip it with magical powers.

Next Dawkins turns to mimicry. Some butterflies escape being eaten by tasting nasty; other nice-tasting butterflies *mimic* the appearance of the nasty tasting ones. But intermediates are not born. Why not?

> By the *unconscious* and *automatic* 'editing' achieved by inversions and other accidental rearrangements of genetic material.
>
> (my italics)

If there can be 'editings' 'unconscious' and 'automatic', leading to achievement, how can there also be 'accidental'? It seems incredible that a scientist can live with such contradictions : implicit in Dawkins here is an innovative printciple he cannot see.

The title of his book, Dawkins tell us, should have been : 'the slightly selfish big bit of chromosome and the even more selfish little bit of chromosome' (but 'this is not a catchy title, so I call the book : *The Selfish Gene*').

The appeal of the book, however, lies even in its poor argument, and its combinations of crude anti-morality combined with evolutionary dogma, a dogma which his own language contradicts (as with the use of the word 'achieved').

> Selfishness is to be expected in any entity which deserves the title of a basic unit of natural selection...

Dawkins even admits that his argument is tautologous :

> What I have done is to define the gene in such a way that I cannot really help being right.

How doubtful it is for Dawkins to apply Darwinist survival theory to genes now becomes apparent, because, as he recognises, the genes, or gene-forms, are *immortal.* Natural selection remains the undefined negative process :

> Natural selection in its most general form means the differential survival of entities. Some entities live and others die, but in order for this selective death to have any impact on the world, each entity must exist in the form of lots of copies, and at least some of the entities must be potentially capable of surviving – in the form of copies, for a significant period of evolutionary time.
>
> p.36.

We may well ask what is a 'significant' period, since sharks and bacteria have not evolved over millions of years, while other species have evolved, at various paces, at various times. However :

> The gene does not grow senile... it leaps from body to body down the generations, *manipulating body after body in its own way and for its own ends.*

Here, surely, he means the *gene form* – for the substance changes? Surely there is confusion here, of the kind Tomlin notes between structure and form? But where, in this, is any vestige of Darwinian theory, of the survival of organisms which have fortuituosly achieved small advantages since the gene in Dawkins' argument merely slips through body after body in endless survival? Even the concept of what is 'alive' begins to dissolve :

> The life of any one DNA molecule is very short... but it could theoretically live on in the form of copies of itself...

Darwinian theory seems here as attenuated as it could be.

From our reading of Marjorie Grene and Tomlin, we may note that Dawkins is discussing achievement again, but his successful genes survive because they are good at making 'survival machines' : this brings the argument back to the species, or Darwinian level :

> a good gene might ensure its survival by tending to endow the successive bodies in which it finds itself with long legs, which help those bodies to escape from predators.

But this argument is absurd, since he is talking about *gene-forms* which inherit bodies which die off, while the genes are immortal. If genes are *immortal*, then it could seem to be irrelevant whether their 'survival machines' function well or not, so long as *some* reproduce themselves. The non-survival (being eaten up) of 'survival machines' can surely have no effect at all on 'immortal' genes – thus there can be no natural selection 'pressure' upon them?

Now his theory seems to be that the best genes can build good survival machines which reproduce before they run away or fall down from the trees, (before they have time to reproduce), because the genes have given them long legs or opposable thumbs, not to be more effective for their own sakes or for the sake of their species. Their real purpose (or 'apparent purpose') is not to produce good jaguars or monkeys or whatever, but simply *more good genes.* In this perverse theory, I suppose, we are with Darwinism of a kind. However the 'pressures' are not acting on the genes but on the animals they 'make' to live in : yet there seems little point in this since the genes are immortal anyway.

Dawkins seems to be translating the Darwinian assumption of a 'ruthless struggle for survival' into the world of genetics where, instead of competition among species we must see competition between the entities of his discipline :

> Genes are competing directly with their alleles for survival, since their alleles in the gene pool are rivals for their slot on the chromosomes of future generations.
>
> (p.39).

Can Darwinian selection theory really be translated into this totally different sphere?

But Dawkins also seems to have taken a leaf out of Kropotkin's book (*Mutual Aid*, 1902), and seems to postulate a new major principle that conflicts totally with his universal selfishness, and the 'struggle'. To make bodies, genes have to cooperate. Genes are not free in their control of embryonic development. 'They *collaborate* and *interact...*' (my italics). 'Building a leg is a multi-gene cooperative enterprise...'. So, it is not all ruthless selfishness, after all.

> Embryonic development is controlled by an interlocking web of relationships so complex that we had best not contemplate it.

While 'cooperation' would seem to be a matter of improvement.

> It is differences which matter in the competitive struggle to survive.

So

> Genes are competitive for places *within the cooperation.*

So, the struggle for survival is now extended to the struggle between genes in seeking to cooperate in developments aimed at advantages to contribute to survival. Some genes which are lethal apparently slip through the net of natural selection simply because they are 'late-acting'.

I do not of course challenge the validity of any of Dawkins' accounts of gene structure. What I do question is whether Darwinian theory can be so played with and stretched, that the gene becomes 'the unit of selection' - meaning the dynamic entity of evolutionary growth.

> It is a consequence of sex and crossing-over that the small genetic unit or gene can be regarded as the nearest thing we have to a fundamental, independent *agent of evolution.*
>
> p.47.

Moreover, there are many doubtful implication in his phrases about a 'bizarre perversion of straightforward replication' while other sentences seemed meaningless ('if crossing-over benefits a gene for crossing-over, that is a sufficient explanation for the existence of crossing-over'). He admits himself that his argument 'comes perilously close to a circular argument'... but 'this book is not the place to pursue the argument'. But what exactly is the purpose of the book?

On page 46, Dawkins asks : '*What is the good of sex*?' but declares he is going to evade the question. However, he does agree with some of the evolutionary problems which arise. The question of *efficiency*, he says (from the individual's point of view) is *irrelevant.* Yet surely, Darwinian natural selection theory is all based on questions of efficiency? If the evolution of sex is explained in terms of the maximisation of the number of genes which survive, say Dawkins, sex actually appears paradoxical because it is an 'inefficient' way for an individual to propagate his genes. Dawkins quotes W.F. Bodner : 'sex facilitates the accumulation in a single individual of advantageous mutations which arose separately in different individuals'.

But if we regard the individual as a survival machine built by a short-lived federation of long-lived genes, efficiency is seen to be irrelevant. Sexual reproduction benefits a gene for sexual reproduction, and that is a sufficient explanation for the existence of sexual reproduction; whether or not it benefits all the rest of an individual's genes is ccomparatively irrelevant. However, what happens to the 'survival' theory if efficiency is thus thrown away?

But now Dawkins takes evolutionary theory right down to DNA itself. 'The true purpose' of DNA he says is 'to survive, no more and no less' (p.97). Discussing the problem of 'surplus' DNA - the existence of which again threatens the 'struggle for existence' theory, he says :

> The simplest way to explain the surplus DNA is to suppose that it is a parasite, or at best a harmless but useless passenger, hitching a ride in the survival machines created by the other DNA.
>
> (p.47)

Discussing opposition to his view, Dawkins says that some argue that it is *individuals* with all their genes who actually live or die. But this, again, is irrelevant :

> The long-term consequences of non-random individual death and reproductive success are manifested in the form of gene- frequencies in the gene pool...

By sex and crossing-over, the pool is kept well stirred and the genes partially shuffled :

> Evolution is the process by which some genes become more numerous and others less numerous in the gene pool.

But to what is this numerousness of certain genes directed? Not, apparently, efficiency in the 'struggle for survival', but simply the eternal persistence of the genes – nothing, it seems, to do with evolution at all : not going on being good for, but simply going on. How, then, can the gene be the 'unit of selection' and the basis of evolution?

The origin and development of life, which is what we are concerned with, is dismissed by Dawkins as :

> Survival machines began as passive receptacles for the genes....
>
> (p.49)

Next came plants which 'started to use sunlight', and then animals 'discovered' how to exploit the plants. The word 'discovered' (like 'apparent purpose') simply evades the problem : *Why should they*? What urge did life follow, to strive towards these achievements?

Efficiency, he had said in the last chapter, is irrelevant. But now it reappears :

> Both main branches of survival machines evolved more and more *ingenious tricks* to increase their efficiency... Sub-branching has given rise to the *immense diversity* of animals and plants which so *impresses* us today....
>
> (p.49) (my italics)

But where did these *ingenious tricks* come from? Where did the evident *ingenuity* of life originate? How can one reconcile the 'achieving' of 'triumphs' with the selfish processes by which *genes* strive to survive? Ingenuity is a telological concept.

Dawkins is in some difficulty now over his 'colonies of genes'. 'We do not know when, why, or how many times independently, animals and plants evolved into many-celled bodies...'.

He likes to think of the body as a colony of *genes*, and of the cell as a convenient 'working unit' for the 'chemical industries of genes'. It all sounds very scientific. But

> bodies have undeniably acquired an individuality of their own. The animal moves as a coordinated while, as a unit. Subjectively I feel like a unit, not a colony...

As Weiss says, I don't feel like an assemblage of molecules – do you? What Dawkins is admitted here is the *primary reality*, which is not the minimals (genes) but *living beings*.

The reason for the autonomy of the living being, however, must still be reduced to the minimals : selection has favoured genes which cooperate with others. Comes the usual incantation :

> In the fierce competition for scarce resources, in the relentless struggle to eat other survival machines, and to avoid being eaten, *there must have been...* a premium on *central coordination* rather than anarchy within the communal body...
>
> (my italics)

(It is, by the way, instructive to note how many of Dawkins' sweeping pronouncements are prefaced by the phrase, 'must have been').

So, having argued that the basic principle in all life is the operation of 'little causal thingummies', he tells us that it would be tedious to go on with this kind of talk.

> In practice, it is usually convenient, as an approximation, to regard the individual body as an *agent* 'trying' to increase the numbers of its genes in future generations. I shall use the language of convenience....

But, incidentally, by his word 'trying' he has given mechanism away, and concedes our whole point, once again.

From time to time, if we look closely, Dawkins gives away his whole position in an important respect : he uses phrases such as 'the individual body as an agent 'trying' to increase the numbers of its genes...' Something 'tries', there are 'plans', and there is 'cunning' – but in what? These dynamics cannot be discussed except in the recognition of complex organisms as *beings* : and so, of a kind of primary consciousness.

E.W.F. Tomlin discusses this question. Life, he says, is consciousness, and consciousness is the immediate self-possession of an organism - the unity of multiplicity. It is seeking to fulfil itself.

To say this is to say that every living being is a *subject* rather than an object, a *true form* rather than a pseudo-form. To Dawkins the whole being is a pseudo-form, a mere shell containing his genes, which alone are allowed to have true form.

Tomlin compares various forms. A cloud or a rock is a purely statistical phenomenon, an assemblage in precarious equilibrium, the product of external forces, without a 'centre', and therefore an object or pseudo-form.

By contrast, a 'being' is a domain of togetherness in dynamic organisation or equilibrium, in which a hierarchy of sub-systems are 'dominated', as a molecule dominates its atomic and sub-atomic constituents or activities : and the more complex the organism, the greater its domain will be'. This dominance is also a dominance of space-time, since the immediate self-possession of a being is precisely the unity or togetherness of a particular 'here' and 'now'. An amoeba's here-and-now-ness, says Tomlin, is part of its subjective unity, which it maintains 'by being an amoeba'.

Reality is the inwardness of subjectivity in an organism, its form; appearance is the external activity whereby we infer or cognise, but never directly perceive, the presence of life.

It is the whole being that must be considered in biology, if scientific explanation is to have any validity : this is clear even if one observes an amoeba feeding or reproducing itself. The genes, while continuous in form, change continually in substance : and while their codes and instructions or messages, their 'control' and 'cooperation' permit, it is the whole amoeba which recognisably and apparently has the continuity of centricity. It will, for example, have a 'character' as do all living creatures, with their uniqueness and intrinsic worth. And it makes sense to talk of the survival of the amoeba, as it does not make sense to talk of the survival of the gene.

It doesn't make sense to talk of the 'selfishness' or otherwise of the amoeba in trying to understand its behaviour. 'Altruism' would seem a distant concept, too anthropomorphic - since altruism is a human capacity belonging to a creature which has the additional dimension of culture, and can consciously contemplate tomorrow as no animal can. Altruism is directed towards the future, and is concerned with outcomes. Darwinian theory recognises a tendency towards the future of some kind (though foresight is, officially, prohibited). But even if we talk of selfishness in genes, we are inevitably employing a posture towards the future, towards outcomes – as in 'surviving', and so admitting directiveness.

'Survival' thus inevitably brings in the contemplation of the future, in time. Ingenuity implies future-time : as indeed, does 'evolution' and 'survival of the fittest' : in this we glimpse Darwinian 'truncated teleology'. These considerations are forced on Darwinism, however mechanistic, by the nature of the world.

Muscles come into existence. 'The gadget', says Dawkins, 'which animals *evolved* to *achieve* rapid movement was the muscle'. In that one sentence are teleological complexities. Why should muscles have been 'achieved'? They represent a leap forward. The muscle is actually a very complicated machine : could such a complex engine simply have been 'achieved' by *chance* mutations? Or must we recognise a development impelled by some other ingenious force – note Dawkins says : 'animals evolved.... muscle'.

Muscle action is complex and (Dawkins says) baffling : 'much more *baffling* is the timing of the operations'.

What follows in Dawkins is a mechanistic comparison of the muscle and its connections with brain and nervous system with engineering functions and computers : the central feature is the neurone. And there is more 'brains' as 'analogous in function to computers'. It is difficult to understand why we enter into a long discussion here of the brain-as-computer in a book on the selfish gene, except in some vague attempt to deal with 'directiveness' and 'apparent purposiveness', by the usual mechanistic false analogue.

Certainly, at this point, the reader becomes confused. What exactly is Richard Dawkins trying to prove? His dealings with neurology and the brain seem simply inadequate. Polanyi's warning is relevant here : thought depends upon tacit integration, and this is indeterminate and unspecificable. So, it can never be reduced (as Dawkins wants to reduce it) to explicit, logically linear, steps :

> We cannot spell this process out in explicit steps, and it is for this reason.. that no 'thinking' machine can ever be adequate as a substitute, or even as a model, for the human mind. Our dwelling in the particulars, the subsidiary clues, results in their synthesis into a focal object only by means of an act of our imagination – a leap of a logical gap : this does not come about by means of a specifiable, explicit, logically operative step.
>
> *Meaning*, p.62.

In Polayni's account we recognise the human mind at work, and we should realise that he indicates modes of thought which cannot be reduced to neurophysical analysis, of a mechanistic kind. But the 'kind of consciousness' the animals display evades this kind of analysis, too. As Straus shows, this is only attempted by those who can be so naive as to ignore the problems like Dawkins :

> To achieve more complex and indirect relationships between the timing of events in the outside world and the timing of muscular contractions, some kind of brain was needed as intermediary... A notable advance was the evolutionary 'invention' of memory....

Dawkins seens to have no sense of the animal's achieved autonomy in its world, or its perception of its world, but must always try to reduce it to simple mechanics ('timing of contractions' rather than whole-body skills). And then we may aks, how and why should 'evolution' 'invent' 'memory'? Such an invention could not possibly have come by 'chance' mutations, evidently, any more than consciousness.

One of the most striking properties of survival machine behaviour, says Dawkins, is 'its apparent purposiveness'. He means, of course, purposiveness – but the dogma has to be preserved, and so this must be called 'apparent'. He doesn't mean that this behaviour 'seems well calculated to help the animal's genes to survive' (we gathered earlier that their survival had nothing to do with the animal's efficiency), 'though of course it is' (of course, everything, we know, is merely directed to that end).

> I am talking about a closer analogy to human purposeful behaviour.

In one animal this has 'evolved' consciousness (can consciousness have ever been the product of chance mutations, giving an immediate advantage – the only grounds for selection)? Dawkins admits he is not philosopher enough to 'discuss what this means', but 'fortunately it does not matter for our present purposes because it is easy to talk about machines which behave *as if* motivated by a purpose, and to leave open the question whether they actually are conscious'. These machines, Dawkins asserts, are basically very simple – 'the classic example is the Watt steam governor'.

It seems hardly worth while going on arguing with a biologist who cannot make a distinction between a Watt steam governor and a conscious being's centricity. As E.W.F. Tomlin says, the problem and the nature of *consciousness* is the fundamental problem to any philosophy of life. He suggests that we distinguish forms of consciousness : organic or psychic consciousness, on the one hand, and perceptive consciousness on the other. It is the former kind of consciousness which in a living creature develops beyond the mere physico-chemical complex, and acquires that unity and self-subsistency which transforms it from an object into a subject. In this process of morphogenetic growth there must be something we must call intelligence.

Dawkins, however, does not find this intelligence, or this consciousness in the living being. After a long discussion of computers programmed to play chess, he declares that in a game 'the computer is on its own... All the programmer can do is to set the computer up *beforehand*, in the best way possible, with a proper balance between lists of specific knowledge, and hints about strategies and techniques...' (p.56)

The genes, too, 'control the behaviour of their survival machines... indirectly like the computer programmer'. But the computer programmer is a man with human intelligence, and the computer is merely fitted with certain human-intelligence *instructions*. What takes the place of this in the genes? The only answer must be that again we have an implicit recognition of what Tomlin calls primary consciousness. As often happens with mechanists (as Erwin Straus makes clear) Dawkins is admitting to a guiding force while denying it, and attributing its operations to something else - mere matter-in-motion.[2]

The cybernetical model employed in this chapter is also extremely misleading. As Marjorie Grene points out, the concepts of cybernetics are unable to take account of the temporal structure of actions. Actions are directed to achievements. But the principles of cybernetics reduce actions to mechanistically conceived causal chains in a monolithic time sequence. Organic time is structured in reference to natural goals.

Secondly, the 'learning' of learning machines is only a pseudo-learning : a computer programmed to play chess is only programmed to avoid bad moves. But a person who learns to play a game stores up information so as to be able to make, flexibly, a number of moves. (*Approaches to a Philosophical Biology*, p.137).

Dawkins goes on to discuss how the genes 'build' a brain : but in all his analogies he fails to distinguish between that initiation of which a human intelligence is capable and the operations of a computer that the human mind creates and programmes.

For instance on p.59 he quotes J.Z. Young as saying that the genes 'have to perform a task akin to prediction'.

> it is the business of genes to programme brains in advance so that on average they take decisions that pay off.
>
> p.59.

But in his model, how can there be anything like this intelligent capacity in genes? Where did these fore-sighted 'genes' come from, that are capable of instructing brains to make decisions that pay off? Have we not here, once again, old Paley's watchmaker learning a bit of his art?

Or, to put it another way, faced with mysteries and inexplicable forms of directiveness and intelligence, doesn't Dawkins fall back on miracles? As Tomlin says,

> it is not the philosophy of organism (to use Whitehead's name for his science) which introduces the irrational and the occult, but the philosophy of mechanism.
>
> *Tokyo Essays, p.215.*

Out of a mere assemblage of physico-chemical elements no organic entity can spring except by a miracle. It is even more miraculous to find that genes can actually make 'predictions' about the future.

This they do, apparently, by a kind of behaviourist programming : nice things will guarantee survival, nasty ones will not. The secret, apparently, is 'simulation' : 'the evolution of the capacity to simulate seems to have culminated in subjective consciousness' – but with true Dawkinian spirit he reduces its role to simulation. Dawkins has abandoned, it seems, his promise to tell us 'how the genes solve the problem of making predictions in rather unpredictable environments...' by 'building up a capacity for learning'. Apparently how the genes do this is by - hey presto - creating a *man*, with his capacity to *simulate* :

> why this should have happened is to me the most profound mystery facing modern biology.

This is self-consciousness : 'the culmination of an evolutionary trend towards the emancipation of survival machines as executive decision-makers from their ultimate masters, the genes'.

Straus declared : 'Man thinks, not the brain', drawing our attention to the activity of consciousness as an activity of the whole being. Dawkins declares 'Genes made the brain and the man in order to think'.

In his argument all the questions are solved by a kind of sleight of hand : *evolution* does this : *selection* does that : the *genes control* and *predict* – and in the end develop the capacity to 'simulate' and to develop consciousness, self-consciousness, by making men and mens' brains to live in. All this is sheer magic, since there is nothing in his account of any dimension in which these intelligent activities could go on. Dawkins has explained nothing, despite the apparent brilliance of his computer analogies, and his scientific talk about

---

[2] Dawkins speaks at this point of something being 'left to the reader's imagination' : is this, too, generated and controlled by the genes?

neurones, models and memory.

The genes, it seems, can plan and design anything. The genes *dictate* 'the way survival machines and their nervous systems are built : but now the nervous systems and brains actually take over'. However, these brains cannot apparently modify behaviour patterns : for each behaviour pattern there is, on the lines of cybernetic theory, a gene like a micro-chip :

> in order for a behaviour pattern – altruistic or selfish – to evolve, it is necessary that a gene 'for' that behaviour should survive in the gene pool more successfully than a rival gene or allele 'for' some different behaviour.
>
> p.69.

How is neo-Darwinian theory brought here into an unholy alliance with neurology? Behaviour patterns are sometimes learnt, and sometimes innate (as with ducks which fly by the stars). But has there ever been any theory put forward in evolution that *behaviour patterns* evolve by small changes brought about by mutation, selected out by 'natural selection' in the survival of the fittest? The theory of evolution relates, surely, to structural changes and richness of genetic material – can it really be extended, as here, to behavioural patterns? And are behaviour patterns determined thus, so that there is a ruthless struggle for existence *between these* (so that - inevitably - 'altruistic' genes are overcome by 'selfish' behaviour)?

It seems at times almost as if, so long as the traditional *language* of evolutionary theory is used, you can talk any old nonsense.

Dawkins confuses, in the coarsest way, behaviour which is gene-influenced; controlled, innate behaviour; instinctual behaviour, and learned behaviour. I am not qualified to discuss particular experiments, but one gathers that zoologists are continually amazed by the results of their experiments, in which they find that some birds will navigate by the stars without ever having had a chance to learn such behaviour, or that some have to learn from their parents : or that a bird will 'learn' a new song it only heard ten months before, or whatever. The secrets of life here are complex, perplexing and fascinating, but the basic subject in biology is to be behaviour, the complex autonomy of a living being, a subject. The geneticist, with his enthusiastic belief in the all-explanatory power of a reduction to minimals, seems content to ignore the complexities of zoology.

Dawkins has only one coarsely applied principle for everything :

> the judge is the ruthless judge of the court of survival.
>
> p.67.

As we have seen, 'survival' for him is not (as for Darwin and the neo-Darwinians) the survival of the species by being more efficient, but the survival of the genes, which is something different. The members of any *species* are but 'survival machines' 'in' which the genes survive. In this, the genes 'programme' for survival – a hypothesis which endows them with a savage kind of foresightful intelligence, which enabled them to 'exploit' the 'court of survival'. Again it would be interesting to know in what this intelligence or cunning resides, since the genes are little more than assemblages of molecules, and in this world nothing is allowed to change except by mistakes in copying, while the large molecules themselves exist almost in defiance of the physical 'rules' – so, in which dynamic (according to reductionism) does this capacity to 'exploit' exist?

Yet, of course, Dawkins cannot put forward such a theme without much philosophical casuistry. So :

> For purposes of argument it will be necessary to speculate about genes 'for' doing all sorts of improbable things. It I speak, for example, of a hypothetical gene 'for saving companions from drowning', and you find such a concept incredible, remember the story of the hygenic bees.
>
> p.66.

In this Dawkins now turns to the problem of 'altruism'.

He relates experiments with bees, by W.C. Rothenbuhler, on foul brood. It may well be that among bees, where work tasks are so divided among specialists, in a complex swarm, alternations in genes may well affect some behaviour. But to extrapolate from this to the problem of human altruism seems evidently absurd. The references to 'purposes of argument', 'speculate' and 'improbable' are meant to prepare the reader for further nonsense in this discussion.

The question is - do genes *determine* human behaviour or not? It looks at one moment as if Dawkins is going to recognise the complexity of the problem :

> Recall that we are not talking about the gene as the sole antecedent cause of all the complex muscular contractions, sensory integrations, and ever-conscious decisions, which are involved in saving someone from drowning.

Well, of course, that would be ridiculous.

> We are saying nothing about the question of whether learning, experience, or environmental influences enter into the development of behaviour.

Was Dawkins sure he was not? What was implied by 'remember the story of the hygenic bees'? Isn't Dawkins whole book devoted to saying that behaviour is *nothing but* the consequence of the genes' selfishness in building survival machines?

This seems to be so, for what follows is this :

> All that you have to concede is that it is possible for a single gene, other things being equal add lots of other essential genes and environmental factors being present, to make a body more likely to save someone from drowning than its allele would. The difference between the two genes may turn out to be at bottom to be a slight difference in some simple quantitative variable. The details of the embryonic process, interesting as they may be, are irrelevant to evolutionary considerations....

yet :

> The genes are master programmers and they are programming for their lives...

'All you have to concede is' apparently, that a *single gene* can 'make a body' which is more likely to be altruistic than another. This may only be a matter of some 'simple quantitative variable', yet it will (isn't the implication?) affect the whole question more radically than 'learning, experience, environmental influences' etc. The gene is the '*cause*' of the complex behaviour in altruism : we are by no means 'saying nothing' about the questions of behaviour, but are explaining it away in terms of the impulses of the minimals to which living beings are here reduced. The ghost of Hobbes is powerfully present!

Dawkins tries to persuade us that the two approaches can live together - those which see life at all levels as complex and autonomous, and those who believe that genes manipulate everything.

Life throws problems at the 'survival machines', and genes are judged by the success by which they have constructed them : the judge is the 'ruthless judge of the court of survival'.

At this point we are told that the details of the embryonic developmental process 'are irrelevant to the evolutionary considerations' - an astonishing statement suddenly to introduce here. How can the problem of the details of embryonic development, one of the greatest puzzles of morphogenesis, not least because it cannot merely be the 'product' of DNA 'coding', be *irrelevant* to evolutionary considerations? If there are 'genes' 'for' altruism, what is there '*for*' the mysterious processes of formative development in a foetus?

Dawkins goes on :

> the obvious first priorities of a survival machine, and of the brain which takes decisions for it, are individual survival and reproduction. All the genes in the 'colony' would agree about these priorities...

Look carefully at the next wild animals that you see, says Dawkins. But surely attention to any living creature should make Dawkins particulate theories seem inadequate?

What we see, if we really do look at a wild animal, is a kind of 'consciousness'. As Tomlin argues, there are two kinds of consciousness, or subjective 'being'. There is the mysterious 'knowing' by which (say) the embryo develops in an adult, or by which (as Crick recognises) a cell 'knows where it is in a cell wall', or by which a damaged organism grows a replacement limb. Dawkins talks about genes growing 'brains' for themselves to 'direct' their 'survival machines'. But, in the light of everything one reads about animals, this is a Noah's Ark kind of myth, to explain morphogenesis, and has no air of serious scientific explanation. The gene alone ('a single gene') cannot manifest the requisite intelligence for that consciousness or subjective being which guides and directs growth and development, nor is it, or could it be, a question of 'genes' 'agreeing'. Dawkins' genes are anthropomorphic – or at least are endowed with those properties observed and found in living beings (like an amoeba) – 'centricity', 'self-directedness'. In his kind of minimalistic thinking, Dawkins cannot conceive of an entity except as a sum of its 'little causal thingummies'. The molecular biologist sees a living being as an aggregation of molecules, and the geneticist as an aggregate of cunning genes. Both are absurd : without denying that a living being is *composed* of molecules or genes, we must say it is unscientific to attribute to the particles qualities which are only observed in the *whole*, functioning as a whole, and operating according to those systems or hierarchical principles which organise the formative patterns of the being, which the components subserve.

The second kind of consciousness is in a different dimension altogether. It is in complex with other functions, as, for example, our capacity for love is in complex with our sexual instincts and our capacities to function at the level of the autonomic nervous system, as in sexual excitement and orgasm. Here one has to take into account not only the various complexities of behaviour, but the whole world of subjective reality, self-consciousness, developmental history, capacities to symbolise and to think, and to behave as a cultured and civilised being. Dawkins persistently fail to distinguish between against (primary consciousness) behaviour; the complex behaviour of the higher animals and the even more complex behaviour (and *experience*) of the animal with culture, man.

Thus his discussion of 'communications' is lame :

> A survival machine may be said to have communicated with another one when it influences its behaviour or the state of its nervous system. This is not a definition I would like to have to defend for very long, but it is good enough for present purposes...
>
> p.67.

Communication, surely has to do with the reading across of meanings. This is true of animals' signs, as when a dog tries to persuade its master to take it for a walk, or when a monkey accepts a human proximity.[3]

There are animal signs which exert what Dawkins calls a causal influence, as when a dog bares its teeth, and a stronger dog hesitates. But Dawkins slips into his list 'human gestures and language' – another example of confusion, since human beings have a culture. He goes on to discuss deceit, or the way creatures tell lies. By the way, he demonstrates an interesting hostility to philosophers : remembering a talk given in America by the Gardners about the 'talking' chimpanzee *Washoe* :

> There were some philosophers in the audience, and in the discussion after the lecture they were much exercised by the question of whether Washoe could tell a lie. I suspect that the Gardners though there were more interesting things to talk about, and I agreed with them.
>
> p.68.

Dawkins declares that he intends to use the words 'deceive' and 'lie' in a much more straightforward sense than those philosophers.

What he actually means is not 'lies' but 'lures' – he discusses deceptions like that of the angler fish, and other forms of behaviour such as the way orchids get bees to copulate with them, and how females of a certain genus of firefly has learnt to lure males of another genus by imitating the light-flashes of their females – not to lure them to copulation, but to be eaten.

These are examples of the remarkable ingenuity of life which suggest that what is at work is no mere process in which chance is predominent, but modes of *intelligence*. It also reveals that scientists are interested in living beings because of their intrinsic interest, not because they are products of 'accident' or the 'ruthless struggle'. But Dawkins (without realising it) often uses anecdotes about animal behaviour with a highly symbolic charge - implicitly bringing them to bear on human ethics.

And he uses his example of intelligence to promote his absurd metaphysic :

> Brought up as we have been on the 'good of the species' view of evolution, we naturally think of liars and deceivers as belonging to different species :

In this he moves without misgiving between the animal kingdom and the human one.

> As we shall see, we must even expect children will deceive their parents, that husbands will cheat on wives, and that brother will lie to brother....
>
> p.70.

---

[3] See Leonard Williams, *Challenge to Survival*, Alison and Busby, 1978

Surely such extrapolations are doubtful, because man lives in a 'mansion of consciousness' (Buber) and a cultural dimension, because of which no communication can be seen as a mere device to deceive or whatever under the pressure of survival. But Dawkins with his extraordinary hubris is content with his own 'powerful ways of thinking' :

The next chapter introduces a powerful way of thinking about conflicts of interest from an evolutionary point of view.

Inevitably, the next chapter is headed 'Aggression' : a topic Dawkins pronounces 'much-misunderstood'. It is indeed misunderstood, as Towers and Lewis argued in their rejection of Naked Apery in *Naked Ape or Homo Sapiens*? The main effect of the misunderstandding is to seem to relieve peoples' responsibilities for their own actions by attributing their aggression to 'natural' impulses. However, we are only to be allowed to consider this sensitive topic within the limits of Dawkins' position :

> we shall continue to treat the individual as a gene machine, programmed to do whatever is best for his genes as a whole :

- so we may, it seems, attribute our ruthlessness to our 'genes'.

To a survival machine, says Dawkins (p.71) another survival machine is a part of his environment, like a rock... It is something that gets in the way. The only difference is that it can hit back.

> This is because it too is a machine which holds its immortal genes in trust for the future, and it too will stop at nothing to preserve them.
>
> p.71.

Dawkins is suspicious of Lorenz's arguments, that animals avoid conflicts because he is a 'good of the species' man. Yet Lorenz's arguments are based on close and painstaking observation of animals, whereas nearly all Dawkins' arguments are based on anecdotal examples, many of which he seems simply to have selected to support his theory.

He goes on to discuss what a certain group of scientists call an *evolutionarily stable strategy* or ESS. (J. Maynard Smith, with G.R. Price and G.A. Parker). This is a pre-programmed behavioural policy (there is no indication as to what intelligence 'pre-programmed' this policy - presumably the genes again).

This strategy is not worked out by the individual : an animal is merely a 'robot survival machine with a pre-programmed computer controlling the muscles'. One wonders what the ethologists make of such reductivism?

This ESS has nothing to do with groups being more successful than others. We have been thinking that a 'ruthless struggle for existence' was the basis of all evolutionary development : but now we find there is postulated a strategy which modifies this (while, since gene survival is the primary principle, efficiency in survival doesn't matter, anyhow). The varieties of competitive dynamic, it seems,are endless. Above Dawkins told us that 'gene machines' simply regard other machines as objects standing in their path : now, p.81, he says :

> crickets do not recognise each other as individuals, but hens and monkeys do.

He also quotes experiments by Tinbergen which show that his earlier absolute statement is nonsense. But we are not to see the interaction between animals as primary: only as manipulations of the genes :

> The lion genes 'want' the meat as food for their survival machine – the antelope genes want the meat as making muscle for their survival machine. The two uses for the meat are mutally incompatible, therefore there is a conflict of interests.
>
> p.89.

So, 'the reason lions do not hunt lions is that it would not be an ESS for them to do so... A cannibal strategy would be unstable'.[4] Clearly, if animals ate one another within the species, it would not be profitable. Yet :

> I have a hunch that we may come to look back on the invention of the ESS concept as one of the most important advances in evolutionary theory since Darwin. It is appliable wherever we find conflict of interest, and that means almost everywhere.
>
> p.90.

---

[4] Yet, I gather, one problem in trout farming is that trout do eat one another, if some are smaller than others.

To some, it may rather look like an attempt to patch up a rather ricketty biological hypothesis : to Dawkins it shows 'clearly how a collection of independent selfish entities can come to resemble a single organised whole'. That is, it may be used to deny the reality of collaboration between members of an animal community, in order to justify the hypothesis that they must *really* be serving the blind 'selfish' purposes of 'gene survival', in the search for 'gene pools' which will become an 'evolutionarily stable set of genes'. In this :

> Most new genes which arise either by mutation or reassortment on immigration, are quickly penalised by natural selection...

We thought 'natural selection' favoured ('ruthlessly') only those organisms, animals or genes, which offered immediate advances in *efficiency*? Now, it seems, by the ESS theory, that natural selection can favour *stability*. But how could it, since its only function can be to allow inefficient consequences to die out : that is all natural selection, by definition, can do?

And what about these other methods of invading the 'gene pool' – reassortment or immigration? Is Dawkins telling us now that there are totally new principles in evolutionary theory? We believed that only tiny mutations, whose effects were selected out by the offspring dying (before they could reproduce) unless there were immediate advantages 'could be the basis of evolutionary development'. Must we now admit to this reassortment and immigration, and effects which are selected out on the basis of *stability*? Perhaps he would tell us by what mechanism?

> There is a transitional period of instability, terminating in a new evolutionary stable set – a little bit of evolution has occured... Progressive evolution may be not so much a steady upward climb as a series of discrete steps from stable plateau to stable plateau.

Selection goes on apparently at the level of the gene :

> Genes are selected on 'merit'. but merit is judged on the basis of performance against the background of the evolutionarily stable set which is the current gene pool.

This seems to suggest something quite different from the struggle-for-survival process which is the basis of classical Darwinism, or even neo-Darwinism.

In chapter 6, page 95, - rather late one might have thought - Richard Dawkins asks : 'What is the selfish gene?' He has already made assertions that evolutionary theory can be applied to this entity, and has made it the determining *factor in all life*. It is not, he declares, 'one single physical bit of DNA' : it is '*all replicas* of a particular bit of DNA distributed throughout the world'.

If we allow ourselves, he declares, 'the licence of talking about genes as if they had conscious aims, always reassuring ourselves that we could translate our sloppy language back into respectable terms if we wanted to, we can ask the question, what is a selfish gene trying to do? The trouble is surely, that Dr. Dawkins does not translate his sloppy language back into scientific respectability, but goes on to make from it metaphysical assertions and moral pronouncements about human life?

Now, it appears, the evolutionary process has not to do with greater efficiency, or stability, but greater numbers of genes :

> It is trying to get more numerous in the gene pool. Basically it does this by helping to programme the bodies in which it finds itself to survive and to reproduce....
>
> p.95.

The Darwinian principle now seems to be so diffused as to have become utterly meaningless. If the key unit of evolution becomes a pattern distributed throughout many individuals, whatever happens to the basic principles? But now, surely, Dawkins is bending his explanation to suit his own dogma?

> The key point of this chapter is that a gene might be able to assist replicas of itself which are sitting in other bodies. It so, this would appear as individual altruism, but it would be brought about by gene selfishness.
>
> p.95.

Just as science in general will enter into any contention, to avoid teleology, so Dawkins will devise any ingenuity to avoid recognising 'altruism'.

Dawkins says 'It has long been clear that this must be why altruism by parents towards their young is so common' – but what has been clear is that this altruism is not real altruism, but an impulse to preserve the gene pool. Whatever the complex mathematical calculations Dawkins examines from the work of W.D. Hamilton, it should be clear to anyone who has read studies of parent - offspring behaviour that it could not be determined by the genetic composition of the individual, but is a complex manifestation of behaviour. And when it comes to human life, there are complex questions of value and meaning. Yet Dawkins seems to believe that even the capacity to save someone from suicide in human beings must be genetically determined. ('A gene for suicidally saving five cousins would not become more numerous in the population... the minimum requirement for a suicidal altruistic gene to be successful is that it should save more than two siblings...'.). But to return to Darwinism, if behaviour were simply determined by the genetic need to preserve certain rare or important genes in other bodies, the concept of the 'struggle for survival' must surely be radically altered? The survival of creatures which have gained small advantages by mutations by chance must give way to a concept of genes 'knowing' how to preserve good gene pool qualities in other bodies as well as in their own, by some primeval sixth sense which now becomes the first principle of life.

Dawkins now makes some mathematical calculations about altruistic behaviour. He quotes Haldane who said that in emergencies life-saving someone from drowning, he had no time for elaborate calculations. Survival machines, says Dawkins, do not have to do the sums consciously in their heads.

> Just as we may use a slide rule without appreciating that we are, in fact, using logarithms, so an animal may be pre-programmed in such a way that it behaves as if it had made a complicated calculation.
>
> p.103.

It is characteristic of the utilitarian and mechanistic tradition to which Dawkins belongs, that he should see the act of rescue determined by a (tacit) calculation inside the 'survival machine' which is scanning the situation, trying to decide whether the genes inside the victim are 11/12 of the necessary rare genes to be preserved in the gene pool, whether the creature in distress is a fifth cousin, or unrelated.

The argument is confused by Dawkins' anthropomorphism and the way he slips back and forth from animal to human altruism. To any ordinary person the problem is clearly one which belongs to the dynamic life world of a human being. Central to our response is the *imagination.* Imagination (as Coleridge knew) is the basis of our capacity to relate to others and to the world. We imagine what it is like inside the other : we act by an impulse of compassionate identification, and if we have a flash of selfishness, this is still modified by imagining what could happen to ourselves. such impulses are a human struggle of anguish, conscience and concern, and emerge from tacit processes as these are integrated within us : and depend how much we can 'find' what is not ourselves.

Dawkins imagines a man catching a ball, and is back again with computers. He seems to have no sense that all our human acts and modes of knowing can only be understood in terms of our subjective life. Only a mechanistic scientist could suppose that a man calculates behind his acts in this way :

> Benefit to self – risk to self – 1/2 benefit to brother – 1/2 risk to Mother + 1/2 benefit to brother...

He seems himself a little doubtful about what 'really happens' :

> What really happens is that the gene pool becomes filled with genes which influence bodies in such a say that they behave as if they had made such calculations.
>
> p.105.

Is parental care altruistic or not? Does it have survival value? He does admit at least that 'the evolutionary advantage of parental care is so obvious... it has been understood ever since Darwin'. But in his accounts both of animal and human behaviour he seems to anxious to preserve various mechanistic dogmas that he completely fails to be able to see the thing as it really is.

So the next chapter, on Family Planning, simply becomes wild. An individual survival machine, he tells us, has to make the kinds of decision, caring decisions and bearing (child-bearing) decisions :

> I use the word decision to mean unconscious strategic move...

– so, these are not 'decisions' at all, but again (automatic) strategies of the genes. Dawkins leaps without inhibition from details of animal off-spring care, to genetic dynamics, to world-population problems and human family planning. So, he offers a number of enormous predictions based on simple mathematics :

> In 1,000 years from now they would be standing on each others' shoulders more than a million deep :

He launches into religious politics :

> It is hard to believe that this simple truth is not understood by those leaders who forbid their followers to use efficient contraceptive methods.
>
> p.119.

From this he goes on to moral pronouncements :

> They express a preference for 'natural' methods of population limitation, and a natural method is exactly what they are going to get. It is called starvation.
>
> p.119.

The truth is, of course that where living standards improve, the birth rate goes down. The problem is urgent, we need not doubt. But the question is whether Dawkins has any real authority to pronounce on it : unless he were to tell us whether, given the theory of natural selection, this population growth problem will be naturally solved : or whether, given his gene survival theory, the genes will take care of it?

How could we, since we are only (survival) vehicles of gene selfishness, and even our most altruistic-seeming acts are gene-controlled, act to change the pattern of rapid population increase?

> Humans [some of them, he says] have the conscious foresight to see ahead the disastrous consequences of over-population.

But since it is 'the basic assumption of this book that survival machines in general are guided by selfish genes, who must certainly cannot be expected to see into the future..., not to have the welfare of the whole species at heart', where is there a place for 'conscious foresight'?

There follows a discussion of animal population trends and their underlying principles. Again, Dawkins must be vigilant against any attribution of altruism which seems directed to preserve the species. A genetic solution is predominant; genes determine behaviour here again. They can even interfere to adjust breeding patterns :

> Genes for having too many children are just not passed on to the next generation in large numbers because few of the children bearing these genes reach adulthood.
>
> p.125.

This is a summary of Lack's work : but we may note that natural selection here takes on another role, not of the survival of the fittest, or preserving the gene pool, but to adjust initial clutch size (litter size, etc.) so as to take minimum advantage of these limited resources.

We go straight on, however, to 'modern civilised man', without a qualification :

> If a husband and wife have more children than they can feed, the State, which means the rest of the population, simply steps in and keeps the surplus children alive and healthy.
>
> p.126.

But surely over-population is a problem in countries where there is no welfare state? In countries where there is welfare state care, population is stable or declining :

> But the welfare state is a very unnatural thing. In nature parents who have more children then they can support do not have many grandchildren, and their genes for *over-indulgence* are not passed on to future generations... Any gene for over-indulgence is promptly punished : the children containing that gene starve...

One would actually like to see the demographic facts on which this is based. It seems yet another social theory based on the mechanistic functional thinking of the nineteenth century.

Can there really be a gene for 'self-indulgence' : and is having many children a 'self-indulgence'? Some strange subjective factor seems to be underlying the argument here.

Discussing the welfare state further, Dawkins declares that 'any altruistic system is inherently unstable, because it is open to abuse by selfish individuals, ready to exploit it'.

Leaving aside the question of whether the welfare state is a cause of population problems or not, we see in these remarks the political dangers of Dawkins' obsessional assault on 'altruism', and the dangers of such a politics supposedly based on 'scientific realism'. It inclines, inevitably (as do the scientific politics of Skinner and E.O. Wilson) towards some kind of control, over human life which has been demonstrated to be unreliable, and unworthy of being free : its fundamental pessimism leads to extremely dangerous and heartless politics.

In chapter 8, we have further examples of Dawkins' sloppy methods of arguing and extrapolation. He reports a finding in Spain by F. Alvarez, L. Arias de Reyna and K. Segura, investigating cuckoo behaviour, that when they introduced a baby swallow into a magpie's nest, it behaved like a cuckoo chick, and heaved out the magpie's eggs. What is the explanation?

Dawkins says 'the true explanation is nothing to do with cuckoos at all. The blood may chill at the thought, but could this be what baby swallows do to each other?'. It doesn't chill my blood, but the question is - what observations or experiments is this assertion based on? The answer is that it is based on none. It is purely a conjectural assertion, offered to explain the Spanish phenomenon. Dawkins has not seen a baby swallow throw out a sibling egg, not, as far as we know, has any other biologist.

Yet this (unfounded) conjecture is translated into gene machine language :

> He can acquire a 1/4 share (of potential investment genes) simply by tipping out the egg... translated into gene language, a gene for fratricide could conceivably spread through the gene pool...

And now (breath-takingly) Dawkins declares that it is 'difficult to believe that nobody would have seen this diabolical behaviour if it really occured'. 'I have no convincing explanation for this...'. It is possible that no-one has ever seen it because it never happened! But next :

> The ruthless behaviour of a baby cuckoo is only an extreme case of what must go on in any family....

Thus, we have a 'battle of the generations...'

'This chapter', says Dawkins, 'could seem horribly cynical, and might even be distressing to human parents, devoted as they are to their children, and to each other'. 'Once again I must emphasise that I am not talking about conscious motives...'. But, as we have observed some of his 'evidence' is no more than fanciful conjecture, starting with his own dogma, and testing nothing against the world. Moreover, under the pressure of his genetic determinism, he pays no attention to all the work that has been done in the phenomenological analysis of whole beings, on sibling rivalry. And yet we come to a conclusion that offers us educational principles :

> when I say something like 'a child should lose no opportunity of cheating... lying, deceiving, exploiting...' I am using the word 'should' in a special way. I am not advocating this kind of behaviour as moral or desirable. I am simply saying that natural selection will tend to favour children who do act in this way... If there is a human moral to be drawn it is that we must teach our children altruism, for we cannot expect it to be part of their biological nature.
>
> p.150.

Here we have demonstrated the appeal of the 'naked ape' and 'selfish gene' kind of approach to human problems. If we ever behave in an immoral way, we can say : 'it is my instincts' or 'it is my genes' : at any rate, we are not naturally moral. Thus, we can evade our responsibility for ourselves. The realism, (which defends itself against being called 'cynical') has the appeal of not recognising, at the centre of human experience, the mysterious truth, that altruism is a primary reality of our natures. Our natural delight in being good derives from those natural impulses towards reparation, which Melanie Klein found, originating in the infants feeling for the mother, and arising from his natural sympathy and imagination : later D.W. Winnicott called this 'the stage of concern'. Philosophical anthropology not only finds man to be a creature with a culture : it finds this culture deeply moral. In all disciplines which seek to answer to Kant's question, 'What is man?' in a phenomenological way, the answer is that *man is naturally moral.*

The mechanists seek to show, by science, that man is not moral. Their politics naturally follow. He has to be controlled, or conditioned to be moral, against his real nature. His real nature is delineated according to the minimals - molecules or DNA or genes - which have their own, and blind, purposes, which are not really purposes either, but mere impulses towards survival.

On such inadequate 'scientific' structures, mechanists like Dawkins would seek to disperse all the findings of the Humanities, and persuade us that all that we know about ourselves is an unreality, that we are not moral, and that our consciousness and conscience are but delusions. Yet it is their fallacious theory, and its erection into a general philosophy of life, that is the real fallacy.

Now, (chapter 9) we turn to the battle of the sexes. Male and female only have a 50 per cent 'genetic share-holding' in the same children.

> Each partner can therefore be thought of as trying to exploit the other, trying to force the other one to invest more.
>
> p.151.

We have usually thought of sexual behaviour, copulation, and courtship as essentially a cooperative venture, undertaken for mutual benefit, or even for the good of the species.

Not so, in Dawkins' selfish gene world, which is essentially an extension into biology of the utilitarian principle of self-interest :

> Ideally, what an individual would 'like' (I don't mean physically, although he might) would be to copulate with as many members of the opposite sex as possible, leaving the partner in each case to bring up the children.
>
> p.151.

Dawkins offers to go back to 'first principles' : of course, again, this means the minimals, the genes in sex. It isn't a question of having a penis or not : frogs don't have penises. So, perhaps then the words, male or female, have no general meaning. They are, after all, 'only words... and we are at liberty to abandon them...' As Donne put it, 'the new philosophy brings all in doubt'.

There is, however, one fundamental difference : male gametes or sex cells are smaller than female gametes. This means (on Malthusian-Adam Smith principles) that the father invests 'less than his fair share.. of resources in the offspring'. The way in which scientific discourse (if this is really an example of it) is conducted is really remarkable :

'Female exploitation begins here', he says.

There are some interesting examples from the animal world in this chapter, but again Dawkins makes no thorough investigation of his hypothesis of animal 'commitment' or the consequences of courtship and copulation : he merely quotes the work of certain scientists and tries to bend it to his gene survival theory in terms of 'genetic pay-off'. And much of the discussion seems anthropomorphically confused, as here :

> A female playing the domestic bliss strategy, who simply looks the males over and tries to *recognise* qualities of fidelity in advance, lays herself open to deception...

Does it happen in the animal world that males try to 'pass themselves off as a good loyal domestic type, who in reality is concealing a strong tendency towards desertion and unfaithfulness', and so 'could have a great advantage' : 'As long as his former wives have any chance of bringing up his children, he strands to pass on more of his genes...'?

I am not concerned with the genetic argument : what I want to know is what the evidence is that female animals 'play the domestic bliss strategy', 'look over males...' 'trying to *recognise* fidelity'. And what evidence is there that animal males 'deceive' females in this respect? If we are discussing human beings, there are tremendously complex issues here, which mean that the genetic questions are seriously complicated. If Dawkins is discussing animals, where is there any evidence? I suggest there is none. Yet, of course, it is all pinned down to Darwinism :

> Natural selection will tend to favour females who become good at seeing through much deception.
>
> p.167.

Where is there any evidence for such a prodigious conjecture?

Males (declares Dawkins) have more to gain from dishonesty from females. But what? Does he mean in sensual satisfaction? It is hardly true of humans, in whom dishonesty seems associated with relational failure, and is likely to cause terrible distress. But in animals it seems likely that mere deceptiveness would be a threat to breeding patterns. Again, one would want to have evidence by way of accounts of actual cases, otherwise one has a kind of simple anthropomorphism, based on a very naive concept of human sexuality.

As for the relation of this argument to evolutionary theory, we have the same invocation of 'natural selection' as an all-powerful force : natural selection now seems a training school in sexual detection :

> natural selection, by sharpening up the ability of each partner to detect dishonesty in the other, has kept large-scale deceit down to a fairly low level...
>
> p.167.

How can 'natural selection' be an educative force?

Dawkins now, however, hits a problem – there are species in which the male does more work in caring for the young than the female. How do we get round this, if we are to preserve our selfish gene theory? Here Dawkins has a neat theory. On land, the female is 'landed' with the fertilised egg or eggs inside her body : so, the male can desert her. In water, where the fertilisation takes place outside bodies, the female has 'a precious few seconds in which to disappear, leaving the male in possession'. Explanations must always, for Dawkins, be mechanical ones. So, 'this theory neatly explains why paternal care is common in water but rare on dry land'. There is no support offered for this theory, from observation or experiment.

Towards the end of the chapter, in which there has been so much sloppy anthropomorphism, Dawkins says : 'I have not talked explicitly about man but inevitable, when we think about evolutionary arguments, as in this chapter, we cannot help reflecting about our own species and *our own experience*' (one may note that there is little or nothing about *experience* in Dawkins' book). While he admits, in the last few paragraphs, that cultural influences affect human sexuality, he makes only rather feeble comments on gender and attractiveness. This in turn makes references to sexual attraction in animals, and here Dawkins seems not to have encountered some of the studies discussed by Marjorie Grene which seem to show that some display is not related to sexual prowess or vitality at all, and so is unlikely to have the evolutionary value attributed to it.

Chapter 10 is about animals living in communities. Here, of course, Dawkins, being highly selective in his use of sources, we hear nothing of Adolf Portmann's *Animals as Social Beings*, but are rather directed to W.D. Hamilton's *Geometry for the Selfish Herd*, by which is meant the 'herd of selfish individuals'.

This simple model, says Dawkins, 'helps us to understand the real world'. There is a discussion of herd behaviour. Here, again, there are puzzles, when we adhere to the selfish gene theme, which must be explained away. Bird alarm calls *seem* like altruistic behaviour, because it has the effect of drawing the predator's attention to the caller. The calls, however, were 'produced by natural selection, *and we know what that means*', says Dawkins. We know that with him more often than not it means that, since natural selection can do anything, a problem is going to be solved by magic.

Bird calls have proved a problem for Darwinian theory, because those birds most likely to call probably get eaten, so natural selection should have selected them out. However, bird calls go on. The answer is simple – even if the calling birds do get eaten *some of their genes still survive in the surviving birds.*

Another theory is that a calling bird calls to cause others in the flock to freeze, so that all remain concealed from a passing hawk (a theory that surely seems a bit too altruistic)? Yet another is that a bird which leaves the flock and goes out on its own is known to be at a disadvantage (as from the hawk). The individual bird thus only calls to upset all the birds so that all fly up into a tree and the individual is safer. This, some researchers, have called 'manipulation', and Dawkins says : 'we have come a long way from pure disinterested altruism' – which seems, in the context, a silly conclusion, since calling in both the latter two examples seems eminently sensible, because the good of the one is bound up with the good of all.

Dawkins next turns to social insects. 'There is a temptation to wax mystical about the social insects', he says, 'but there is really no need for this'.

Anyone who has read the account of the worker wasp (in *Approaches to a Philosophical Biology*) or who has watched bees dancing, then flying out on their bearings exactly to a food source, must be struck with awe and mystery. But for the mechanist there must be no mystery : selfish gene theory will explain away all and we can lapse into a happy scepticism.

To Dawkins, however, ants and bees are not ants and bees, but gene machines : it isn't these social insects who organise their life on fantastic patterns but :

> Genes trying to manipulate the world through queen bodies (who) are out - manoeuvred by genes manipulating the world through worker bodies...
>
> p.191.

This book should have been called *Genes Behind The Scenes*, since every manifestation in nature turns out in the last analysis to be a manipulation 'by genes'.

Yet this, and 'natural selection' seem baffled by internal conflict. Suppose genes attempt to disguise male eggs, making them smell like female ones :

> Natural selection will normally favour any tendency by workers to 'see through' the disguise. We may picture an evolutionary battle in which queens continually 'change the code' and workers 'break the code'.

This is a new light on the 'evolutionary battle' - that there can be war between the *same* species :

> The war will be won by whoever manages to get more of her genes into the next generation...

Later in the chapter there is more play with genetic theory, freed (as it were) from the chains of Darwinian theory.

> Now that we have eschewed the 'good of the species' view of evolution, there seems no logical reason to distinguish associations between members of different species as things apart from associations between members of the same species...
>
> p.197.

This follows the hypothetical explanation ofthe possibility that 'viruses may be genes that have broken loose from 'colonies' such as ourselves, so that we might just as well regard ourselves as colonies of viruses. Or collections of molecules? 'These are speculations for the future', says Dawkins. There is nothing wrong with speculation, so long as it is presented as speculation – but if speculation is extended thus, what happens to evolutionary theory, which is a basic paradigm in established biology? You cannot abolish its basic tenets (struggle for survival between species) with one hand, and continually invoke it as a fundamental principle with the other, which is what Dawkins does.

Dawkins ends his chapter 10 with a discussion of what seems to be altruistic mutual grooming, in symbiosis, and elements of deception in such processes. Dawkins puts one ESS theory through the computer to test it.

No computer, however, could save him from his sudden lurch, from discussing 'cheating' in symbiosis, into discussing altruism in man.

Trivers ('The evolution of reciprocal altruism', *Q Rev Biol.*, 46,35-37, 1971) is quoted as suggesting that : 'many of our psychological characteristics - envy, guilt, gratitude, sympathy etc. – have been shaped by natural selection for improved ability to cheat, to detect cheats, and to avoid being though a cheat'.

These psychological faculties could only have been 'shaped' by 'natural selection' if they had given immediate advantage : what then could have been the advantage of our mixture of envy and gratitude? In any case, natural selection only selects consequences of mutation : it is not appropriate to apply natural selection to complex psychological factors (which arise, according to psychoanalysis, from complex inter-subjectivity with the mother). But there seems no limit to Dawkins' absurd interpolations :

> It is even possible that man's swollen brain, and his predisposition to reason mathematically, evolved as a *mechanism* of ever more devious *cheating*, and ever more penetrating detection of cheating in others. Money is a formal token of delayed reciprocal altruism...

'There is no end to the fascinating speculation...' he says again. Yet this is presented as a 'scientific' work.

Defending Richard Dawkins against the criticisms of Mary Midgeley, Paul Seabright said that Dr. Dawkins 'attempts no moral philosophy', and that his 'remarks about selfishness... are a metaphor that serves to make a purely biological point'.

This is not so. The whole of the last chapter discusses questions of human altruism, God, culture and the survival of the identity. At the end, he says :

Even if we look on the dark side and assume that individual man is fundamentally selfish, our conscious foresight could save us from the worst excesses of the blond replicators... We have the power left to defy the selfish genes of our birth... we can even discuss ways of deliberately cultivating and nurturing pure, disinterested altruism – something that has no place in nature.

p.215.

What are these remarks but moral philosophy? They insist that altruism is not natural, that our basic fabric (the 'replicators') is selfish, and that any attempt to make man altruistic must necessarily go against the grain of his nature. Moreover, he implies that 'science says' this is so.

At the very beginning of his chapter Dawkins admits that he has not 'deliberately excluded man', and, indeed, he has often extrapolated into anthropology. He includes us in 'survival machines', and what he has said applies, (he admits) to any evolved being.

However, our species is to be exempted because we are unique. We have 'cultural transmission' and such phenomena as language have 'evolved' by non-genetic means, at a rate that is orders of magnitude faster than genetic evolution. Even in birds, Dawkins admits, 'cultural mutations', say, can evolve by non-genetic means.

Dawkins admits that he is not satisfied with any explanation that tries to fit culture into evolutionary theory. Even such theories as those around 'reciprocal altruism' do not 'begin to square up to the formidable challenge of explaining culture, cultural evolution, and the immense differences between human cultures....' (p.205).

For an understanding of the evolution of modern man, says Dawkins, we must begin by throwing out the gene as the sole basis of our ideas on evolution.

I am an ethusiastic Darwinian, but I think Darwinism is too big a theory to be confined to the narrow context of the gene. The gene will enter my thesis as an analogy, nothing more.

p.205.

This is a welcome turnabout : but it comes oddly at the end of a book which has attempted to reduce all life to gene selfishness, in terms of gene-machine survival, irrespective of species survival, on doubtful Darwinian principles.

Dawkins then turns to ask if there is one universal principle in biology. There is, says Dawkins :

This is the law that all life evolves by the differential survival of replicating entities. The gene, the DNA molecules, happens to be the replicating entity which prevails on our own planet... the basis for an evolutionary process.

p.206.

The genes, then, are not an 'analogy', but the fundamental minimal, however narrow their context.

There are two things to say at this point, in the light of Dawkins' book. First, if the genes 'manipulate', as he suggests, 'building survival machines' and 'brains', scheming to ensure gene-pool survival', this 'intelligence' or 'dynamic' or directiveness *cannot be the mere product of the laws of the physics and chemistry.*

Secondly, if this is so, what we must look for is a way to understand the morphogenetic dynamics in terms of some 'primary consciousness' such as Tomlin postulates.

Dawkins, however, has found something else : human culture, and he must now submit this to his reductionist mechanism. Having found the gene as the unit of natural selection, he now (thinking minimally) looks for the 'unit of cultural transmission'. Culture must be allowed to be a function of a whole complex creature, of consciousness and imagination. It must be, too, reduced to a biological 'little causal thingummy'. This is the *meme.* (It should be pronounced to rhyme with 'cream').

Dawkins leaps into our sphere, of the subjective disciplines, with a sickening lurch. These memes are to culture as DNA is to life :

examples of memes are tunes, ideas, catch-phrases, clothes, fashions, ways of making pots or building arches...

Whatever Dr Dawkins' qualifications in the Scientific world, we in the Arts and Humanities who study the nature of human culture and its transmission can only say that on this subject he writes maive nonsense :

> Just as genes propagate themselves by leaping from body to body via sperm or eggs, so memes propagate themselves in the meme pool by leaping from brain to brain via a process which, in the broad sense, can be called imitation.

We have seen that imitation is as far as the mechanist can go, in recognising imagination and mind. The scientist has no idea of the complexity of thought, symbolism, and the meeting of minds. Dawkins quotes a colleague :

> When you plant a fertile meme in my mind you literally parasitise my brain...

('literally'?)

> turning it into a vehicle for the meme's propagation in *just the way*

('*just the way*'?)

> that a virus may parasitise the genetic mechanism of a host cell. And this isn't just a way of talking...

Undaunted, Dawkins goes on to tackle belief in life after death, and the idea of God ('we did not know how it arose in the meme pool').

Now Dawkins is in 'our' area, of culture and psychology, it becomes more clear that he confuses levels of discourse. The analogy he makes between 'memes' and DNA is absurd. Memes, apparently, are not products of the human mind and nurture, between union and separateness : they float about in a brain-soup :

> The old gene-selected evolution, by making brains, provided the 'soup' in which the first memes arose...

To memes, too, apparently, the qualities which provided for high survival value in 'natural selection' (longevity, fecundity and copying-fidelity) also apply to memes.

What it amounts to is a pathetically inept attempt to find and discuss mind ; but, to the mechanist, mind itself can only be a container for bits. The body is not a body in its own being, a being-in-the-world - it is simply a 'survival machine' built by, and used by, the genes. The mind, in Dawkins, is equated with 'the brain' and in the brain float bits - of music, fashion or idea, the memes.

There is a dim sense in Dawkins that behind that knowledge of the world is a man : so, there is something else besides bits in the world. But even so, his conception of his own ideas is like bits in a computer. His reference to Darwinism here is extraordinarily revealing : even Darwinism becomes a bit :

> When we say that all biologists nowadays believes in Darwin's theory, we do not mean that every biologist has, graven in his brain, an identical copy of the exact words of Charles Darwin. Each individual has his own way of interpreting Darwin's ideas... He probably learned them, not from Darwin's writings but from more recent authors. Much of what Darwin said is, in detail, wrong. Darwin, if he read this book, would scarcely recognise his own original theory in it. Yet, in spite of all this, there is something, some essence of Darwinism, which is present in the head of every individual who understands the theory.

What people do have in their heads is a Darwin 'meme' :

> The meme of Darwin's theory if therefore that essential basis of the idea which is held in common by all brains who understand the theory.

The 'meme' now seems to have a meaning something like the 'Spirit of God', since clearly the ideas of Darwinism are variously interpreted, and altered, according to the scientists' inclination.

Genes, Dawkins declares, have not been shown in his book as 'conscious, purposeful agents'. Yet they, have, as we have shown, been described as controlling, building, manipulating, towards the goal of the gene survival : they have definitely been portrayed as having an *end*. But this is not to be admitted :

> Blind natural selection, however, makes them behave as if they were purposeful... and it has been convenient... to refer to genes in the language of purpose.
>
> p.211.

As we have seen, the Darwinians always want it both ways – an end which is survival, but only blind movement towards it, a directiveness, but this only 'apparent purpose', moving aimlessly towards its aim.

> For example, when we say 'genes are trying to increase their number in future gene pools' what we really mean is 'those genes which behave in such a way as to

increase their numbers in future gene pools tend to be the genes whose effects we see in the world...'.

But in the body of the book this was not a matter of convenience, because to discuss all questions of altruism, Dawkins declared that what the genes were 'really' *trying* to do was – an argument that wouldn't work if he put the matter in a non-convenient way. To say that genes are 'selfish' and 'ruthless' implies something quite different from what is implied in the second way of putting it.

But now Dawkins (who is a master of extravagent postulations) goes on to suggest that cultural *memes* are as ruthless and selfish in the same way!

Culture might seem to be the last bastion of altruism : as Dawkins has admitted, culture has no evolutionary value in the battle for survival. There must be, however, - pursuing his ridiculous analogy - competition in the *brain*, between memes :

> If a meme is to dominate the attention of a human brain, it must do so at the expense of 'rival' memes...

We might suppose that, having discovered human culture, Dawkins might have allowed that the choice of celibacy was a complex issue of meaning, choice and behaviour. But no – if there cannot really be a *gene* for celibacy, there must be a *meme* for it.

> The meme for celibacy is transmitted by priest to young boys who have not yet decided what to do with their lives...

'I have been a bit negative about memes', says Dawkins. What he has really done has been to try to explain away human consciousness and culture by applying to this sphere the simple mechanistic and competitive principles of Darwinism.

What about the way culture survives, however? Physically,

> we were built as gene machines, *created to pass on our genes...*

– only DNA making more DNA. But we can seek immortality also through our memes :

> But if you contribute to the world's culture... the memes of Socrates, Leonardo, Copernicus and Marconi are still going strong...

Whose survival are we talking about? A cultural trait (a meme) may have evolved as it has because of its *advantage to itself.* Memes will evolve which exploit the capacity to the full...

Dawkins surely has no idea how a cultural artefact exists? It can exist of course only in a human consciousness. A poem, picture or symphony only exists in the 'criss-cross of utterance between us'. It cannot exist outside men and women : it is utterly absurd to suggest that memes might take on a survival life of their own, except of course in the physical artefacts.

In his last chapter, Dawkins tries to find mind and freedom. But such manifestations are *against* our 'real' nature. 'Man has a capacity for 'conscious foresight'. But he is composed of selfish genes which have none : 'they are unconscious, blind replicators'.

> they will tend towards the evolution of qualities which, in the special sense of this book, can be called selfish...

'Natural selection' is always bound to favour the ESS. And if we strive to be moral, this must always be against nature and our own nature :

> we are built as gene machines and *cultivated as meme machines*, but we have the power to turn against our creators. We, alone on earth, can rebel against the tyranny of the selfish replicators.
>
> p.215.

Surely, whatever the value of Dawkins reports on scientific observations, this is not justified by the observed facts of life. There is no 'tyranny' in the molecular and biological processes : the animals observed by biologists are not victims of tyranny, but in various degrees triumphant achievements, capable of amazing forms of social life, ingenuity, intelligence and prowess. Man has an additional cultural dimension, besides other capacities such as upright posture and consciousness. His culture is the product of encounter, and arose out of his faculty of imagination, based on symbolism which emerged from the nature of infancy. It is not at all composed of autonomous fragments of sense-things which, in some way, have a (competitive) life of their own : we achieved our culture, and we were not built as 'meme machines'. We came into being as automonous living creatures, and we were not 'built' by our genes, or by our memes : we can only be understood as living beings, having emerged out of the developing flux of life.

Certainly, in the behaviour of such a complex creature as man, must not the activities of this creature be examined, as the humanities examine them, in terms of the whole being? That is, are not 'scientific' incursions of such subjects as microbiology and sociology into cultural and ethical matters really absurd? They would certainly seem so, to an important stream of European philosophical anthropology, to say nothing of common sense.

Such pronouncements as Dr. Dawkins' on the relationships between genes and human altruism woulsd seem not only ridiculously naive, but also menacing to the Humanities in general, because of the implication that only the 'little bits of goo' are 'real', a point Mary Midgeley takes up in here *Beast and Man.*

Not only does Dr. Dawkins go in for moral philosophy : there is also implicit nihilistic metaphysic which comes over to us from his book, and it is summarised in the sensational blurb (I have found leading scientists and medical men in Cambridge who take the book on the lines of that blurb as an important statement of scientific truth about life).

But as I hope I have shown, it is really an absurd extrapolation into the realm of philosophical anthropology, from reductionist microbiology, in which a highly tendentious and confused use is made of Darwinist principles which are doubtful anyway.

In conclusion, it must be said that Richard Dawkins' thesis about us being 'throwaway gene machines' can only be accepted if the following things are true :

1. The minimals of any living being determine its behaviour.
2. Natural selection is the directing force in evolution, devoted solely to ruthless survival.
3. Anything can happen in evolution, given long enough time.
4. There is no directiveness in evolution.
5. Darwinism has offered a fully satisfying account of the origin and development of life.

In fact, none of these are true. Yet among the company of distinguished scientists in Cambridge I have found Dawkins' absurd thesis treated with respect - a respect which, I have to report, has combined with an aggressive disdain for my own kind of 'humanities' concern with human values and being so persuasive in his simple-minded scepticism.

Yet the slightest acquaintance with philosophical argument reveals vast fallacies in this dangerous book. It assumes that life is to be understood by being broken down into its smallest components : it confuses the level at which it is appropriate to discuss molecular and cell organisation with the level at which it is appropriate to discuss the operations of and behaviour of higher animals, and of man, with his consciousness and culture; it assumes that hypotheses about the former must necessarily determine all problems in the latter dimension : it dodges, fudges and confuses arguments hopelessly, transferring neo-Darwinist theories of evolution from the consideration of plant and animal life to the molecular and gene level, confusing in the process what is meant by 'survival' : it uses terms life 'natural selection' like an incantation, often endowing this purely negative force or principle with creative powers and foresight : it assumes that is is proven beyond doubt that all evolution depends upon accidental or chance mutations, and 'mistakes in copying', without any examination of the residual problems - certainly the problem of why life is here at all, and the realities of prodigious and complex leaps forward; and it confuses the dimensions in which morality and the nature of things (what is and what ought to be) may be properly discussed.

I have spent too much energy and space on Dawkins. But my encounter with his book was the spur to write the present work. Of course, by the time my criticisms are published, opinion among the scientists will have moved on. Whatever happens, the important thing for them is to uphold the dogma - and it was no doubt the Darwinian dogma which enlisted such sympathy with this ridiculous book. So, no doubt, the scientist will close ranks - even against the growing tide of doubt. Yet surely there must come a day when Darwinism can no longer provide a ticket to talk endless nonsense about man, offered with a 'scientific' tag?

10 The Concept of Life: the work of E. W. F. Tomlin

Some light may be cast on the new modes of thought we require by the work of E.W.F. Tomlin in the *Concept of Life*, a still unpublished work from which some chapters have appeared in print.[1] Tomlin is concerned to 'restore the vital' to our account of life. In the attempt to reduce biology to physical science, biologists in this century have divorced nature from everything to do with life, and the reductive tendencies of all mechanism culminated in a *reductio ad absurdum*. First the organism was represented as a more than usually complicated machine. Then the attempt was made to discover what it was that lent it dynamism. It must be put into movement in time.

But this is not to be achieved, as it were, by adding a 'T' for time to the diagrams and formulae. The developed organism is a structure in space, but it has undergone *morphogenesis*, development, in a process in which the form arrives phase by phase. Before, there was no structure : now a structure is there.

Tomlin examines various inadequate attempts to solve the problem. Often it involves some way of shoving a vital spark or entelechy of some kind into the mechanism. Some solutions are occult and irrational : requiring, virtually, a belief in magic.

Tomlin believes the solution must lie in some kind of bringing together of biology and psychology. He refers to the philosophy of organism, of which Whitehead laid the foundations, and we have seen Sheldrake's examination of this approach. Tomlin invokes Husserl's concept of the *Lebenswelt*, the living world.

Many scientists feel that the 'world of science' represents the 'real' world. But the world of science, Tomlin points out, exists only within the Lebenswelt itself. Where biology is concerned, we must think of societies as in the organic thought of Whitehead, or of *beings*.

> Instead of consisting of groups of bodies, colliding blindly in space or in the void - which, whatever their disclaimers and whatever double-think the mechanist biologists employ – nature is a community or society of significant beings.

Presumably by 'significant' Tomlin means having intrinsic interest.

Life is consciousness : this indicates the solution to the Cartesian problem. By consciousness here we do not mean merely the highly aware and self-conscious consciousness of man, but :

> in a primary sense as the immediate self-possession of an organism, and in the secondary sense, as that which, mediated through the brain, proceeds to the ordering of the external world...

Every living being is a *subject* rather than a object, a true rather than a pseudo-form. To comprehend life we need to bring together philosophy and (as Sheldrake indicates)

---

1 e.g. *The Concept of Life*, *The Heythrop Journal*, July 1977, Vol XVIII no.3 : *Biology and Metaphysics*, *Books and Issues*, Vol 1 No.2, 1979.

metaphysics, for we must know where we stand and try to state this systematically, while also developing new modes of observing and understanding the world. For instance, we need to find life and consciousness and consciousness as the self-possession of an organism - the unity of a multiplicity, so that we may see that :

> a being is a domain of togetherness in dynamic organisation, or equilibrium in which a hierarchy of sub-systems are 'dominated', as a molecule dominates its atomic and sub-atomic constituents or activities; and the more complex the organism, the greater its domain will be. This dominance indeed is also a dominance of space-time, since the immediate self-possession is precisely the unity or togetherness of a particular 'here' and 'now'. An amoeba's here-and-nowness is part of its subjective unity, which it maintains by 'being an amoeba', and this maintenance is observable externally to the biologist by its structural activities... The development of the human brain, with a nervous system of unprecedented complexity, implies a much greater domination of space-time. Here, then, we have a new version of the distinction between appearance and reality. Reality is the inwardness or subjectivity of an organism, its form; appearance is the external activity whereby we infer or cognise, but never directly percceive, the presence of life.

As Tomlin says, it is the divorce from science that has driven philosophy to extremes. But conversely it seems true that science divorced from philosophy is also driven to extremes : as we have seen, the incapacity to subject their arguements to philosophical scrutiny leads scientists to utter metaphysical nonsense. The truth is that there is an urgent need to restore metaphysical disciplines to scientific discourse : Tomlin quotes F.I.G. Rawlins on Max Planck :

> here was indeed a true philosopher, ready to acknowledge that metaphysics may well advance if common cause is made with natural science.

It is essentially philosophy of the metaphysical kind which Marjorie Grene and Michael Polanyi restored to science. At our universities the kind of philosophical debate which Marjorie Grene conducted at Boston, as collected into *The Understanding of Nature*, should be obligatory. It is a question of making our thoughts more at home in the world, and restoring the awareness that 'the things we call living are more real than other things' (Marjorie Grene), and seeing that 'knowing is a form of life' : restoring the scientist's awareness of his own thinking as a form of that cconsciousness which *is* life.

And as far as evolution is concerned, we must recognise that directiveness, that gradient, so that we can see that, looking at the 'fossil record', 'the fact that we can speak of them is one of the surprising results of the process that they record'. (Marjorie Grene).

At the end of the *Knower and The Known*, Marjorie Grene writes :

> evolution as macro-evolution, as the emergence of life and of higher forms of life, out-runs both the concepts of gene-substitution, and of improvement in relation to environment. *It makes sense only as an achievement* - an achievement for which statistical methods can measure the necessary, but not the sufficient, conditions.
>
> p.266.

What can 'achievement' mean? It points to a response to the origin and development of life that leads to none of the pessimistic, meaning-destroying and value-destroying attitudes with which we in English and the Humanities have to contend.

And we should not have to contend with these, if science and biology put their own house in order. Philosophy, Tomlin points out, tends to assume that 'the highest form of anything is that which is most real'...

The question of what is 'real' seems to lie behind such a debate as the present day one already mentioned on abortion. To some, the newly conceived foetus is endowed with an immortal soul : to others it is just a bundle of molecules until in some way it displays 'human' aspects. However, in a world which lives within the Cartesian paradigm, the 'objective' can never become integrated with the 'subjective' : as Tomlin puts it, meaning and value are assigned to mind while physical and mechanical causality are confined to nature. Science deals exclusively with a quantitative world of extension and movement (*res extensa*) while philosophy or metaphysics deals exclusively with a world of thought (*res mensa*). Thus insulated in an 'endless psychomachy' as Tomlin calls it, Descartes never engaged with nature, despite his hypothetical link between mind and body via the pineal gland. In this situation, the sphere of life became marooned : 'it was neither subjective nor objective, but something in between, a tertium quid, or, as Galileo said, an inexplicable intrusion'.

> In due course life was arbitrarily annexed to the empire of physics, from which it has still not succeeded in liberating itself.
>
> *Heythrop Journal* XVIII, No.3, p.290.

This kind of problem lies behind the failure of the modern mind to engage with many urgent issues which concern life. The problem, to use a phrase from Tomlin, is the '*omission of the vital*'.

By the way here, I should like to point to one particular aspect of the Evolution debate, which is relevant here. There is one evident reality in Nature, which is that of a gradient : the movement, by directiveness, towards higher and more complex forms.

Darwin conceived of the progress towards new forms in terms of minute, slow and gradual changes :

> Natural selection acts solely by accumulating slight, successive, favourable variations; it can produce no great or sudden modifications; it can act only by short and slow steps.
>
> 1888 Edition, p.413.

Nature is a 'niggard in innovation' : yet, as we have seen, there are innovations like the eye which can only have come into existence by sensitive and complex coordinated innovations which cannot be explained on Darwinism principles.

There is thus, evidently, an innovatory, directive dynamic - and books like Rupert Sheldrake's are an attempt to bring this dimension into biological thought.

The mechanistic scientist 'solves' this problem by mystification. Marjorie Grene refers us to Schindewolf, who in his *Grundfragen der Paläontologie* (p.413, and pp.430-31) says that 'nothing could be more mystifying than the prophetic vision of selection which the neo-Darwinians implicitly assume : the idea that selection selects what will be advantageous in another ten million years, and thus foresees advantages that will become to the remote descendents of the forms selected'.

Here is one blankness crucial to the present debate. There is in life a movement forward towards new potentialities in time, which is not recognised in conventional thinking - except by magic or mystified dodges such as that to which Schindewolf draws attention. This is the kind of question scientists avoid by talking about 'apparent purpose' - thus suspending the teleological questions neatly.

Another trick to side-step the problems is that of solving all questions of development in time by declaring that evolutionary periods are so long (See Sir Julian Huxley in *Evolution as a Process*, 1957, p.3). But, as Marjorie Grene 'Can an incomprehensible happening become comprehensible by lasting a long time?' Doesn't 'selection' in conventional biology stand for some mysterious force which controls the whole process? Schindewolf, as against this highly abstract and hypothetical Darwinian theory, suggests that *life can originate novelty*. This is a preferable assumption because it is simple :

> He does not pretend to 'explain' this proposition, and in that sense it may be 'mysterious', but no more mysterious, he says, than physical concepts like 'force' or 'gravitation'...

And not nearly as mysterious as the way Darwinians use the concept of 'natural selection' as a creative force, when, in their terms, it cannot be (see *The Understanding of Nature*, p.131).

The second attribute of life, then, is that *it can initiate novelty*.

Cartesian dualism, as we shall see, persists in fundamental attitudes to such an organism as the human foetus. At what point does it have, as it were, injected into it a 'mind' and become a 'human being'? Recent developments in electronic information technology are taken as proof, as Tomlin puts it, that organic and cerebral inventive activity is itself mechanistic. It is rather a further revelation, he insists, of the vital inventiveness which produced the organism with its own automtisms - its *esse*, or its dynamics as a form-in- itself

> A flower is not a complicated laboratory workshop in which the process of photo-synthesis takes place; it is a form-in-itself, that is to say, its network of photo-synthesizing and hormonal activity is 'contained' by its primary

consciousness. A flower's *esse* is to 'floralize'

Heythrop article, p.302

Tomlin then takes a more specific example of a life-form :

> A single cell is an organism of the highest complexity. Like each of the millions of cells of higher orders, it is engaged in incessant micro-activity according to strict thematism (which may cover a constellation of themes)... a protozoan... is not a mere physico-chemical complex, or a drop of fluid maintained by surface tension. As Jennings first demonstrated in 1904, it is a creature which manifests definite behaviour... it takes in nourishment by means of its pseudo-podia ('false feet'); it breathes, though lacking lungs and gills; it responds to stimuli, though lacking nerves ; and, by dividing itself in two, it engages in reproduction. Grimstone remarks that 'for the most part the problems of morpho-genesis and differentiation in protozoa cannot yet be formulated adequately in physico-chemical terms', but it would be nearer the truth to say that they 'cannot be formulated in physico-chemical terms'.

Tomlin insists that such creatures must be described as individuals, beings or subjects. This creature, like a paramecium, *does not have a brain or a nervous system.* Many commentators, some of whom Tomlin quotes, see mind/thought and consciousness - as products of the human brain. But in lowly creatures such as the amoeba, declares Tomlin, we must recognise a *form of consciousness* anterior to, and presupposed by, cerebral consciousness. Moreover, discussing how the nervous system itself comes into being, Tomlin suggests that there is something called *primary consciousness*,

> The fact that certain cells destined for one part of the embryo should develop into nerve cells associated with cerebral consciousness is explicable only on the assumption that embryonic development, as a unified and coordinated activity, is throughout informed with what we have called primary consciousness.

p.303

When we take account of man, we must take account of what Tomlin calls the *Lebenswelt* (Husserl's concept) : but even with the lowest forms of life, we must recognise that we are dealing not with mere 'molecules', but with *life-worlds*.

This is not some new kind of mysticism, or religion. Indeed, to believe that 'mind' is, in some mysterious way, suddenly 'introduced' into a living organism is the mystifying explanation. We ought to know, from biology, that the following things are true about life :

1. That a living thing is alive is the most real thing about it : there must be no 'omission of the vital' in an approach to biological forms.
2. In living things there is a *directiveness*, towards higher forms.
3. Life can initiate *novelty* : and evolution is only understandable by taking account of some such dynamic.
4. Each living creature displays intentionality : is trying to become itself, to realise its potentialities is a form in itself, and manifests its *esse*.
5. Even the simplest life-forms manifest a kind of *primal consciousness* and so must be approached as beings.

May one humbly suggest that these simply take account of realities and so make good biology. It would seem to me as a Humanities man, that any biology worthy of the name must recognise those special qualities that belong to 'life'.

It is interesting to turn back to Darwin from Waddington's suggestion that 'learning and innovation' would be a better phrase than 'chance and necessity'. Most mechanistic scientists are so anxious to eradicate God and Design that they have put themselves in a position of scotomising (that is, turning a blind spot upon) evident realities.

Darwin himself never really solved the problem. In his *Autobiography*, he says :

> Although I did not think much about the existence of a personal God until a considerably later period of my life, I will here give the vague conclusions to which I have been driven. The old argument about design in nature, as given by Paley, which formerly seemed to me so conclusive, fails, now that the law of natural selection has been discovered. We can no longer argue that, for instance, the beautiful hinge of a bivalve shell must have been made by an intelligent being, like

> the hinge of a door by man. There seems to be no more design in the variability of organic beings and in the action of natural selection, than in the course in which the winds blow. Everything in nature is the result of fixed laws.
>
> p.87.

Concealed in this are two fallacies : the negative survive-or-not processes of natural selection cannot *sui generis* generate design : nor can the mere accidental processes of mutation design a beautiful hinge. The 'learning and innovation', the 'ingenuity' that life displays demand some other explanation, and that we do not yet have. There may not be an 'intelligent being' behind the beautiful ingenuities of life, but there is something that can only be called intelligence, or primary consciousness in the beings themselves.

Secondly, everything in nature cannot be 'the result of fixed laws', since the more complex molecules and organisms exist in some defiance of the laws. All mechanisms which purposefully, although temporarily, reverse 'noise' and entropy seem to be the product of intelligent orchestration. Codified direction or constraint of the degree necessary to engender the plans and processes of life is never observed in non-living matter. And, as many have observed, there is an impulse (call it 'apparent purpose' though we may) which is not 'the result of fixed laws' and cannot be.

The enigma may never be resolved : but it is not resolved by mechanism, nor is it ended by ignoring it in favour of dogmas, as scientists try to ignore it.

E.W.F. Tomlin has followed Darwin's confusion throughout his *Letters* over this question. As we have seen, in his *Autobiography*, Darwin declared himself a theist, because of the 'extreme difficulty, or rather impossibility, of conceiving this immense and wonderful universe, including man with his cpacity for looking far backwards and far into futurity, as the result of blind chance and necessity'. (*Autobiography*, p.92, *Letters*, Vol.I. p.312). In a letter to Asa Gray, of 22 May, 1860, he wrote that he could not see God's design and beneficence on all sides : there seemed to him too much misery in the world :

> I cannot perseuade myself that a beneficent and omnipotent God would have designedly created the Ichneumonidae with the express intention of their feeding within the living bodies of caterpillars....
>
> Vol II, p.312.

But in flinching from this variation in the pathetic fallacy, Darwin goes on : 'Not believing this, I see no necessity in the belief that the eye was expressly designed'. The truth is that the eye cannot be explained except in terms of a number of complex features all coming into being in some coordinated way, as a manifestation of 'learning and innovation' : it certainly cannot have come into being by small chance changes which gave no immediate advantage in the struggle for existence. But Darwin evades the problem by finding himself in the metaphysical predicament of supposing that, if intelligence is involved in the 'design' of the eye, it must be intelligence of a God who ought not to have invented cats who play cruelly with mice. Yet Darwin goes on :

> On the other hand, I cannot anyhow be contented to view this wonderful universe, and especially the nature of man, and to conclude that everything is the result of brute force. I am inclined to look at everything as resulting from designed laws, with the details, whether good or bad, left to the working out of what we may call chance. Not that this notion *at all* satisfies me.
>
> p.312.

In such discourse we can see that Darwin's best value was in positing and investigating profound questions about the nature of life : in Waddington's comment on chance we can see that they remain puzzling questions. The disaster is that, before the ideas were fully examined, they became dogma, of the kind into which Monod elevates 'chance' – even though Darwin himself found it far from satisfactory.

'I am conscious (he wrote to Asa Gray on 26 November 1860) that I am in a hopeless muddle'. :

> I cannot think that the world as we see it is the result of chance; and yet I cannot look at each separate thing as the result of Design...
>
> Vol. II., p.313.

And to the same a month later :

with respect to Design, I feel more inclined to show a white flag than to fire my usual long-range shot. If anything is designed, certainly man must be; one's 'inner consciousness' (though a false guide) tells one so, yet I cannot admit that man's rudimentary mammal... were designed... You say you are in a haze; I am in thick mud... yet I cannot keep out of the question...

Vol.II., p.382.

Tomlin returned to the attack on Darwinian dogma in a review-article in *Universities Quarterly*, Winter 1983/4, vol 38 No.1, 'The dialogue of evolution by natural selection'. This was a review of *The Great Evolution Mystery* by Gordon Rattray Taylor, also taking account of *The Descent of Darwin* by Brian Leith.

Tomlin sums up Taylor's argument by pointing out tht there are three possible attitudes to Evolution (a) Creationism (b) Evolution by Natural Selection and (c) a new synthesis based on post-Darwinian scientific developments, especially in the field of genetics.

Evolution cannot be denied : but Creationism is (says Tomlin) an irrational and mythical response to it. Gordon Rattray Taylor argues that natural selection cannot be the 'main' explanation of evolutionary change. Taylor has assembled a full and clear summing of the doubts. Leith in his chapters 9 'How strong is natural selection?' 'and 11 'Darwin in Descent' offers likewise excellent summaries.

The essential question, according to Tomlin, is whether a concept of directiveness can, without being providentialist, throw light on the nature of morphogenesis, or whether morphogenesis is intelligible in the absence of what Whitehead called 'aim'. Taylor admits or suggests that 'the world is ordered, but that the order springs from an inner necessity'. And this poses the question 'Is there some principle at work alongside natural selection?' But, according to Tomlin, Taylor does not answer this question because he is afraid of teleology - he tries to reconcile 'inner directiveness' with 'blind impulsion'.

One of the problems, which we have glimpsed, is that the closely coordinated programmes we find in life can scarcely be believed to be products of chance variation, according to the theories of evolution by natural selection. Take, says Tomlin, the case of *Hydra* :

> an animal about 1/2 inch long, which lives in fresh water. This creature consists of a semi-transparent tube, with an opening at one end surrounded by trailing or searching tentacles. These tentacles capture minute protozoa for sustenance. Attaching themselves to underwater plants, *Hydra* swim about in the water, occasionally turning somersaults. Certain *Hydra* possess stinging cells; and from these thread cells tiny harpoons or barbs called nematocysts, responding to stimuli, are propelled, each barb containing poison. The mechanism is so delicate and swift in its operation that its function cannot easily be detected. Now the structure of *Hydra* is complicated enough : but the use to which it is put by the planarian flatworm called *Microstomum*, a smaller creature, is one of the most extraordinary phenomena in nature. *Microstomum*, whose mouth is half-way down its body, devours the stinging *Hydra* ; but instead of digesting the nematocysts, it passes them through its body and, having done so, employs them as artillery on the surface of its skin. Then when the approaching prey touches its surface - or sometimes without waiting for it to do so - it releases its sting harpoons precisely as the *Hydra* had done. Having discharged all its weapons, poison and all, the worm then proceeds to rearm itself by devouring another *Hydra*. And so on.

This instinctive routine has established itself with the most rudimentory sensory apparatus : yet such creatures manifest a recognisable *behaviour*. To attribute such behaviour to chance or accident or a combination of chances or accidents is to defy reason. In the migration of the Hydra we seem to see a purposiveness of the kind which Darwinists refuse to believe in, says Taylor.

Discussing evolution in general, over Taylor's book, Tomlin points out that it has not taken place over the whole of nature as we sometimes suppose. Many species have faded out altogether. Nine of the thirty or so phyla recorded have become extinct and several, though surviving, have undergone only slight change. The Coelacanth is a notorious example. Bees preserved from the Tertiary Age 2 million years ago (preserved in amber) have been shown to differ hardly at all from those of today. And as for bacteria, they seem indifferent to Evolution : 'The survival of the fittest would therefore seem to have occured in the absence, rather than the presence of Evolution'.

The concept of 'ruthless competition' as the basis of the 'battle for survival' is also doubtful. Lorenz has demonstrated that fishes of different species can feed in proximity on a coral reef, without displaying any rivalry. Conflict seems to be a concept popular in Darwin's time (and ours) : Prince Kropokin's *Mutual Aid* (subtitled 'A factor of Evolution', 1902) which powerfully stated the opposite case, has never been popular.

Tomlin next discusses Taylor's views on Lamarck. His views are not to be dismissed as absurd : they are based on plain facts. For instance, the ostrich has calluses or callosities on its rump, breast and pubis. It is born with them, and they occur in those places where the animal, in sitting, presses on the ground. The warthog has calluses on its knees.

The implication of Lamarck's view is that the genome is capable of being altered or modified by experience. The Darwinian view is that no such development can happen unless a mutation comes along *by accident* at the right moment. But while a skin-thickening gene is conceivable, is a callous-creating gene in selected spots? According to the neo-Darwinians,

> Every single change of structure and function throughout the history of life must have been due to a gene or a group of genes having by fortuitous mutation come into being at the 'right' moment, in order that species should have been formed in the way they have.

Of course, the virtually insuperable difficulty of accepting this does not prove Lamarck right.

Yet there have been experiments which seem to suggest possibilities that Lamarck's ideas may have been correct. Alan Durrant at the University College of Wales, Aberystwyth, and J. Hill at the Welsh Plant Breeding Station have done experiments which seem to show that changes of fertilizer may induce trends which, persisting for generations, implied genetic changes. A young scientist, Howard Temin, of Wisconsin disproved the assumption that information could travel from DNA to protein but not from protein to DNA. Temin proved that viruses could carry genetic material into host cells and embedded it in the host DNA, where in due course it would give rise to more viruses, using the host cell machinery for the purpose of making and assembling the raw materials. They manufactured a special enzyme for the purpose : Temin called it 'reverse transcriptase'. (If we combine this result with Gabriel Dover's concept of 'molecular drive' we may suspect that the systems under discussion are more flexible and open to change than established theory allows).

Another new concept discussed by Taylor is that which has caused much public discussion : that of 'Saltation'. One puzzle in the evolutionary picture is the absence of intermediate forms. Changes in evolution do not seem to have come gradually : they come by leaps. And some developments suggest anything but gradual adaptations : for example there is the mysterious evolution of birds, whose feathers evolved *before* flight. This is shown clearly in *Archaeopteryx* fossils, where stiff flying feathers are superimposed upon the downy heat-conserving ones.[2]

Taylor's objections to orthodox Darwinism are summarised by Tomlin thus :

1. The manifest existence of orthogenesis, i.e. definite and recognizable trends which continue for millions of years.
2. The repeated occurence of the same evolutionary process, or the same kind of evolutionary process in species widely divergent.
3. The appearance of structures involving intricate coordination long before the need for them arises.
4. The development of organs such as the eye which, entailing equally complex coordinations, could not conceivably have taken place by chance mutation of genes, unless the concept of 'chance' is so processed as to be made no different from pure fortuitousness. (Leith's view concurs with this).

Tomlin rejects mystical explanations, such as Sheldrake's : only by plodding rationalism can we ever outline a more positive view.

---

[2] Doubt has been cast on Archaeopteryx, the supposed link between reptiles and birds. Sir Fred Hoyle, Professor Wickransinghe and Dr Lee Spetner have tried to show that the beautifully preserved feathers were added to the fossil to make a 'missing link'. *British Journal of Photography*, No.10, March 8, 1985.

The main encumbrance to this is the concept of the fortuitous genetic mutation, or a series of mutations. The origin and development of organs of extreme complexity, and of *species*, cannot be accounted for in that way. The length of time taken to fabricate a single molecule by chance collocation of atoms, even given that these atoms are already 'lined up', would be of the order of 10^{243} milliards of years. (calculated by Lecomte de Noüy, *L'Homme devant la Science*). Pure chance cannot have caused the appearance of life on earth.

All examples of chance selection also depend upon a surreptitious retention or consolidation of such order as has already been established, which further presupposes a phantom consciousness presiding in the wings, says Tomlin. For in a world of *pure chance*, there could be as such retention. 'Pure fluctuations imply pure flux'. Since there is order in the world, there cannot be such a chaos.

The theory of natural selection implies series of retentions and consolidations, stage by stage, governed by orthogenetic themes. For that reason, although it may exploit chance, it cannot operate by chance. It implies a phantom selector, as we have seen, stationed 'above' the process and 'choosing' and conserving the 'good' mutations - and anticipating future choices. In other words (Tomlin points out) natural selection *implies the very consciousness it was devised to eliminate.*

Chance, randomness and hazard have no meaning in or by themselves : they acquire meaning only from a pre-existing order. The organisms in nature exhibit a structure and function which may be altered by chance, but which cannot be a product of chance. The idea of chance as a factor, or (as in Darwinism) a duel factor in organogenesis is quite untenable.

'The mechanism of Evolution is the process of natural selection' was a pronouncement of *The Times* (7 January 1982). This is to claim for chance what could operate only by some form of directiveness. If natural selection is *the* only effective agency in evolution, there would have been no evolution at all.

As a philosopher of science Tomlin declares that however brilliant or experimental scientists are, insofar as they continue to subscribe to Darwinism, they are 'amateur thinkers beguiled by a myth'.

11 'Chance': has genetics altered the picture? Dobzhansky's view of natural selection as 'engineer'

The complexities of genetics are now so daunting that probably many people feel they would never be able to understand them : therefore they simply defer to the experts, and except it as 'proven' by molecular biology that Monod is right to say :

> Pure chance, absolutely free but blind, at the very root of the stupendous edifice of evolution; this central concept of modern biology is no longer one among other possible or even conceivable hypotheses. It is the sole conceivable hypothesis, the only one that squares with observed and testable fact.

Moreover :

> Man knows at last that he is alone in the universe's unfeeling immensity, out of which he emerged only by chance.

So, while this may arouse an intuitive protest 'from the intensely telemonomic creatures that we are', we have to sit down under the 'facts of science' and the metaphysical conclusions which are the only ones to be drawn from these.

However, in an interesting essay on *Chance and Creativity in Evolution*, Theodosius Dobzhansky says : 'I do not think that the modern biological theory of evolution is based on 'chance' as much as Auden fears or Monod affirms'. In the end (p.309) Dobzhansky talks in a very different way : 'evolution resembles artistic creation; its greatest masterpiece is man'. (p.336). And he seems to give away the whole mechanistic argument at points :

> A gene is a macromolecule; it is also an organic system that carries within itself the billions of years of its evolutionary history. The mutational repertory of a gene is a function of its structure, and hence of the billions of years of its evolution. Chance is, therefore, brought under restraint, both at the level of the origin of mutations, and of their retention or loss in the population.
>
> p.315.

The question to be asked is what is this 'history' embodied *in*? Where does the 'repertory' reside, and from whence is the 'restraint' exerted? This, again, cannot be *merely* the laws of physics and chemistry, or even the 'DNA code' : is there then an intelligence, or 'primary consciousness', or 'morphological resonance' to exercise that 'function'?

But before we examine the stages of Dobzhanksy's argument, let us examine his neo-neo-Darwinian additions to evolutionary theory. In his view, natural selection is a 'cybernetic servo mechanism that channels the flow of information from the environment to the gene pool'. We have looked elsewhere at the cybernetic model, used as an analogy instead of 'intelligence'.

But the question is important, because it seems to suggest a way of transforming the 'sieve' model of natural selection, which, because of its negative nature, could not really explain anything about development. Has genetic theory really shown that 'natural selection' can be an innovative force?

Here we may first arm ourselves with Marjorie Grene's philosophical analysis of the concept of 'fitness' in genetic arguments about evolution, especially in Sir Ronald Fisher's

Genetical Theory of Natural Selection. This theory has a mathematical core, so it is extremely difficult for us to follow : and it seems to present us with conclusions from the exact sciences : Fisher says that 'the vigour of the demonstration requires that the terms employed be used strictly as defined'.

As we have seen, Darwinism and neo-Darwinian theories hold that evolution is the result of the joint action of random variation by mutation and 'environmental pressure' or natural selection. A corollary of this is that the variations brought about by mutation which are preserved by natural selection give an immediate advantage.

In neo-Darwinian theory, there are two important concepts : adaptation and improvement. An organism is regarded as adapted, if one can imagine its condition as an improvement over another possible condition that would be slightly less favourable.

What Fisher tries to do is to establish tables to establish what he calls the 'Malthusian parameter of population increase', and to measure *fitness* by 'the objective fact of representation in future generations'. He arrives at a fundamental theorem of natural selection, which is that 'the rate of increase of fitness of any organism at any time is equal to its genetic variance of fitness at that time'.

But this, as Marjorie Grene shows, is not really a fundamental theorem of natural selection : it is a statistical device for recording population changes. This is not corrected by calling it 'natural selection'. Darwinian selection has to do with a causal connection between environment and improvement : given the adaptive nature of organisms, variation, inheritance, and Malthusian population trends, natural selection follows. To simply show that changes in reproductive value occur does not demonstrate natural selection : as Marjorie Grene points out, it could perhaps be shown that red hair is increasing, and so the probability of there being large numbers of redheads in the future is increasing : but this tells us nothing about the usefulness of red hair.

When examined closely, it appears, some scientific arguments do not at all prove what they are assembled to prove. Thus, in an earlier paper, R.A. Fisher had argued that very slight trends in the frequencies of characters in the populations may be recorded if they persist over a long period of time. Were these the result of 'natural selection'? Fisher argued that such trends were not the result of mutations, since mutations are infrequent, and advantageous mutations still more infrequent. Therefore it must be natural selection which directs the trends his statistics describe. But this is so only if mutation and natural selection are the only two possible courses of evolutionary change – and this is what the theory is supposed to prove, not to pre-suppose. This kind of argument is frequent in the evolution debate, in which we find many circular arguments.

Marjorie next analyses some ambiguities around the words 'fitness', 'improvement' and 'progress', which lie behind evolutionary thinking. In Darwinian thinking, natural selection controls the chance of death : moths which show up get eaten, so those which are by (genetic) chance less visible survive. This is the *immediate advantage* natural selection gives. Of course, moths which are eaten cannot reproduce. But reproduction as the continuation of the species is a matter that is indirect and inferred : 'it is staying alive that is the immediate benefit'.

This basic notion in evolutionary theory is complicated by certain principles in genetic theory. Here comes into account not only the probability of leaving descendents, but the summation of such probabilities itemised in the gene pool of the relevant population. To say that the rate of increase in fitness is due to changes in the gene ratio is to assert a fundamental belief of modern genetic theory. The problem is to decide in genetic theory what causes what, and often it is assumed that a process determines a consequence, proving Darwinism, when all that has been said really only means that 'wherever there are characters there are some genes that cause them'.

What Marjorie Grene points to is the ambiguity in the term 'fitness', involving three kinds of selection. There is genetical selection which is future-directed but not really selection at all. There is Darwinian selection for now, giving immediate advantage as with the sooty moth that is not eaten. And there is Darwinian selection for later, in reference to future advantage (but this, of course, must not have any teleological connection with 'trends', which would be orthogenetic and finalistic – devoted towards ends).

Crept into this ambiguity are the notions to be found in genetical selection - the future-directed character of the evolution of genetic dominance, in which there is the 'chance of leaving a remote posterity', by storing up characters that will some day be useful in some distant future environment. In this, since there is no question of immediate advantage, we have something which is surely outside traditional Darwinian theory, and requires a quite different type of thinking.

As Marjorie Grene points out, Fisher sees that in trying to formulate principles of evolutionary theory based on mathematical analysis of fitness in population, he observes that such a theorem must differ from a law like the second law of thermo-dynamics, because, while physical systems run down, evolution tends 'to produce progressively higher organisation of the organic world'. This recognises a kind of *progress* which escapes all Darwinian considerations.

Marjorie Grene calls the ambiguities in the argument 'self-enclosing'. She takes the concept of 'improvement'.

First, there is the strict statistical meaning of the 'Malthusian parameter' : the increase in the probability that organisms possessing certain characters will leave offspring. This has meaning only in time, and what it says about organisms in time is either tautologous (what has survived has survived), or as the retrospective appraisal of an achievement (what has survived is not what has failed to survive, but what has succeeded).

But this says nothing about *why* this may have happened, and so says nothing about adaptation or relationship to the environment.

There is a second kind of concept of improvement which is very different – 'biological improvement' 'in relation to the environment' which is 'progress determined by Natural Selection'. Organisms are constantly becoming better adapted to their niches in nature as natural selection weeds out the imperfectly adapted. This refers to characters or functions which are genuinely advantageous in relation to the environment, and these improvements are effected by Darwinian selection.

Here, the concepts pre-supposed are quite different from those supporting the statistical concept. While statistical improvement entails the conception of the organism as an aggregate of gene effects, Darwinian improvement entails, in addition, the conception of the organism as a machine with parts adapted to the performance of their specific functions :

> The improvement that is effected here is that of increasing adaptation in the sense of specialisation, of fitting in better and better to a special niche in nature.
>
> Grene, p.263.

As will appear, I believe Stephen Dawkins and Dobzhansky confuse these two concepts of *improvement.* As against genetical improvement, which is meaningful only as an assessment of a tend over a lapse of time, this second kind of improvement is short run. It is not intelligible in terms of gene ratios at all.

This observation of Marjorie Grene's as a philosopher is of great importance, for arguments like Dawkins' and Dobzhansky's discussions of chance, rest upon this confusion. Dawkins sees whole organisms as being at the mercy of the genes : Dobzhansky tries to persuade us that the 'probability-that-organisms–possessing-certain-characters will leave offspring' aspect of genes means that 'natural selection' is a positive force, and not merely a sieve. But, as Professor Grene goes on to say, the second kind of improvement (which is the essence of evolutionary theory) depends for its assessment on :

> the recognition of phenotypes, as wholes, not as aggregates of genes and on the recognition of the relation of such wholes to their environment, to predators, to climate, and so on and so forth. It is neither future moths, nor gene pools, to whom it is advantageous not to be taken by a bird, but this black moth on this tree trunk today. And the evolutionary trend which establishes and maintains a phenomenon like melanism expresses the accumulation of millions of such individual escapes and individual disasters, not to genes or gene pools, but to moths, whether today or yesterday or... a million years ago.
>
> p.264.

Marjorie Grene turns her attention to a third kind of improvement – that which develops very slowly, and which does not appear to benefit the possessor at the beginning of the development. This refers to characters and to functions which will be 'better' for future phenotypes in future environments. This kind of development R.A. Fisher excludes : he attacks the idea of the 'benefit of the species' as teleological and irrational – and here, of course, Stephen Dawkins follows him.

But what about the emergence of new forms? Did all ultimately useful characters have in some way something useful in their minute beginnings? It has been argued that even feathers, primitive photoreceptors and electric organs could have had. But what about Fisher's own theory of the evolution of dominance : we have recessives, either useless or even harmful, hidden away in a population over a long period of time – up to the moment of a new environment situation which makes them advantageous, and so calls forth modifiers that turn them into dominants. But this story, as Professor Grene points out, requires the idea of 'remote posterity'. The modifiers must be lurking in the population ready to leap into action when the environment demands, and 'thus the future directedness of the whole procedure is simply transferred to them'. (Grene, p.265).

This is the idea, I believe, that Dobzhansky picks up. One scientist even writes of how the genes are 'gifted with automatic foresight' (C.D. Darlington : *Evolution of Genetic Systems*, 2nd Ed., Oliver and Boyd, 1958, p.239).

But what are the implications of this for the Darwinian evolutionists? Here Marjorie Grene brings the three concepts of improvement against one another. Improvement I, statistical selection, is the measure of Improvement 3, long-run adaptation. But long-run adaptation is *adaptation*, and if we accept Darwinian theory, there must be improvement 2 : Darwin has proved that this is how adaptation is produced.

Improvement 3 cannot be explained in Darwinian (2) terms because this depends upon *immediate advantage*, and improvement 3 is a matter of remote benefit. Yet Improvement 3 is included by the Darwinians, as when they measure improvement by the techniques for measuring Improvement I :

> Moreover, Improvement I is expressed by a deferential, which can be interpreted as a summation of gene changes *now* - short-run trends; but at the same time, sine it is the one statement (the fundamental theorem) it must express one relationship, and since in Darwinian terms improvement 3 would be nonsense, this one relationship must be the situation covered by improvement 2.
>
> p.265.

This confusion, as we shall see, enables Dobzhansky to declare that natural selection is not a sieve but a positively creative force. He takes it from the confusion that Improvement 3, which cannot be explained by Darwinian theory (2), is explained by Improvement I, as a product of gene increase, which he can take (as a consequence of the confusion) as a product of improvement 2. The confusion also enables Fisher as Marjorie Grene points out to believe that selection can do so much 'because it has so long'. By a certain kind of measurement, irregular variation in the rate of increase in fitness is more apparent in a single generation than over a longer period, and therefore very small 'selective values' are sufficiently strong to establish themselves *over a long enough time*. This statistical observation is used against those who object to natural selection theory because of the difficulty of accounting for the first beginnings of what will ultimately be useful traits.

Finally, Marjorie Grene discusses a fourth kind of improvement which is truly an advance to higher forms of life : the appearance of genuine novelty at a higher level of richness or complexity. As so often, Fisher implicitly recognises this. But it is a kind of improvement which neither statistical genetics nor selectionist biology can explain, since it is neither quantitive nor adaptive. Yet these innovations, these sudden leaps forward, - and the most remarkable of them, (as Professor Grene puts it), is that we can speak of them - we, as conscious intelligent knowers who have come from the improvement process :

> evolution as macro-evolution, as the emergence of life and of higher forms of life, outruns both the concepts of gene-substitution, and of improvement in relation to environment. It makes sense only as an achievement for which statistical methods can measure the necessary, but not the sufficient, conditions.
>
> p.266.

It will be useful to return from this final paragraph of the *Knower and the Known* to Dobzhansky's essay *Chance and Creativity in Evolution* in *Studies in the Philosophy of Biology*.

As we have seen, he declares that 'The mutation process is not synonymous with evolution; it is only the source of the raw materials for evolutionary change'. (p.315). To him, however, all genes may well be the 'products of divergent evolution from the primeval self-copying that arose from inorganic nature some three billion years ago'. 'In their chemical aspects, the mutations that were taking place in this entity and in its descendents were copy errors, accidents, chances'. The vast majority of these errors fell by the wayside. Those that were retained where those which 'proved helpful, or at least neutral, with respect to survival and reproduction'. (p.315).

Anyone turning from Marjorie Grene's analysis can see that here there is nothing to account for improvement 4, or, indeed, improvement 3. There is no reason why the 'primeval self-copying entity' should have come into being in the first place, or why it should be a self-copying entity. The terms, 'error', 'accident', and 'chance' imply divergencies from a self-copying 'normality' that is taken as read, but for which no explanation is offered. Nor is there any explanation for the remark that the gene is an 'organic system that carries within itself the billions of years of its evolutionary history' – though this implies a systems organisation which is not the same as 'chemical aspects', but governs them ('The mutational repertory of a gene is a function of its structure' indicates dynamics not accounted for in the account of how life originated and evolved here).

But now Dobzhansky delineates another source of development. First, *gene recombination*, 'a source of evolutionary raw materials second in importance only to mutations'. Evolution is not simply a mutation giving rise to a new clone differing from its ancestor in a single gene alteration. Gene recombinations occur owing to processes of transformation, transduction, parasexuality and fully-fledged sexuality. 'More or less complex networks of developmental reactions intervene between the genes in the sex cells and the characteristics of adult organisms'. We are here in the realm of Improvement I : but this does not prove evolutionary theory. Are these states merely throws of genetic dice? Asks Dobzhansky. 'The evolutionary dice are loaded; the loading comes from natural selection'.

It must be stressed, says Dobzhansky, that 'selection puts a restraint on chance and makes evolution directional'. It increases the adaptation of the population to tis environments. So, we are concerned with Improvement 2 – the way in which changes giving immediate advantage are selected out. But then Dobzhansky adds :

> It is responsible for the internal teleology (Ayala, 1968), so strikingly apparent in living organisms.

This seems a recognition of improvements 4 and 3.

> The turmoil of mutation and recombination is curbed and channelled in the direction of adaptedness. This does not quite make evolution orthogenetic, because natural selection depends on the environment and the environment does no change in a constant direction.
>
> p.317.

But, in fact, the first sentence, which refers to Improvement 2, the Darwinian process, does not explain Improvements 3 and 4. The great leaps forward, and the potentialities which seem to be 'gifted with automatic foresight', with forward reference, are not explained by the relationship between natural selection and the environment : evolution *is* orthogenetic.

It is the nature of the environment that it may bring out the latent possibilities in Improvement 3 : but the impulse to take advantage of possibilities must be there in the first place, as must in the first place have been the impulse for life to begin. These immense pregnancies in the world are what are taken for granted, and hidden, behind the mechanistic Darwinian 'explanations', of Improvements 3 and 4. And it is these the lurk behind the anti-teleological, anti-directive terms 'error' and 'accident'.

Dobzhansky does his best, in his allegiance to the denial of these positive drives to improvement, to show that 'natural selection' is a positive force.

Natural selection, he tells us, constitutes a bond between the gene pool of a species and the environment :

> It may be compared with a servo-mechanism in a cybernetic system formed by the species and its environment.
>
> p.317.

He says that 'somewhat metaphorically' 'it can be said that the information about the states of the environment is passed to and stored in the gene pool as a whole and in particular genes'. He uses Toynbee's phrase 'challenge and response' to apply to the relationship between evolution and environment.

As we have seen, improvement 3 seems to be derived from the presence in the gene pools of potentialities which may be fulfilled in certain environmental conditions. However metaphorical Donzhansky's account may be, it is still not possible to accept that these potentialities are explained on the basis of Darwinian theory (Improvement 2) for the reasons given by Marjorie Grene.

Dobzhansky is reverting to the confusion she analyses, by going on immediately to say, 'selection is, as a rule, directed towards maintenance or enhancement of the Darwinian fitness. Darwinian fitness is reproductive fitness' :

> It is quantifiable as the rate of transmission to the next generation of the components of a given genotype...

– which is further to confuse what is found by statistics, quantitatively, with the qualitative changes in organisms on which evolutionary theory fixes its attention – that is, in whole creatures, where it is not a matter of reproductive fitness, but of life or death : of survival in that sense.

Natural selection, Dobzhansky goes on to say, is an impersonal, and, by itself, purposeless process, 'which nevertheless conduces as a rule toward internal teleology of living beings'. It is, he says, perfectly correct to stress the mechanical, automatic nature of selection, but this may be misleading. One exaggeration is to see selection as a sieve, 'or screener of unalterably deleterious or useful genetic variations generated by mutation'. (p.319).

But, whatever Dobzhansky may say, this is the only way selection can work, according to the theory. Dobzhansky tries to get round this by invoking Improvement 3 again :

> The Darwinian fitness is not an intrinsic property of a genetic variant arising by mutation, but an *emergent* product of its interactions with the environment and with the rest of the genotypic system. The sieve model becomes therefore downright misleading.
>
> p.320.

The word 'emergent' which I have underlined allows into the argument improvements 3 and 4. But since the only Darwinian explanation of these (2) is that of mutations and a sieve-like natural selection, what is Dobzhansky's? *A fortiori*, a sieve is the only thing natural *selection* can be : the whole foundation is based on the Malthusian concept that, in the 'struggle for survival', fitness alone can live, by small immediate advantages. If this is not the way things work, if natural selection is *not* a sieve, then the whole hypothesis breaks down, and some other hypothesis must be put forward.

In Dobzhansky, however, as in Dawkins, the process transfers itself somehow to the gene pool, and it becomes, instead of the survival of the fittest', *normalising* :

> normalising selection sweeps the gene pool clear of unconditionally deleterious genetic variants...

Is it really legitimate to extrapolate Darwin's Malthus-population theory of selection to this sphere?

Dobzhansky opts for an 'engineer' model of natural selection. 'Selection does more than merely permit the rare beneficial mutants to reproduce'. He goes on to discuss a complex problem of pigmentation : and then evolutionary convergence – the way in which, according to theory, the bodies of whales and dolphins may have been adapted to function in water.

But, as Dobzhansky recognises, such complex changes could not have come about according to the theories of natural selection : 'It would, however, be naive to imagine that the fish-like body shape appeared one fine day as a lucky number'. There must have been millions of gene changes :

> The point is that these fish-like characteristics were built up gradually by selection from genetic variants and their recombination, some of which were arising by mutation - perhaps also in human ancestry, but were not utilised by selection as building blocks for the contruction of adaptive gene patterns.

These genetic variations :

> Were quickly compounded or arranged into adaptively coherent patterns, which went through millions of years and of generations of responses to the challenges of the environment. Viewed in the perspective of time, the process cannot meaningfully be attributed to the play of chance, any more than the construction of

the Parthenon... what is fundamental in all these cases is that the construction process was meaningful.

p.323.

Dobzhansky is discussing, clearly, Improvement 4 : implicit in his argument, too, is the richness of gene potential, recognised in Improvement 3. He recognises clearly the directiveness of life, and that there is a movement towards higher and more complex being.

He does not admit that there is nothing in Darwinian theory that can account for this. He does not admit that the processes of Improvement 2 cannot explain it, though he implies the theory is 'naive'. He does not conclude that natural selection is not an adequate concept to explain innovation. He ignores the question of coordination and supposes 'meaningful construction' can come by linear steps.

So, instead, he tries to use natural selection as a term to mean that innovative, shaping power which *does* produce Improvements 3 and 4. What he should do is declare the concept 'natural selection' inadequate, obsolete – no longer capable of fitting the facts. Instead, in true Darwinian spirit, he deifies it, and makes it capable of doing anything. But, of course, while the original term 'natural selection' was offered by Darwin as an explanation, in the way Dobzhansky uses it there is no explanation at all.

Dogma imposes an extraordinary compromise : there is meaning and purpose but generated by blind and dumb chaos :

> The meaning, the internal teleology, is imposed upon the evolutionary process by the blind and dumb engineer, natural selection. The 'meaning' in living creatures is as simple as it is basic – it is life instead of death.

p.323.

This is little more than rhetoric, but it does allow one important theme into the argument : Dobzshanky has been trying to show that life and evolution is not all 'chance' : there is also 'natural selection'. Natural selection is blind and dumb : but at least it *governs* chance – and has to do with *meaning* and *life*. This enables Dobzhansky to go on in his chapter to discuss how the origins of life on the earth and the origins of man are turning points in the history of the earth and the cosmos. Evolution, to him, resembles artistic creation, and its 'greatest work is man'.

But, for all this, there is nothing in Dobzhansky's essay to offer any explanation as an alternative either to Darwinian mechanism, or to vitalism. He is forced to glimpse the impulse to improvement in life : he recognises the principle of innovation, but cannot say where it resides.

So he drags his Darwinian dogma like a ball and chain ('Has natural selection implanted in man his self-awareness and death-awareness'?), and sinks from time to time into the reductionist's mechanism :

> Except on the human level, evolution is a blind, if you wish, mechanical process. It cannot plan for the future, conceive purposes, or strive for their realisation. How can a purposeless process transcend itself?...

p.335.

He evens refers to 'the unmistakeable resemblance of biological evolution to human creativity' only to sink back into mechanism : Natural selection, again, is a 'cybernetic servomechanism that channels the flow of information from the environment to the gene pool' – a statement which seems both misleading and based on a false analogy.

What is lacking in Dobzhansky's essay is the capacity to find mind and consciousness : to find that primary consciousness of which Tomlin speaks. It is true Dobzhansky says at the end : 'life arose from life lacking self-awareness'. But this is rather by way of rejecting the view that some shadowy rudiments of mind are built into all matter. The problem is one of overcoming our Cartesian modes of thought, and really searching into the nature of life which, as Tomlin argues, from the simplest level, displays something that must be called mind. Dobzhansky's word 'cybernetic' clings too much to the scientists' conception of such dimensions having to do with 'information' and 'codes'. The cell 'knows', and the amoeba 'knows'. There is, in modes of improvement 3 and 4, a factor which is *striving to achieve.* This factor is often called 'natural selection', but the use of the term nails it down to a concept of mechanistic preservation of fitness (Improvement 2), which cannot explain it. What we have to speak of is 'intelligence' or the 'psychic' or 'primary consciousness', developing in the elementary combinations that first became living organisms. The concept of 'automatic foreknowledge' in the gene pools, or Dobzhansky's 'interactive properties in the genetic system' do not at all offer a satisfactory explanation. There is still something else to be found behind the innovation, the ingenuity, the improvement and mystery of life; and for both scientist and poet this offers as marvellous a challenge as charm and quarks, or the black holes and pulsars, or the new theories of a 'third force' offer in contemporary physical science.

12 Does recent science endorse Darwinism? Michael Denton's doubts

Michael Denton is a molecular biologist, and his criticisms of Darwinism come from one trained in the reductionist paradigm, well aware of the latest mechanistic theories of molecular behaviour.[1] He came to feel that there is 'something fundamentally wrong with the currently accepted view of evolution'. He tries to show that 'the problems are too severe and too intractable to offer any hope of resolution in terms of the orthodox Darwinian framework' (p.16).

If biologists cannot substantiate the claims of Darwinism, 'upon which rest so much of the fabric of twentieth century thought', 'then clearly the intellectual and philosophical implications are immense' (p.16).

Before Darwin the established view of the basic order of nature was static and discontinuous. The model of nature was typological : it seemed self-evident that species bred to type generation after generation. From ideas influenced by Plato and Aristotle, all individual entities were seen as physical expressions of a finite number of ideal unchanging forms. This view had a theological backing, and all the forms were seen as evidence for the existence of a Creator and the grandeur of His design. The overwhelming consensus among biologists favoured castastrophism and a recent Earth, and the special creation of each and every species as a fundamentally immutable entity.

Darwin's liberating experience was *The Voyage of the Beagle.* Reflections on the rivers of Patagonia and the mountains of the Andes brought him to question Genesis and the 6,000 year time scale. His examination of coral reefs introduced into his mind a new geological time scale. The specimens he collected began to undermine the doctrine of the fixity of species. It was in the Galapagos Islands that he felt himself confronted with irrefutable evidence that the species was not an immutable entity. One of the main features of the rich natural life of the Galapagos Islands was the way many of the organisms (tortoises, iguanas, mocking thrushes and many of the plants) varied from island to island - some of the variations being so marked that they seemed like quite distinct species.

Could these variations have evolved from an original ancestor? The finches were the most suggestive example. The idea that they were all related to a common ancestor seemed irresistable. At the same time Lyell and Alfred Russell Wallace had come to parallel doubts. The main influences were the recognition of the immensely long geological past and the evident variation which was not due to any supernatural catastrophe or intervention : biology was on the way to coming under the fundamental aim of science to reduce wherever possible all phenomena to purely natural explanations.

The idea that living things originated as the result of chance and selection, Denton says, has a long history : it is clearly exposed in the Ionian nature philosophers like Empedocles, in

---

[1] *Evolution: A Theory in Crisis,* Michael Denton, Hutchinson, 1985.

Democritus and Epicurus, and had come down through Hume. Denton discusses the various attempts to explain life made by Empedocles, Hume and Lamarck. An important influence on Darwin was Malthus, and his account of the 'struggle for existence'. This enabled Darwin to exclude any concept of an 'inner drive'. The theory at first was elegant and simple : it involved premises which were self-evident organisms varied; the variations could be inherited ; all organisms were subject to a struggle for existence which was bound to favour the preservation by Natural Selection of beneficial variations.

> given variations, given that they could be inherited, and given Natural Selection, then evolution and adaptive change were inevitable. Darwin had an evolutionary theory that was entirely materialistic and mechanistic.
>
> p.42.

The mechanism which produced change by genetic variation, Darwin believed (and this is still believed) was *entirely blind to the adaptive needs and requirements of the organism.* As Monod put it, 'Pure chance, absolutely free but blind, at the very root of the stupendous edifice of evolution...'

Here Denton distinguishes between the *special theory* and *the general theory.* The first merely proposes that new races and species arise in nature by the agency of Natural Selection. The general theory is much more radical : it proposes that the special theory applies universally. The vast variety of life on earth can be explained by extrapolation of the processes such as generates the variations among the Galapagos finches.

An important element, of course, is the long geological time-span now introduced. Darwin examines the changes introduced by breeders and concludes

> I can see no limit to the amount of change, to the beauty and complexity of the co-adaptations between all organic beings, one with another and with their physical conditions of life, which may have been affected in the long course of time through nature's power of selection, that is by the survival of the fittest.
>
> *Origin of Species*, p114-5,
> 6th ed 1962, quoted by Denton p.48.

Another important element in Darwin's theory comes from comparative anatomy. Here there are two forms of resemblance. Homology is the resemblence between structures like skeletons, where elements have been modified for different ends. There is also analogous resemblence, where in fundamentally different structure has been modified or adapted to similar ends (like the flipper of a whale and the fin of a fish). Darwin saw in the former a justification of his theory, since these forms of homology suggested a common ancestor.

Comparative anatomy also provided Darwin with the basis for a hierarchical pattern of nature. Instead of categories representing expressions of a basic theme in God's mind, the varieties of species arose from common ancestors by descent with modification.

Paleontology also provided evidence that evolution had occurred. The fossil record revealed that the history of life on earth was overall one of progress from simple to more complex forms of life. Darwin also postulated that life could have originated spontaneously from 'Some warm little pond' (Denton, p.53).

As Denton sees, to postulate such a mechanistic origin, followed by chance development, meant 'the elimination of meaning and purpose from human existence' : and he shows how glimpses of this impact disturbed Darwin's wife while he himself suffered much conflict of mind over this question.

But this was all developing on the basis of an *entirely theoretical edifice. 'By its very nature evolution cannot be substantiated in the way that is usual in science by experiment and direct observation*' (Denton p,55).

> Neither Darwin nor any subsequent biologist has ever witnessed the evolution of one new species as it actually occurs. Outside of direct observations the only means of providing decisive evidence for evolution is in the demonstration of unambiguous sequential arrangements in nature.
>
> p.55.

Denton discusses the difficulties of such proof. Of course, there seem to be clear-cut sequential patterns and many have taken these as virtually irrefutable evidence for natural evolutionary transformations. But the one great problem is evolutionary theory is the *absence of intermediate forms.* Darwin himself admitted

> the distinctness of specific forms and their not being blended together by innumerable transitional links is a very obvious difficulty.

Besides this, there were also problems of what *hypothetical* paths evolution might have passed. This is particularly true of various *highly specialised organs.* Darwin argues that the imperfection of the fossil record was to blame : but this is a circular argument, because what is missing are the intermediates around which the whole problem centres!

> *Archaeopteryx* was problably the best intermediate that Darwin was tc name, yet between reptiles and *Archaeopteryx* there was still a very obvious gap[2]
>
> Denton, p.58.

Darwin's theory requires 'innumerable' transitional forms and the fossil record provides us evidence for believing this infinitude of connecting links ever existed : even Darwin saw this as the gravest and most obvious of all the many objections which might be urged against his views.

Yet Darwin persisted in his belief, which required *gradualism* : Natura non facit saltum. Even if he had been able to provide clear evidence for slow continuity he would still have had to prove that evolution happened by the purely random processes of natural selection. *Has there been time for this*? *The Origin* provides no quantitative evidence with mathematical backing, to show how any one major evolutionary transformation would have been possible according to his theory.

Moreover, it is one thing to show such a development is *possible* : another to show it is *probable.* Denton takes the case of the eye :

> Even if Darwin had been able to demonstrate the existence of a continuous sequence of increasingly complex organs of sight, leading in tiny evolutionary steps from the simplest imaginable photo-sensitive spot to the perfection of the vertebrate camera eye in a single phylogenetic line (in fact no such series exists in any known lineage) and even if he had been able to show by quantitatise estimates that the immense number of mutational steps could have occurred and been substituted by natural selection in the time available, this would only have meant that evolution by natural selection was *possible.* It would not have meant it was *probable.*
>
> p.61.

Darwin's theory postulates a 'gigantic random search' : can we believe this is the basis of life's achievement? Commonsense alone, Denton suggests, finds it incredible that such complex and ingenious devices as the camera eye, the feather, the organ of corti or the mammalian kidney could have come into being by such a 'blind undirected search mechanism'. Darwin himself, of course, was often prone to doubt. Yet nowhere in *The Origin* does he supply any quantitative evidence that chance and chance alone could lead to such improbable ends. Time itself is no answer - it tells us nothing about the probability of achieving any sort of goal unless the complexity of the search can be quantified.

There are other problems of which Darwin was aware. All his evidence was circumstantial, and ' nowhere was Darwin able to point to one *bona fide* case of natural selection having actually generated revolutionary change in nature, let alone having been responsible for the creation of a new species'. His theory of inheritance was erroneous and to explain it he was forced into an almost Lamarckian position, having to toy with the idea of some sort of directional bias in the occurence of variation and mutation.

---

[2] And now the Archaeopteryx fossil comes under suspicion of fake. See *The Journal of Photography* 8 March 1985.10.p.269.

Darwin was also aware that in breeding there were always limits to what could be achieved. If this were so, how could one species evolve into another? How could he extrapolate from micro to macro evolution?

So, evolutionary theory was *extremely speculative*, on many counts. Darwin was not dogmatic or fanatical - he was a man of great common sense, always aware of the hypothetical nature of his theory. His was a masterful work, persuasively written : it was a sensational best-seller from the start. And it produced a crisis in the philosophy of being :

> Instead of being the pinnacle and end of creation, humanity was to be viewed ultimately as a cosmic accident, a product of a random process no more significant than anyone of the myriads of other species on earth.
>
> p.66

Darwin set man adrift in a cosmos without purpose or end.

Denton goes on to say that it is assumed at large today that recent research confirms Darwinism : *nothing could be further from the truth.* The evidence was so patchy one hundred years ago that even Darwin himself had increasing doubts as to the validity of his views. Not even the most trivial type of evolution had ever actually been observed directly in nature : Darwin was quite unable to demonstrate the 'infinitude of connecting links' that was crucial to his theory.

Yet Darwin's theory was elevated from a highly speculative hypothesis into a unchallenged dogma within twenty years. The reasons for this were due to the spirit of the age - the Victorian belief in gradualism, and the *inevitability of progress.* At the same time there seemed to be an analogy between the competitive spirit of the free market economy and selection as the driving force behind evolution.

In science itself there was the drive as in the new physics, to find *first and natural causes.* God retreated as a first cause, the architect of a *clockwork universe* which continued to operate automatically without any divine intervention.

It seemed to most educated people that all past phenomena would prove explicable in terms of presently operating processes and that the universe had gradually developed from a few elementary particles into the present state though the operation of the basic laws of physics and chemistry. Darwin extended his ambition to living things.

There was no scientific alternative, and special creationism seemed not only unscientific but also discredited as the species barrier seemed to have been breached by perfectly natural processes. So, Darwinism has an immediate appeal - despite all the lack of evidence and the lacunae in the theory.

Modern scientists, because of this, misjudge the nature of Darwinism. Denton quotes Julian Huxley in 1959

> The first point to make about Darwin's theory is that it is no longer a *theory* but a *fact.*

Stephen Dawkins declares that 'the theory is about as much in doubt as the earth goes round the sun'. But, says Denton '*such claims are simply nonsense*'. Darwinism is very much in doubt when it comes to macro-evolutionary phenomena. It is impossible to verify by experiment or direct observation. Theorists of science like Sir Karl Popper have doubted whether evolutionary claims, because they cannot be falsified, can be classed as truly scientific hypotheses. Moreover, evolution deals with a series of *unique* events (the origins of life, the origins of mind) and these are unrepeatable and cannot be examined experimentally. Such events offer scope for fascinating speculation, but the ones here are beynd validation. And while the evolutionary hypothesis is *incapable of proof by normal scientific means* the evidence is *far from compelling.*

Despite these objections, any resistance to Darwinism is fiercely condemned in the scientific world : dissent becomes by definition irrational ('and hence especially irritating if the dissenters claim to be presenting a rational critique' p.76). 'Once a theory has become petrified into a metaphysical dogma it always holds enormous explanatory power for the community of belief'.

Because Darwinism has been elevated to the status of a *self-evident axiom* the very real problems have become invisible. It is believed that the theory was proved a hundred years ago, and that evidence since from paleontology, zoology, genetics and molecular biology have provided ever-increasing support for Darwinian ideas. But the truth is that the general theory, *that all life on earth has originated and evolved by a gradual successive accumulation of fortuitous mutations remains a highly speculative hypothesis entirely without factual support* and is *far from self-evident.*

Denton now turns to specific problems.

The origin of a species from another species has never been observed.

Half a century after Darwin Kettlewell demonstrated that natural selection operates, and since his work on peppered moths other cases of natural selection have been observed. But, however impressive, the sort of evolution observed by Kettlewell is *relatively trivial. 'It falls far short of the evolution of a new species, if we accept the usual definition of a species as a reproductively isolated population of organisms*' (p.81).

There are demonstrations of the reality of specification in nature, as, for intance in observations of gulls, and the insects and birds of Hawaii and elsewhere. From all this work it is clear that Darwin's *special theory* was largely correct : 'it is now possible to explain in great detail the exact sequence of events that lead to species formation'. Both natural selection and processes like genetic drift play important parts in the process.

The question remains whether it is acceptable to extrapolate from the species theory to the general theory : can macro-evolution be explained by microevolutionary processes?

As Mayr points out, there are here a number of doubters : Goldschmidt the geneticist; Schindewolf the paleontologist; the zoologists Jeannel, Cuenot and Cannon. As Denton point out, there are many instances outside biology where such an extrapolation is clearly invalid

> where large scale 'macro' changes can only be accounted for by involving radically different sorts of processes from those responsible for more limited 'micro' types of change.
>
> p.87.

With complex systems their function arises from the integrated activity of a number of coadapted components - like Polanyi, Denton takes the example of a watch. If a system of this kind was to change this 'necessitates a relatively massive reorganisation, involving the redesign or respecification of all or most of the interacting component subsystems' (p.91). When one thinks of the changes from lizard to snake, or flight, or the appearance of mind, obviously a succession of minor changes in the component structure could not have brought these innovations : such macro-evolution requires a sudden 'saltational' change : and simultaneous and coadaptive changes, so that components of the organism could still function together in a coherent and integrated way.

Darwin as we have seen said that Nature does not take leaps - 'this canon if we look to the present inhabitants of the world is not strictly speaking correct' (Denton p.93). In his next chapter Denton discusses the typological perception of nature. Typology contrasts completely with the idea of organic evolution : it believed that there were absolute discontinuities between each class of organism, so life was a *discontinuous phenomenon* and sequential arrangements should be totally absent from the entire realm of nature. To the typologist the eidos or type was real, and the variation an illusion : for the evolutionist the type is an abstraction and only the variation is real. Biological variation can never be radical or directional to the typologist.

As Denton points out, all the founders of modern biology were typologists : Linnaeus, Cuvier, Agassiz, Richard Owen (who invented the term 'dinosaur') Charles Lyell who suggested the terms Eocene, Miocene, Pliocene and Pleistocene. It is generally assumed that their typology was bound up with their religious views, but this is not so. Too often (says D.L. Hull in *Darwin and his Critics*) opponents of evolutionary theory are lumped together and their persistence explained away as religious bigotry. Typologists like Agassiz believed their position was empirically based, and that it was the evolutionists who were prejudiced by *a priori* concepts and were chasing a phantom.

If we examine fossil remains they point out, *there are no intermediate forms.*

> Why have not the bowels of the earth preserved the monuments of so remarkable a geneology, unless it be that the species of former ages were as constant as our own?
>
> (Cuvier)

To observers like Cuvier, the coordination of the limits of a mammalian creature was so complex and marvellous that it seemed evident that any major functional transformation would require *simultaneous coherent coaptive changes in all the component structures* : this is so vastly improbable that it seemed to preclude any kind of evolutionary transmutation. To suppose that such changes could come about by chance or accident was inconceivable (M.J.S. Rudwick, *The Meaning of Fossils*, 1972). Moreover, experience in breeding had shown *distinct limits* beyond which change was impossible.

Even in Darwin's time it was appreciated that evolutionary theory seemed to offer the best explanation of how organised beings developed in previous epochs : yet it seemed impossible because inconsistent with the observed fact - such as the crucial absence of intermediary forms. But while scientists did not turn to the Creationist argument, they clung to Darwinism because there was no *conceivable naturalistic alternative*. This is the problem Norman Macbeth examines, of course.

The distinctness of classes is as typology implied. Each class of organism possesses a number of unique characteristics by which it can be defined. These occur in a fundamentally unvariant form in all species of that class and *are not found even in rudimentary form in any species outside that class*. With mammals, for instance, they have :

> a hairy integument, each hair being a complex structure consisting of a keratinized cuticle, a cortex and a central medulla; mammary glands exhibiting alveoli surrounded by a network of myoepithelial cells responsive to the hormone oxytocin producing milk, a nutricious secretion containing fat globules and sugars ; specialized sweat glands in the skin ; a four-chambered heart with left venticle delivering aereated blood to the aorta; discrete and reniform kidneys, with nephron form and function specialized to generate a concentrated urine containing a high concentration of urea; a large cerebral cortex with distinctive six layers of cells; a diaphragm, a special muscle used by mammals for respiration ; three highly specialized ear ossicles - a mallus, incus and stapes conducting vibrations across the middle ear; the organ of corti, a specialized organ for reception and analysis of sound.
>
> p.105.

If we take merely the complex structure and function of hair, there are no structures which may in any sense be considered *transitional between hair and any other vertebrate dermal structure*. When we examine the brain, there is nothing like it in any non-mammalian vertebrate, and it is not led up to gradually through any sequence of less complex neurological structures in any known group of organisms.

Denton points out that if we were to list all the identifiable groups of organisms in such a way as to satisfy the axioms of typology, we would have to name every significant characteristic of every living thing on earth - from the wing of the bat to the vertical column of vertebrates and the spinnaret of the spider.

There are also the essentially invariant forms the species in which they occur which are not led up to gradually though a sequence of intermediate structures, like the cilium, the tiny microscopic hairs on the surface of cells. Their structure is universally based on the 9-2 pattern : every cilium so far examined prossesses the the same basic structure. There is no halfway structure in this complex molecular organisation.

Nor, of course, is there any variation in the universal pattern of the DNA structure and code : this is not led up to gradually through a sequence of transitional forms. Of all the millions of living species known to biology, only a handful may be considered as intermediate : the lungfish, and the monotremes, like the duck-billed platypus, for instance. But these cannot be seen as transitional between two types, they are a combination of the characteristics of fish and mammal or reptile and mammal. These forms may be anomalous in terms of typology, but they *do not provide evidence for believing that one type of organism has ever gradually converted into another.*

Most existing taxa are *remarkably well defined and strikingly isolated,* say Denton.

> Not only are *bona fide* intermediates virtually unknown, it is impossible to allude to any more than a handful of cases where the pattern of nature seems to exhibit something of a sequential arrangement.

One of the supposed sequential arrangements is the vertebrate series from the cyclostomes through fish, amphibia and reptiles to mammals. *Every school text teaches the story of vertebrate evolution as a series of successive transformations.* The evolutionists claim morphological evidence for this - how things branch out, as between the different kinds of fish, and so on. But when the morphology is studied in detail the evidence is far less convincing.

Denton points to problems over the aortic arches in the heart, where there are fundamental differences between mammals, reptiles and fishes (lungfish) : the lungfish heart is *not* intermediate between fish and amphibia.

The most convincing 'sequence' of all, from lemur to monkey and then via ape to man, is far less convincing when critically examined. It is not clear that any living primate population can be taken to represent truly transitional forms or new forms which led to the next most advanced stages.

To refute typology and securely to validate evolutionary claims would require hundreds of not thousands of different species all unambiguously indeterminate in terms of their overall biology and in the physiology and anatomy of all their organ systems. Darwin said of one; 'imagination must fill up the very wide blanks'. But the evidence does not exist in the world, however much the imaginative appeal of Darwinism projects it.

Louis Agassiz (*Methods of Study in Natural History*, 1863) asserted that sequence was essentially absent from nature : a significant number of modern biologists are asserting that 'no species can be considered ancestral to any other'. These include the cladists, to whom Denton now turns.

This problem has to be approached by examining our methods of classification. Denton discusses the sytems used by Aristotle and Linnaeus. A diagram of classification can imply that each class is perfectly distinct and totally inclusive or exclusive of other classes. But it is easy to draw a diagram which implies sequential relationships. In the nineteenth century scientists drew hierarchic schemes : when Haeckel drew his famous tree of life, he probably regarded the groups as cousins to one another rather than as ancestors and descendent. But since evolutionary theory scientists have tried to draw evolutionary trees. Great difficulties arise, however, over how groups should be placed in the trees.

Cladistics is a school of taxonomy and it tends to depict nature in non-sequential terms. It takes no account of evolutionary claims regarding the geneology or derivation of any particular species or group. It tries only to depict the distribution pattern of unique shared characteristics (homologies) among a group of organisms. The resulting procedures, like earlier forms of classification, result in striking heirarchic patterns.

Evolutionary biologists claim that the hierachic pattern of nature provides support for the idea of organic evolution. Yet direct evidence for evolution only resides in the existence of *unambiguous sequential arrangements*, and these are never present in ordered hierachic schemes.

In many diagrams, the 'links' represented are often acknowledged as *ideals*, as *purely theoretical.* Yet this, of course, is not the impression given by 'evolutionary trees' in school text books. Such trees suggest descent.

The puzzle is that when such classifiction diagrams are made, what does become clear is the *highly ordered pattern* of nature, and this puzzled Darwin. The question inevitably arises - 'how a random evolutionary process could have generated such a highly ordered pattern'. If an evolutionary tree is postulated, one feature of it must be that acquired traits are *conserved* : they must not be lost or transformed : so, evolution must be a conservative process 'such that each phylogenetic lineage gains a succession of what are essentially immutable character traits'. Can the existence of invariant character traits be compatible with the notion of evolution as a random radical process of change? Why should evolution depend upon gradual random process, but then become immune to such change - especially considering that many character traits are of dubious adaptive significance?

If a hierachic pattern is to result as the end product of an evolutionary process, no ancestral or transitional forms can be permitted to survive. Darwin saw extinction as playing an important part here ('many forms of life have been utterly lost... through which the early progenitors of birds were formally connected with the early progenitors of other... classes' - but where are their remains?). But surely (says Denton) no purely random process of extinction would have eliminated so effectively all ancestral and transitional forms, all evidence of the trunk and branches?

The hierarchic pattern is nothing like the straightforward witness for organic evolution it is commonly assumed to be. Cladism, meanwhile, seems closer to typology. Many in biology are disturbed (and *Nature* attacked the cladists at the Natural History Museum (292 :295.6)) Cladism has rejected the assumption that the nodes in cladograms represent ancestral species : it has rejected dependence on evolutionary theory. Cladism concentrates on 'sister group relationships' and is 'non-evolutionary classification'. Some (like Patterson) see much of today's explanation of nature in neo-Darwinian terms as empty rhetoric : the 'ancestors' 'exist not in nature but in the mind of the taxonomist'. Cladism is making explicit the lurking sense in classification schemes that '*nature's order is not sequential*'.

Denton quotes Darwin at length on homology, as his argument is presented with great clarity - homology being one of the most powerful lines of evidence for organic evolution. Morphology, declares Darwin, is the soul of natural history. Where we see conformity of type, is it not powerfully suggestive of true relationship, of inheritance from a common ancestor? He rejected the creationist argument, that it pleased the Creator to construct on a uniform plan as 'not a scientific explanation'.

However, one problem which arises at once is that homologous structures are often specified by non-homologous genetic systems, while it also seems difficult to establish an embryological basis for homology. If we take homologous organisms, the earlier events in the first stages of development are not identical : while gastrulation and the gastrula may be homologous in all vertebrates, the way the gastrula is formed and the positions of the cells in the blastula are markedly different.

Homologous structures are assisted at by different routes. For instance the vertebrate alimentary canal is formed from the roof of the embryonic gut cavity in the sharks, from the floor in the lamprey, from roof and floor in frogs, and from the lower layer of the embryonic disc, the blastoderm, in bird, and reptiles (Gavin de Beer, *Homology : An Unsolved Problem*, Oxford, 1971). There are many other such problems, suggesting that correspondence between homologous structures cannot be pressed back to similarity of position of the cells of the embryo or the parts of the egg out of which these structures are ultimately differentiated. Moreover, homologous structures can owe their origin to 'different organiser - induction processes'.

It seems that Darwin's use of the term 'homology', defined as 'relationship between the parts which results from their development from corresponding embryonic parts 'is *just what homology is not.*

When Denton turns to the way in which genes specify apparently homologous structures, it appears that these are specified by quite different genes in different species. 'Homologous structures need not be controlled by homologous genes and homology of phenotypes does not imply similarity of genotype' says de Beer. Denton adds :

> with the demise of any sort of straightforward explanation for homology one of the major pillars of evolutionary theory has become so weakened that its value as evidence for evolution is greatly diminished.
>
> p.151.

Nor can this be dismissed as a trivial problem : Denton quotes Sir Alister Hardy as saying that the concept of homology is 'absolutely fundamental to what we are talking about when we speak of evolution'.

Denton also refers to a number of other puzzles : adaptions of great complexity which exhibit very close resemblance in their design but which must have arisen entirely independently. He discusses the design of fore- and hind-limbs, and their supposed descent from the pectoral and pelvic fins of fish. They have arisen independently, but are evidently constructed on corresponding principles. The profound resemblance cannot be explained in terms of a theory of descent : how can the independent origins of structures so incredibly similar be explained in terms of the *random* accumulation of tiny advantageous mutations? It seems as if a complex and seemingly arbitrary pattern was assured at twice independently in the course of evolution. There are many such examples.

Possibly some such apparently homologous structures are really only analogous, like the fish-like shape of the whale. But if more and more forms are discovered to be analogous, and the phenomenon of homology was found to be less significant, there would be little need for a theory of descent with modification.

In this sphere, the failure of homology to substantiate evolutionary claims has not been widely publicised. But it is yet another sphere, declares Denton, in which advances in knowledge do not support the Darwinian paradigm. Nor, even when homology is likely, does this tell us anything about how it came about : homology is really only circumstantial evidence.

Denton then turns to the fossil record. Here the central problem is that there are no intermediates or transitional forms to bridge the enormous gaps. Without these the concept of evolution must still seem an outrageous hypothesis.

In the decades following the publication of *The Origins of Species* it was widely believed that eventually the missing links would be found, thus confirming the theory of evolution. Since then there has continued to be public excitement, of the Conan Doyle *Lost World* kind. Yet despite intense activity and excitement no missing links have been found which are intermediate or ancestral. The Poganophora, a new phylum found in Indonesia, a marine worm (sometimes six or seven feet in length) turned out to be a most unusual and highly specialised organism with no mouth or digestive tract : but it has to be placed on 'an extremely distal twig of the hypothetical evolutionary tree' (p.160).

What about fossils? In Darwin's time huge tracts of land were still unexplored. Today 99.9% of all paleontological work has been done since 1860. But virtually all the new fossil species discovered since Darwin's time have been either related to known forms or like the marine worm Poganophora, unique types of unknown affinity.

There have been discoveries of genera that have defied all efforts to link them with known phyla, like the Burgess shale fauna and the Ediacara fauna in South Australia, among which was the Tribrachidium, a three-legged organism. But what the rocks have never yielded is any of Darwin's 'myriads of transitional forms'.

The first representatives of all the major classes of known organisms are already highly characteristic of their class when they make their initial appearance in the fossil record. Neither the phyla nor their main sub-divisions are linked by transitional forms.

A curious feature is that the earliest representatives of most of the major invertebrate phyla appear over a relatively short space of time, about 600m years ago in the Cambrian period. The strata laid down in the hundreds of millions of years before that era are almost completely empty of animal fossils.

The same is true of plants. Like the sudden appearance of the first animal groups in the Cambrian rocks, the angiosperms suddenly appear in the Cretaceous era - a phenomenon that startled Darwin. The angiosperms transformed the world's vegetation.

A similar pattern is true of vertebrate fossils. The first members appear abruptly, unlinked to other groups by transitional or intermediate forms. As they appear, they are already well differentiated and characteristic. The same is true of the amphibia.

At the Darwin Centenary Symposium in 1919, G.G. Simpson pointed out that it is one of the most striking features of the fossil record that most new kinds of organism appear abruptly :

> They are not, as a rule, led up to by a sequence of almost unperceptibly changing forerunners such as Darwin believed should be usual in evolution.
>
> p.165.

The fossils provide none of the crucial transitional forms required by evolution. The phenomenon of sudden appearance becomes more universal and more intense as the hierarchy of categories is ascended. The incorporation of fossil taxa into the Systema Naturae leaves the whole orderly hierarchic scheme intact.

As paleontology progresses, says Denton, the numbers of hypothetical phylogenetic relationships increases, while profound and undoubted discontinuities appear in the fossil record. Again, to close these gaps there would have to be great numbers of transitional forms : as Darwin himself said, their number must have been 'inconceivably great'.

Denton attempts to show an evolutionary tree of the development of the whale. It requires, to complete the 'flow', inventing ten hypothetical species on the direct path and an additional fifty-three hypothetical species on collateral branches. Is this necessary? Yes, because if we decide that if a hypothetical sequence of species did not contain such collateral branches, this would be tantamount to postulating an external unknown directive influence in evolution which would be quite foreign to the spirit of Darwinian theory and defeat the whole aim of attempting to provide a natural explanation for evolution. So, if one sticks to strict Darwinism, and thinks of all the elements which require evolution (tail flukes, streamlining, bringing nostrils to top of skull, specialised nipples for feeding under water) one must think of possibly hundreds, even thousands, of transitional species. *But there are no remains of these.*

Denton goes on to discuss the exceptional case of *Archaeopteryx*, though this is now perhaps clouded by the allegation that feathers were added to the fossil. An article in *Nature*, however, argues that it was as capable of powered flight as a modern bird (Olson L.S. and Feduccia, A. 'Flight Capability and the Pectoral Girdle of *Archaeopteryx*', *Nature*, 278 :247-8, 1979). Even if Archaeopteryx is not a fake, however, it is not led up by a series of transitional forms from an ordinary terrestrial reptile through a number of gliding types with increasingly developed feathers until the avian condition is reached. The development of *flight*, as Marjorie Grene declared, is a major step forward whose evolution raises enormous problems. (Denton discusses this later).

Considerable problems are also caused, of course, by the fact that there is little or no clue in fossils to the nature of the soft parts of an organism. The whole question of the transition from birds to reptiles is bound up with this problem, so the extent to which Archaeopteryx was avian in its major organ systems must remain conjectural.

Besides the problem of lost soft organs, there is that of convergence : similarities in forms, like that between the forelimbs of a mole and a molecricket, or the similarity of the design of the eye in vertebrates and cephalopods, *do not imply any close biological relationship.*

Denton gives us an example of both problems, of judging overall biology on skeletal grounds, and that of convergence, by referring to the rhipidistian fishes. These were considered to be amphibian ancestors because of the pattern of their skull bones, the structure of their teeth and vertebral columns and patterns of bones in their fins, in all of which they resembled the earliest known amphibians. It was assumed that their soft biology would also be transitional between that of typical fish and amphibia.

But the discovery of a living coelacanth in 1938 changed all this. Much of its soft anatomy, particularly that of the heart, intestine, and brain was not what was expected of a tetrapod ancestor :

> the modern Coelacanth shows no evidence of having internal organs for use in a terrestial environment.

Because of such difficulties, the status of even the most convincing intermediates is bound to be insecure. In S. Stanley; *Macroevolution* (1979) he declares :

> The known fossil record fails to document a single example of phyletic (gradual) evolution accomplishing a major morphological transition and hence offers no evidence that the gradualistic model can be valid.
>
> (p.39 : Denton p.182)

What about the well-known evolutionary series of the evolution of the horse? The leading evolutionist Simpson says that the description of the 'gradual reduction of the side toes' is fictitious. And though the horse series is the most convincing, it requires little change. Of another well-known sequence in biology text-books, that of the elephant, when the fossils are subjected to detailed analysis, 'it requires elasticity of the imagination to see anything more than a superficial resemblance between the available parts of the skeletons' (Sylvia Sykes, *The Natural History of the African Elephant.*) Denton says :

> Considering that the total number of known fossil species is newly one hundred thousand, the fact that the only relatively convincing morphological sequences are a handful of cases like the horse, which do not involve a great deal of change, and which in many cases like the elephant may not even represent phylogenetic sequences at all, series to emphasize the remarkable lack of any direct evidence for major evolutionary transformations in the fossil record.
>
> p.185.

The fossil record, then, does not confirm Darwinism : all the myriads of life forms that should be there have mysteriously vanished. Is it that the search has been insufficient? Denton shows that this is not so. Is it that the record is imperfect? But now fossils of unicellular and bacterial species have been found in rocks dating back thousands of millions of years before the Cambrian era : new fossils have been found in Cambrian and late pre-Cambrian strata, yet none can be considered as ancestral to any specific living phylum. Why is there still a *mystifying absence of primitive transitional forms*? Even if for some reason organisms did not form hard relics, their after-death imprints would be found.

Of the living families of vertebrates from 79% to 97% have been found as fossils : imperfection of the fossil record does not seem to be a strong case.

Two American paleontologists, Niles Eldredge and Stephen Jay Gould, see the imperfection of the record as a record of the mechanism of evolution itself ('Punctuated Equilibria : An Alternative to Phyletic Gradualism' in *Models of Paleontology*, ed. T.J.M. Schopf, 1973). They view evolution as an episodic process occurring in fits and starts, interspaced with long periods of stasis. New species evolve rapidly in localized geographical areas, and because of this the chance of the fossilization of transitional forms is small.

This may explain the gaps between species. But, Denton believes, it does not explain the larger systematic gaps - as between (say) a primitive terrestial mammal and a whale. It is incredible, that in such transformations there should have been thousands or even millions of intermediate forms which have disappeared. And while Gould and Eldridge wanted to help fill the gaps, what they have done is actually to have drawn attention to the fact that the intermediate forms are 'missing believed non-existent' - the trade secret of paleontology, and the way this questions the grounds for believing that the phenomenon of life conforms to a continuous pattern.

Denton now turns to the possibility of imagining or hypothesising intermediate forms - and this raises fundamental problems in Darwinism.

> Any change... which on the surface may at first appear quite trivial, on closer examination would inevitably necessitate extensive reorganisation of the entire anatomy and physiology of the organism.
>
> p.202

I look at this problem below, over Michael Pitman's discussion of the eye.

Denton now turns to one of biology's most beautiful and amazing developments - the flight feather.

> Each feather consists of a central shaft carrying a series of barbs which are positioned at right angles to form the vane. The barbs which make up the vane are held together by rows of barbules. From the anterior barbules, hooks project downward and these interlock with ridges on the posterior barbules. Altogether, in the flight feather of a large bird, about a million barbules cooperate to bind the barbs into an impervious vane...

> In addition to its lightness and strength the feather has also permitted the exploitation of a number of sophisticated aerodynamic principles in the design of the bird's wing.
>
> p.202

The arrangement of the feathers enable the bird to deal with turbulence, and to vary the shape and aerodynamic properties of the wing at take-off, landing and for various sorts of flight - flapping, gliding, soaring.

Birds are supposed to have envolved from reptiles. But how did they evolve? Flying from the ground up? Or from the trees down? By gliding? Did the friction of the air gradually 'fray' the scales, until these horny appurtenances changed into feathers? But this theory would surely bear no strict mathematical aerodynamic examination, involving calculations to do with wing/weight ratios. Moreover, as Denton points out, there are serious doubts about the transition from gliding to powered flight. For this the physical adaptations are *in opposition to those of gliding flight.* Here the arguments are very complex, but it seems that experiments have shown that flight is a far more complex *achievement* than any evolutionary approaches have envisaged - not least the problem of the *feather* itself.

It is not easy to see how a reptile scale could be converted gradually into an impervious feather without passing through an intermediate state in which it would be weak, easily deformed and permeable to air. The feather for flight needs an exquisite coadaption of components, and the arrangement looked at above, to make it an aerodynamic tool. To this we must add what Denton omits - the development of light hollow bones, certain kinds of muscle, and, of course, the whole capacity for flying and seeing in the flying way.

Denton does indicate the very special characteristics of the *avian lung*, in which there is a undirectional flow of air. How could such an utterly different respiratory system have evolved from the vertebrate design - not least since the slightest malfunction would mean death in minutes?

> Just as the feather cannot function as an organ of flight until the hooks and barbules are coadapted to fit together perfectly, so the avian lung cannot function as an organ of respiration until the parabronchi system which permeates it and the air sac system which quarantees the parabronchi their air supply are both highly developed, and able to function together in a perfectly integrated manner.
>
> p.212.

Denton quotes Darwin again : 'if it could be demonstrated that any complex organ existed which could not be formed by numerous, successive, slight modifications, any theory would absolutely break down'. But surely it must, over flight - since those aspects we have already discussed are not the end of it? The design of the heart and gastrointestinal system are all of a special kind related to the problem of flight. *Can reptile really have been converted into bird?*

Parallel problems arise over the structure of the bat and bat flight, in relation to the gliding lemur dermoptera (*Galeopithecus*) : and also with the morphological development of sea-creatures. There are great differences between the amphibian egg and the amniotic eggs : there are hardly two eggs in the whole animal kingdom which differ so fundamentally.

The amniotic eggs made genuinely terrestrial life possible : this required at least eight innovations. The amphibian solved the problems in quite different way. Attention to the details, as of heart and aortic arches of the amphibian, when trying to develop an evolutionary theory of how they evolved to the reptilian and mammalian condition are daunting.

So, too, it seems impossible to provide evolutionary explanations for other complex phenomena : the mating flight of the dragon fly. Its copulating organ is one of the great mysteries of nature, unique among organs in the animal kingdom.

Metamorphosis, of which there are many examples, is another process difficult to explain, as are many forms of migration and other behaviour patterns. Forms of symbiosis seem like interspecific altruism (as with the larva of the wood wasp *Sirex* and its complex relationship with its predator, the parasitic wasp *Ibalia*, Denton p.223).

One of Denton's, most spectacular examples of a feature in Nature which it seems impossible to consider as a feature developed by chance mutation is the Rotary Motor of the Bacterial flagellum. It is the only structure in the entire animal kingdom which exhibits a true rotary motion.

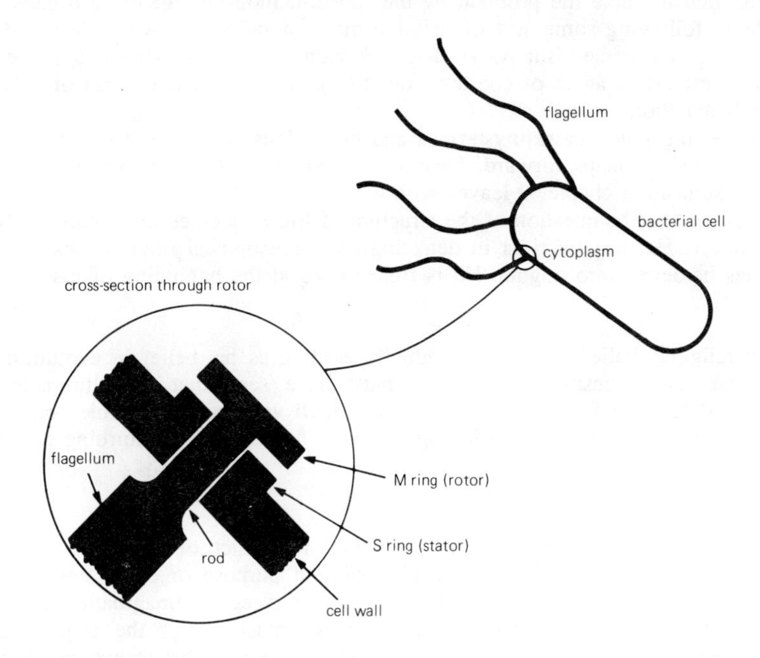

The Rotary Motor of the Bacterial Flagellum. This is the only rotary device known in nature. According to Howard Berg the torque is generated by a translocation of ions through the M ring (a disc mounted rigidly on the rod of the flagellum and free to rotate in the cytoplasm), where they interact with charges on the surface of the S ring (which is mounted on the cell wall), imparting a rotary motion to the flagellum.

from Berg, H, 'How Bacteria Swim',
Scientific American, 233(2) : 36-44, fig p.44, 1975.

It is actually driven by a rotary motor impelled by the active translocation of ions.

Denton offers other examples from the world of flowers and insects. His examples confirm my intuition that biology will save us : that the facts of the mystery of life's ingenuity must force us into new ways of thinking. But Darwin side-stepped this problem by suggesting that our trouble is that our imaginations are not as good as 'Nature', :

> what Darwin is saying... is that though we cannot imagine exactly how the gap were bridged in any particular case, this is became our imagination is relatively crude alongside the ingenuity of nature. Thus the problem of providing detailed reconstructions of credible sequences of transitional forms is avoided and we are asked instead to wonder at the beautiful creativity of nature.
>
> p.228.

But, surely, the 'chance mutation' theory implicity denies the existence of the ingenuity we indicate when we say 'life strives' or 'life innovates'? Denton sees the strategy as one by which evolutionists hide their inability to confront the problem of the gaps : and it is tautologous. If gradual evolution is true, then the gaps must have been closed gradually, even if we can't imagine how it happened!

Now, however, gradualism is being questioned, as by Stephen Jay Gould in *The Panda's Thumb* : gradualism cannot account for the major innovations.

In physics and chemistry many phenomena are discontinuous. So there is no coherent reason for evolution to be continous. Goldschmidt in *The Material Bais of Evolution* 1940, as we have seen, tried to solve the problem by the 'hopeful monster' theory : organs and types emerge suddenly following some sort of massive macromutation. Darwin considered anything of the kind to be a miracle. But Mayr declared such genetic monstrosities to be *hopeless freaks*. The problem arises again of coordination : to give a thrush the wings of a falcon does not make it a better flier.

There is something much more mysterious and miraculous : the sense of an intelligence in life which generates the leaps forward. Denton does not say as much, but it is the sense of this which his astonishing chapter 9 leaves with us.

Denton turns next to the question of the structure of life molecules in relation to the general theory of evolution. He is quite right in detecting the philosophical implications of Darwinian theory as it has hardened into dogma. He is right to see at the beginning of his book that (of Darwin)

> His own religious beliefs had been gradually eroding as his belief in evolution had grown, and, as a sensitive person, he must have seen that the elimination of meaning and purpose from human existence, which was the inescapable conclusion of his position, was for many including his wife, a profoundly disturbing reality to accept.
>
> p.54.

'humanity has to be viewed as a cosmic accident, a product of a random process.... So 'Darwinian theory... set man adrift in a cosmos without purpose or end'. Yet this happened because of a *speculation* : Darwin's great speculative piece of imagination hardened into dogma. Today we take it as proven a fact that we must accept the implications for a philosophly of being - yet, at least as Denton argues it is not true that recent research confirms Darwin : 'nothing could be further from the truth':

> the facts of comparative anatomy and the pattern of nature they reveal provide nothing like the overwhelming testimony that is often claimed. Simpson's claim that 'the facts simply do not make sense unless evolution is true' or Dobzhansky's that 'nothing in biology makes sense except in the light of evolution' is simply not true if by the term evolution we mean a gradual process of biology change directed by natural selection.
>
> p.154.

Moreover, molecular biology does not provide, as many mistakenly suppose, further endorsement of Darwinian mechanism and its consequent metaphysical bleakness.

Michael Denton points out that there has been a rapid and fundamental growth in an understanding of the molecular basis of life since 1950, at which time hardly anything was known. A series of dramatic discoveries transformed the biological sciences. In 1953 Crick and Watson published their famous paper, and the phrases 'double helix' and 'DNA' are now household terms.

This momentous discovery says Denton, solved the Centuries-old puzzle of heredity, revealing its chemical basis. In 1955 Sanger reported the first complete chemical structure of a protein, insulin, after ten years work. In 1957 X-ray crystallogaphic studies of sperm whale myoglobin provided the first 3-D picture of a protein, and in 1959 Perutz announced the 3-D structure of horse oxyhaemoglobin. In the next few years rapid advances in many different areas of biochemistry began to reveal the *structure* and *function* of all the main molecular components of the cell. Protein molecules are the ultimate stuff if life :

> If we think of the cell as being analogous to a factory, then the proteins can be

> thought of as analogous to the machines on the factory floor which carry out individually or in groups all the essential activities on which the life of the cell depends.
>
> p.234.

So far, so good : or, rather, not so far, so good. There is something seriously wrong with this analogy.

In the way it is offered there is an implicity assumption - that this 'factory' (the cell) can be explained (as the 'ultimate stuff of life') by such an analogy.

But in a 'factory' the arrangements of the machines are designed for a purpose : the function of each machine is designed for a purpose which is not to be understood in terms of the composition and structure of that machine : the factory was set up by planning minds. DNA is only analogous to the metal of the machines, their spacing on the floor, their arrangement of pulleys. What the factory serves, and how it is organised and controlled is still not explained.

Michael Denton describes the cells as 'tiny molecular machines' that consist 'fundamentally of a long chain-like module, or polymer, made up of a linear sequence of simple organic compounds called animo-acids'. We have looked at the structure of life-molecules above.

Michael Denton speaks of the 'backbone' which is formed by the linkage of the amino acid carboxylic acid groups and is identical throughout the molecule. It is the unique side groups which jut out from the backbone which confer different chemical properties to different regions of the amino acid chain. The linear sequence of amino acids in a protein can be

> thought of as a sentence made up of a long combination of the twenty amino acid letters
>
> p.235.

Of course, Denton's is a simplified account for the layman. But for the moment, let us attend to his word 'sentence'. It is clear from every account we read of the structure of life molecules that there is a perception that the structure is related to the question of *information.* The basic life-molecules, as we shall see, have a way of *replication*, reproducing themselves, by a marvellous process of 'reading off' forms and what seem to be 'codes'. To the modern scientist, what seems particularly fascinating about these processes is that they resemble certain techniques such as are used in present-day 'information technology', as in the use of tapes to store information which is read by reading heads, and so forth.

However, we need to step with great care in this argument, and we need to bring up all our own understanding of what 'communication' is, when a scientist uses a word like 'sentence'. He may speak as if a sentence were simply produced by a mechanistic process, exists as a diagram of 'information content' which is then 'decoded' by a 'reading head'. This kind of process is indeed used when I push my bank card into a cash machine. But even in this, there is an intelligence which has composed the system and processed the card - and the reading machine. And when we speak of a *sentence* we may take ourselves into a very complex process of one consciousness speaking to another through a ineffable system of cultural symbolism - in which it is *not true* that all we are concerned with is 'information content', but a process in which there are many unseen and inexplicit dynamics whose 'tacit' existence is the very basis of the communication. For instance, when I speak a sentence to a student or one of my family, we are each in readiness to talk and hear : we are alert to exchange like that. We are educated (it is no good speaking a sentence to a baby) : we have a whole grasp of reality (the 'living principle') to which the sentence relates. And we are in time with intentionality as conscious 'beings-in-the-world'.

It is important not to confuse one kind of 'communication' with another : when we talk of 'codes' or 'sentences' in molecular biology, we must continue to ask where is the *intelligence*? Where is the *knowing*? (An analogous problem is that of the way in which a life-molecule 'knows' its own stuff, and 'knows' what is not self).

To follow Denton's account further, he now turns to the all-important question of the *folding* of the chain of amino acids. This happens *automatically*, by which he presumably means according to physical forces : but is it not possible that some kind of *memory* is involved here? By such folding the molecule assumes its 'tertiary structure'. The shape, Denton tells us, is known as the *minimum energy conformation*, and is dictated by the amino acid sequence, and occurs in such a way as to bring about the maximum number of favourable atomic interactions between the various constituent amino acids'. It also has to do with electrical charges. What it is important to note here is that the form is presented as being the inevitable *manifestation of the laws of physics and chemistry* : if this is so, the form cannot be the embodiment of any kind of information, as we have seen looking at Polanyi's arguments. Yet it is just that, so other principles must be involved in its 'design' because without these all we would have would be noise.

Most proteins, Denton tells us, consist of some several thousand atoms folded into an immensely complex spatial arrrangement. Proteins which perform different functions have completely different overall 3-D structures and functional properties. The proteins which form the keratin in hair and nails are long and thin and intertwined round each other like the fibres in a length of wool. Enzymes tend to be rounded in shape and have a special region known as the *active site*. This is generally a cleft-like structure which extends from the surface into the interior and fits the compound upon which the enzyme acts (the substrate). Proteins carry out structural and catalytic functions and transport and logistic functions. Although Denton does not discuss it, the question remains : by what overall controlling dynamic are these functions regulated and '*where*' does this control exist?

We now come to replication. The proteins, despite their amazing versatility, cannot assemble themselves without the help of the nucleic acids. Denton has called them the 'machines' : the nucleic acid molecules he says 'can be thought of as playing the role of the *library* or *memory bank* containing all the *information* necessary for the *construction* of all the various *machines* (proteins) on the factory floor... we can think of the nucleic acids as a *series of blueprints*, each one containing the *specification* for the construction of a particular protein in the cell' (my italics). It is the words in italics that conceal, I believe, a confusion.

> we can think of RNA molecules as photocopies of the master blueprint (DNA) which are carried to the factory floor where the technicians and engineers convert the abstract information of the blueprint (RNA) into the concrete form of the machine (protein)
>
> p.239.

Denton turns now to the fundamental dogma of molecular biology. Information in living systems travels thus :

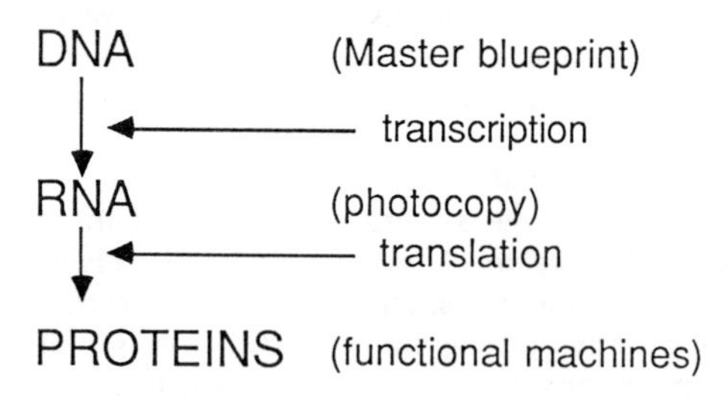

'Just as in the factory', says Denton, 'the information in the blueprint flows via the photocopy and into a manufactured article on the factory floor' (p.239).

Again, we have to be alert to the analogy. The RNA is not itself a product : it simply acts to generate the replication as a template. But there are unanswered questions raised by the analogy : *what controls the process overall*? And while the process of the manufacture of proteins is clear, in what lies the '*information*' guiding the development of some cells into nerve cells, some into skin and bone, some into brain cells, some into eyes, and so on. This kind of 'guidance' cannot be a product of the pattern and reading processes of DNA - RNA replication : so, where is it? There is always a search in science for the one single fundamental principle : if everything is reduced to its miminals, we shall find the answer, it is assumed. So, once the basic process of replication has been analysed, it is assumed we have found he secret of life. But this is not so - the operation of the nervous system, for example, is not explained by DNA theory. But nor, too, are the many functions of the cell, especially its international dynamics, between memory and present knowledge, in becoming a fully developed whole being.

As it is explained, DNA - RNA theory is a product of chemical and physical laws : the diagrams reinforce this impression.

However there are some key terms in Denton; passages which we need to ponder : 'the RNA *moves... to the actual site of translation* : *attaches itself* : *can recognise*' : and, of course, we should note the whole way of talking : '*carried out*', '*where the decoding takes place*'. It seems to me that in the way of talking here we have a tacit acknowledgement that we are no longer talking about the way (say) a molecule of salt dissolves in water, or sulphur combines with iron to form H_2S. We have the acknowledgement of *behaviour.*

> During the process of translation the mRNA passes through the ribosome just as the magnetic tape passes the recording head on a tape recorder. As each triplet reaches the reading head, it associates loosely with its appropriate tRNA which is also carrying the appropriate amino acid. Special proteins in the ribosome remove the amino acid from the tRNA and hence the amino acide chain is gradually assembled amino acid by amino acid as successive tRNA bring their attached amino acids to the reading heads... when the amino acid chain is completed it is detached from the tRNA and folds automatically into its correct 3-D functional conformation.
>
> p.245.

All this Denton sees as the product of an intimate relationship between one class of molecules - the proteins - and another quite different class of molecules - the nucleic acids.

> The nucleic acid contains the information for the construction of proteins, but it is the proteins which extract and utilize that information at all stages as it flows through this intricate series of transformation.

Another important process which goes on in the cell is that of DNA replication. Because DNA can encode for all the proteins necessary for transcription, translation, *and its own replication*, the cell system can replicate itself.

The whole system is presented as if it were self-operating, automatic, and a mere machine. Denton himself gives a description of a cell as a vast factory, on his analogy of machines running in a great work house of lots of apparatus, operating according to the laws of physics and chemistry.

The replication, the transcription and decoding and the rest, however, take place within either the complex processes of a bacterial cell, or the even more complex processes of a higher organism. They belong to subtly governed processes. If we could see it all enlarged,

> we would wonder at the level of *control* implicit in the movement of so many objects down so many seemingly endless conduits, all in *perfect unison...* we would wonder even more as we watched the strangely purposeful activities of these weird molecular machines... the life of the cell depends on the *integrated* activities of thousands, certainly tens, and probably hundreds of thousands of different protein molecules...
>
> p.329 (my italics)

The cell contains equivalents of our

> artifical languages and their decoding systems, memory banks for information storage and retrieval, elegant control systems... error fail-safe and proof-reading devices... assembly processes... an immense automated factory...
>
> p.329 (my italics)

If we turn back to our detailed account of DNA-RNA replication, there is nothing in it to account for any of these processes - except the engagement of molecule structures like cogs. It is true that the transference of the 'message' from DNA to mRNA, and the transcription by tDNA into more proteins is a fantastic process, but basically the explanation we are offered does no more than show how complex molecules interact *according to the laws of physics and chemistry* - that is, until the tRNA moves into the reading site and *recognises.*

There is also the question of the gene's intervention :

> most genes are about one thousand nucleotides long (this being the length of DNA necessary to specify the average protein), each mRNA molecule, being merely a copy of a gene, consists of a chain about one thousand nucleotides in length :

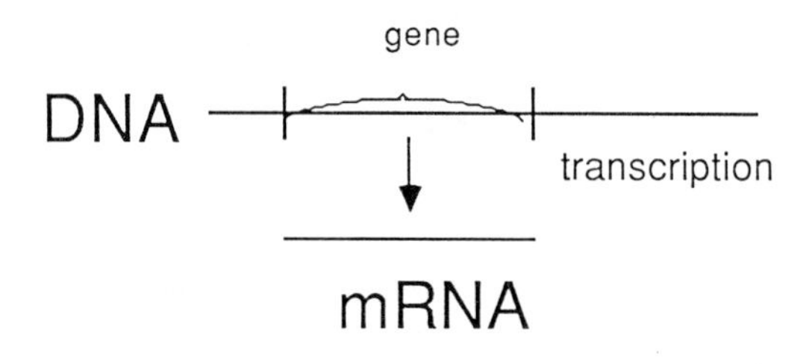

p.242

The gene 'asks' for a certain trancription to be made, into further proteins. But in all the processes so far described in replication, while it is possible to see how the molecules interact *according to the laws of physics and chemistry*, and how that kind of 'information', as to whether the proess is to come up with AGU CGA UUG ACA and thus produce SER-ARG-LEU-THR, is communicated, this is all still within the laws of physics and chemistry.

The questions begin to arise when (as we have seen) there are 'reading heads' which 'recognise' - and here we begin to enter into the invisible problem recognised by Crick when he pointed out that a cell 'knows where it is in the cell wall'. All these processes, marvellous enough, are *employed* (as machines in a factory are *employed*) in some overall purpose (and this cannot be called 'apparent purpose' since life goes on and 'works') - controlled by forces which must also be related to memory, time, intention and the rest. *The analysis of life-molecules, their structure and processes, still gives us no indication of where and in what dimension these controlling principles are to be found.*

There may be a belief within the mechanistic dogma, in its particulate ambition, that the overall controlling principles are merely a product of the processes of replication - but they cannot be. For the 'heads' that read and produce the new long strings of protein, indeed, the long strands of DNA and RNA themselves are clearly *instruments* of some larger organisation. The 'information' they contain is only information of the kind a set of complex gears bring in engagement with another set if complex gears : the 'decoding' is only a decoding of complex instructions, such as I fit into my bank cash machine.

A logical confusion thus centres in the words 'information', 'message' and 'decoding' in the discussion of molecular biology. The use of such terms seems to imply that the 'instructions' or 'information content' for the creation and maintenance of a living organism is to be witnessed there in the coding and decoding of the composition of the long life molecules. *But this is not so. Those molecule transcriptions only follow the laws of physics and chemistry* - until we reach the question of what asks for proteins here and proteins there : what requires proteins here to make hair, here to make flesh, here to develop the embryos, here to bring about adolescence (and where is the I factor, or intentionality, later to become adolescent or menopausal 'stored'?)

Here we need to turn again to Polanyi. DNA may determine the boundary of a biological system. But the form and function of the resulting biology system cannot be explained by the laws governing its parts.

Polanyi as we have seen compares the organism with a machine. The machine is irreducible. Its design, shape and operation are comprehensive features not due to physical and chemical forces. A description of a machine purely in terms of its physical and chemical composition would yield a topography of atoms and molecules unique to the subject. If could not identify the machine as belonging to a class of machines based on certain operational principles.

Biological systems, like machines, have forms and functions inexplicable by chemical and physical lawns. The argument that the DNA molecule determines genetic processes in living systems does not indicate that life can be explained by reduction. A DNA molecule essentially transmits information to a developing cell. But the transmission of the information cannot be represented in terms of physical and chemical principles. The DNA process, of course, can be understood in one dimension as being explicable in physical and chemical forms, where that kind of information (as of the relationship between groups of atoms) is concerned : but when we examine the 'information' the process *subserves* - need for this kind of protein, stop producing this protein, introduce protein for hair, prepare for this enzyme reaction &c - *this information cannot be a product of the laws of physics and chemistry*, since it is somewhere inherent and embodied in a *living system* which, like the machine, has overall principles and (though this is not Polanyi's point but one I am introducing) a controlling 'intelligence' or 'primary consciousness' - such as controls and prompts the 'reading heads' of the molecular system to function. (Just as, one might say, my kidneys or liver are not the mere product of physical or chemical laws, but serve my body).

The angle Polanyi uses is that of a description used to *patent* a watch - of this were based on physical and chemical analysis, it would require only one atom to be changed, for the patent to be circumvented. *A watch must be defined according to the principles underlying its operation.*

The implication of this is that any description of DNA mechanism simply based on physical and chemical analysis does not enable us to understand it : there remains the problem of the operative principles, as it would if we were trying to patent DNA. This becomes evident when (as in Denton's account) we use terms like 'reading' codes, and discuss 'controls' : DNA could only be patented if we brought into the definition on account of its operative principles, and the exploration of these raises the questions of what controls the whole process, and the degree to which we have here a higher operative principle by which a cell 'knows' that it needs to produce, and how to organise and deploy its forces. There is nothing in the physics and chemistry to determine these controlling overall operative principles any more than the principles of cog operation determine the purpose or function of a factory : the supposition that they are dealt with by adequately such analysis is false.

When we turn to artefacts that convey information, the problem becomes much more complex. Nothing, says Polanyi, is said about a book by a physical and chemical analysis of it. All objects conveying information are irreducible to the terms of physics and chemistry. Analogies are made, as we have seen, between the activities of DNA molecules and magnetic tape, and the use of these analogies suggests strongly that replication is a process by which information is conveyed : but for information to be conveyed, as opposed to 'noise', or the mere interaction of structures as cog-wheels, there must be an *informative intervention*, that is, an overall *controlling principle*. While reductionist approaches are blind to the necessity of recognising this, the logical need for such recognition, as we must see, is essential.

As Polanyi goes on to point out, if we look at higher animals and processes in them such as breathing and thermal regulation, the way the parts of the anatomy perform their functions and operate, it is clear that these conceptions cannot be defined in terms of physics and chemistry. The workings may be analysed in terms of these but the functions are not defined by such an account.

Here Polanyi considers Ernest Nagel's attempt to dispel any idea that physiological functions display purpose : they are mere events that *happen to be* beneficial to the organism, but they do not purposely serve this benefit. But, Polanyi consists, 'a process can be regarded as a biological function only if it does benefit the organism... *This remains its essence*, as much as it is the essence of a machine to serve a purpose acknowledged by its designer'. (my italics). We acknowledge this when we think about disorders interfering with these functions, in machines and living organisms. Moreover, Nagel uses the biological names for vital functions, and so implicity admits their irreducibilty.

In this there is an unseen element Polanyi has examined elsewhere. 'The mere shape of a living being defeats any physical-chemical definition and this is true throughout the anatomical features of life'. The material of the machine is subject to the laws of physics and chemistry (and DNA theory shows how this is so). But the 'shape and consequent working of the machine are controlled by its structural and operational principles'. This is also shown by

DNA theory as when it tells how long protein chains '*automatically*' assume a folded shape. It is when we move into this sphere, however, that we begin to get into the second perspective - for how is it that the molecule turns into this shape rather than that? As others have suggested, some kind of memory must be involved, because the shape has meaning to do with a higher function, which is to contribute to a process which is under some kind of control (in the cell - producing this substace or that to serve the overall functions of the 'factory' and its contributions to the overall economy of the organism).

'No biological process ever takes place in an unstructured medium' declares Polanyi's : the diagrams and accounts we have in molecular biology often ignore this or deliberately pretend the physics and chemistry can be independent of the structure : yet the operations examined are meaningless without that overall structure.

We have seen that Polanyi invokes the concept of *boundary conditions*, which form the frame of physics : without that gift physics is dumb. Physics is a matter of using differential equations which describe systems with a set of fixed conditions. If you describe the sun as a sphere or the path of a planet as elliptical you offer a differential equation with the postulated findings on the one hand and a base or zero point on the other : 'the arrangement arises under the dual control of a differential equation working within the bounds of a particular set of conditions'.

While such boundary conditions are indispensable in physics and chemistry, they are not highly significant. They are 'fixed conditions'. But with machines it is different for here the boundary conditions are the *structural and operational principles of the machine* :

> we say therefore that the laws of inanimate nature operate in a machine *under the control of operational principles that constitute* (or determine) *its boundaries.*
>
> p.61

A machine is thus under dual control. And if we think of DNA theory, it is clearly under the control of the laws of physics and chemistry (bonds &c) : but what is missing from much thinking about it, not least in molecular biology, is the other part of the dual control - its subservience to the control of operational principles that constitute (or determine) its boundaries - that it 'serves life'.

The relationship between the two controls - the devices of engineering and the laws of natural science - is not symmetrical. The machine operates by the principles of engineering. But the laws of physics and chemistry are indifferent to these - they would go on working even if the machines were smashed. They serve the machine while it lasts : and so the machine always relies on the laws of physics and chemistry.

Now Polanyi turns to living beings. To speak of life as something to be explained by physics and chemistry is absurd for physics and chemistry by themselves do not determine any finite system.

> biochemistry and biophysics are always concerned with processes *that have a bearing on an existing organism.*

Any chemical or physical study of living things that is irrelevant to the working of the organism is no part of biology. Similarly, the chemical and physical study of a machine must bear on the way the machine works, if it is to serve engineering.

The fallacy in DNA theory is that it examines the physical and chemical interactions and some of the 'engineering' processes, but leaves out of the account the overall boundary conditions - the *bearing on an existing organism.* It is assumed that the particulate analysis of the physical and chemical processes accounts for the overall total picture.

> But it is not conceivable that an organism developing from a fertilised cell might shape the boundary conditions of the developed organism, without itself being subject to such boundaries?

Polanyi's reply is that a machine that manufactures machines for a factory produces them within its own boundaries, as set to the machine by its operational principles.

The embryo producing the biological boundaries of maturity works likewise within its own embryonic boundaries.

No biological process ever takes place in an unstructured medium.

Polanyi assumes next the currently prevailing view that DNA determines altogether the outcome of embryonic growth, including the design of the final organism. Does this not face us with the fact that a pure chemical compound controls supreme biological functions? Where is then the boundary condition which controls this chemical effect?

Polanyi gives one reply, though he admits this is not to the point. A DNA molecule produces nothing by itself, its genetic programme being initiated within the richly structured framework of a fertilized cell : subsequently DNA controls morphogenesis within a steadily developing framework.

But it is DNA itself that introduces within its chemical structure a pattern that acts as a controlling framework to the ensuring generative process. DNA controls the genetic development if an organism by trasmitting to its cells a quantity of information that induces in them an equivalent amount of organic differentiation. Where then is the boundary condition?

Polanyi returns to his analogy with the book. The analogy is with the *pattern* in DNA, the pattern which forms part of an organism and transmits information through this pattern : such a pattern is the *morphological feature* of the organism and this cannot be reduced to physics and chemistry. The pattern by which DNA spreads information is not part of its chemical properties.

Its functional pattern must be recognised as a boundary condition located within the DNA molecule.

That is, there is a third dimension which we must learn to apprehend. The basic account of the composition of DNA and RNA molecules makes it seem as if the only factors involved are such as can be reduced to chemistry (C,H and O molecules) and to physics (links, energy requirements &c). But as we have seen, there are other factors such as the way each protein molecule is folded, the dynamics of replication, and the way certain groups of molecules 'read off' others. Moreover, in the cell, as Denton indicates, there is a huge organisation of molecules and processes, like a factory under subtle and complex *control*. This 'control' must be 'there' in the assemblages of molecules, and if we speak of 'codes', 'reading' and 'control' we are, willy-nilly, admitting another *dimension* altogether.

The boundary conditions of a machine have two interrelated aspects : they consist in a *distinctive structure* sustaining a *purposive operation*. The various functions of a living organism are similarly sustained by its structure known as its morphology.

A written text transmits information by its structure alone : it acts passively by being read. The living parts of organisms also pass on information by being 'read' : some force determines red hair, blue eyes; devotion of cells to this part of growth or that ; stages of growth ; etc. While these may be products of the DNA replication system,

If DNA is regarded as bearing a pattern that forms part of an organism and as transmitting information through this pattern, then such a pattern is to be classed likewise as a morphological feature of the organism and hence irreducible to physics and chemistry.

This is not to ask for any entelechy or 'soul-stuff' : it is to admit that our logic in this area is deficient.

There is no 'code' in the mere raw chemicals of inorganic chemistry : in all organic chemistry there are codes by which behaviour is controlled and outcomes sought. The very orderliness of atoms precludes their capacity to convey information :

All chemical compounds consist of atoms linked in an orderly manner by the energy of chemical bonds. But the links of a compound forming a code are peculiar. A code is a linear series of items which are composed, in the case of a chemical code, of groups of atoms forming a chemical substituent. In the case of DNA, each item of the series consists of one out of four alternative substituents. In

> an ideally functioning chemical code... each alternative substituent forming a possible item of the series must have the same mathematical chance of appearing at any point in the series. Any difference of alternative chances would reduce the amount of information trasmitted, and if there were a chemical law which determined that the constituents can be aligned only on one particular arrangement, this arrangement could transmit no information. Thus in an ideal code, all alternative sequences being equally probable, its sequence is unaffected by chemical law, and is an arithmetical or geometrical design, not explicable in chemical terms.

Polanyi here is difficult to follow : but perhaps the fundamental distinction he is making is not difficult. All the sequences one can diagram as in the DNA theory above can be reduced to physics and chemistry : but if they only obeyed physical and chemical laws the molecules could not do anything biologically functional or purposive. In doing these things they serve other laws, too. He considers the theory advanced by neurophysicists,

> that the nervous system registers the memory of a habit acquired by an organism in the structure of its RNA molecules. This is called the fixation of experience by RNA in the manner of a tape recording... The information imparted to a molecule of this type is received and held by it in a way similar to that in which a tape recorder would do this. The pattern of its traces is the pattern of the impacts in which the message was embodied. The pattern can no more be derived from the laws of physics and chemistry when engraved in a RNA molecule than it can when inscribed on a tape...

For this overall pattern which controls the system Polanyi invents the phrase 'a profoundly informative intervention' which puts the system under a *non-physical-chemical* principle.

But as Polanyi puts it, we do not yet have a quantitative measure of the 'negantropy' (as Schrodinger called it) of a living thing.

Polanyi refers to the analogy made by W. Ostwald between a living being and a flame : its identity is not defined by its physical or chemical topography but the operational principles which sustain it. It is a fundamental property of open systems that they stabilize any improbably event which serves to elicit them.

> The first beginning of life must have... stabilized the highly improbably fluctuation of inanimate matter which initiated life.
>
> p.389

As we have noted elsewhere, the first thing to note about life on earth is its *improbable* nature in this sense, and the need to recognise that for such a thing to come about requires a powerful operational principle to initiate it. The same is true of evolutionary developments : Polanyi declares

> I deny that any entirely accidental advantages can ever add up to the evolution if a new set of operational principles, as it is not in their nature to do so.
>
> p.385

Once we recognise that such operational principles cannot be explained in terms of reduction to physics and chemistry, it becomes absurd to see complex advances such as flight or seeing, and especially thinking, as being the product of 'chance and necessity' on reductionist lines. Moreover, at the heart of the evolutionary process

> the deepening of sentience and the use of thought are the most conspicuous [among the changes tending towards higher levels of organisation]

And here such changes cannot be determined merely by *their adaptive advantage*,

> since these advantages can form part of such progress only in so far as they prove adaptive in a peculiar way, namely on the lines of a continuous evolutionary

achievement.

The rise of human consciousness is a case in point : that it should be explained on the basis of 'adaption', arising from differential reproductive advantage, is absurd. It evidently emerges from some latent '*persistent creative trend*' and it is this that is overlooked or denied by the theory of natural selection. Mutation and selection may *release* and *sustain* the action of evolutionary principles : but all major evolutionary achievements are defined by evolutionary principles. Or, to put it another way, the conditions in which evolution takes place do not control or initiate it : we must seek some other dynamic in another dimension which does that.

Polanyi next turns to man and declares that man is the 'most precious fruit of creation'. The knowledge of this fact lies outside natural science, and to declare it in the context of natural science today seems uncomfortable - even while science assumes passionately that it is itself a primary expression of humanness and consciousness. It is absurd that science turns against itself by denying, in its 'objectivity', that man has a supreme position among all known creatures, and it is a profound flaw in its metaphysical position (not least in its denial that it has a metaphysical dimension).

Polanyi traces what he calls our anthropogenesis : the history of man going back to the origins of life, as a continual stream of germ plasm. This historical process, to which we all belong, must have been governed by an operational principle

> From a seed of submicroscopic living particles - and from inanimate beginnings lying beyond these - we see emerging a race of sentient, responsible and creative beings. The spontaneous rise of such incomparably higher forms of being testifies directly to the operations of an orderly innovating principle.
>
> p.387

Polanyi develops an existential argument. The original primitive beings developed a 'self-controlled shape and structure, and the physiological functions serving its survival, set up a centre of elf-interest against the world-wide drift of meaningless happenings'. Polanyi traces the development of beings, through the protozon to multicellular organisms, sexuality, and further developments towads what he calls 'personhood'. (It is interesting to note that sexuality brings tragedy, for while more simple organisms divide and go on living the sexual reproduction involves the death of individuals, even as their progeny go on living).

Stages on the way include the simplest original learning processes, perception, creativity and understanding.

> A whole firmament of self-set standards was prefigured here and soon the faint thrills of intellectual joy appeared in the emotional life of the animal. And it became also liable to puzzlement and frustration.
>
> p.388

50,000 years ago a new stage developed : what Teilhard de Chardin called the *noogenesis*. Men formed societies, invented language and created by it a lasting articulate framework of throught : the *noosphere*. 'This was the second major revolution against meaningless inanimate being'. The reductionist mechanistic approach to existence which leads to nihilistic philosophies of being denies all these revolutionary achievements. Polanyi's account differs : the noogenetic achievement offers a fabric of life *not* centred on individuals and transcending the natural death of individuals :

> When man participates in this life his body ceases to be merely an instrument of self-indulgence and becomes a condition of his calling. The inarticulate mental capacities developed in our body by the process of evolution become then the tacit coefficients of articulate thought... these tacit powers kindle a multitude of new intellectual passions. They set in motion heuristic endeavours. They make us love human greatness and accept as our guides those who have achieved it. By accepting such teaching man testifies to the existence of grounds on which he can claim freedom.
>
> p.389.

To conclude, Polanyi asserts that the use of man can be accounted for only by other principles than those of physics and chemistry.

> If this be vitalism, then vitalism is mere common sense, which can be ignored only by a truculently bigoted mechanistic outlook.
>
> p.390.

Darwinism has diverted attention for a century from the descent of man by investigating the *conditions* of evolution and overlooking its *action*. 'Evolution can be understood only as a feat of emergence'.

To return to Michael Denton, he follows his discussion of molecular theory with a contemplation of the enigma of the origin of life. Darwin made the suggestion that 'in some warm little pond' with 'all sorts of ammonia and phosphoric salt', a 'protein compound' might be chemically formed, ready to undergo still more complex changes...

Is there a distinct break between life and the inorganic world? Molecular biology, Denton declares, has emphasised the hugeness of the gap between the physical and biological worlds. Indeed, it represents one of the most dramatic and fundamental of all the discontinuities of nature. Even the simplest bacterium is so complicated as to be without parallel in the non-living world.

While molecular biology has shown that the basic design of the cell system is essentially the same in all living systems on earth from bacteria to mammals, here, as in the rest of biology :

> no living system can be thought of as being primitive or ancestral with respect to any other system, *nor is there the slightest link of an evolutionary sequence among all the incredibly diverse cells on earth,*
>
> p.250.

While Darwin made no claim that his theory could be extended to explain the origin of life, some like Thomas Huxley believed it might be extended to show how living protoplasm evolved from not-living matter. Today evolutionary biologists believe the process began with a primitive self-replicating molecule which slowly accumulates beneficial mutations that enabled it to reproduce more efficiently. After eons if time it grew into a more complex self-replicating object. Life is now widely viewed as the inevitable product of any planetary surface which has the right conditions. Immense sums of money and effort are devoted today to finding life that must have initially originated elsewhere in the universe. Sums as great are devoted to listening for possible messages from beings in outer space. Experiments on the surface of Mars sought for evidence of biological activity. No signs of life were found. Some scientists, after giving much thought to the problem, have been forced to the conclusion that 'the evolution of life is a much less probable event than the molecular biologists would have us believe'.

Denton next turns to the 'prebiotic soup'. The existence of this is crucial to the theory of life's spontaneous origins : yet no traces of abiotically produced organic compounds have been found in the rocks of great antiquity, such as the 'dawn rocks' of Western Greenland (3,900m years old). Paleontology here has failed to support evolutionary presumptions once more. '*There is absolutely no positive evidence*' for the existence of prebiotic soup.

There are other problems, as of the presence of oxygen in the atmosphere : the question of the provision of protection against ultra-violet light by the ozone layer, and the fact that nucleic acid compounds are particularly sensitive to ultra-violet radiation damage and mutation.

The time-scale has also been reduced by recent research. Remains of simple types of algae have been found in rocks as old as 3,500m years,

so the time for life to form, between the origins of the earth (several thousand million years) and the origins of life has been reduced to a few hundred million.

Even if there had been a prebiotic soup, it is only *possible* that life would evolve : it has yet to be shown as probably. The American biochemist Harold Morowitz[3] has speculated on the minumum requirements for a self-replicating cell : one is a bilayered lipid cell membrane and this requires a minimum of five proteins. Eight proteins would be required for a very simplified form of energy metabolism. A minimum of ten proteins would be required for DNA synthesis and so on. The total minimum requirements are about 80 proteins. Even this tiniest of cell capable of self-replicating could never have been thrown together suddenly by some kind of freakish, vastly improbable, event. 'Such an occurence would be indistinquishable from a miracle'.

If one postulates simple primitive cell forms, the trouble (declares this molecular biologist) is that their translational apparatus would be less than perfect. Francis Crick has spoken of 'very crudely made proteins', but

> everything we have learned about protein structure and function over the past thirty years implies that the function of a protein depends upon it being very accurately manufactured and processing highly specific configurations.
>
> p.265

So, we seem to be asking for the basic axioms of modern biolchemistry to be contradicted, Denton declares. How could a cell prone to error ever do its functions and be viable?

But, again, Denton finds that extraordinary kind of argument which follows from the need to preserve evolutionary theory at all costs : C. Woese declares:

> Nevertheless it is essentially a certainty that at an early enough stage in evolution such cells as these did exist and some had to be viable.
>
> 'On the origins of the genetic code', *Proc.Nath.Acad.Sci. U.S.* 54; 1546-52; p.1549.

Evolutionary biologists, Denton shows, have striven to imagine a cell which was primitive and didn't work as efficiently as the cells we know today. But, as Denton points out, this is to imagine a hypothetical cell *in which efficient proteins evolved in a system basically incapable of their manufacture.*

Here, again, molecular biology does not support evolutionary theory : it exposes the reality, that highly organised forms could never have come from chance and chaos : they could only have come from some organising principle. If translation were inaccurate, this would have introduced further errors : to improve itself, such a system would have had to 'overcome its fundamental tendency to accumulate errors in exponential fashion', and this is incredible.

Modern organisms have a low rate of mutation : if this is raised, then this may lead to an accumulation of errors, and this may prove lethal - it can lead to autodestruction. Each cycle increases the 'noise' and erases crucial information. This is an empirical observation, and so it becomes clear that the overall principle of organisation in living things depends for its continuing existence on a highly complex integration of processes which are extremely accurate, complex, and coordinated : and could only have come into existence on some highly sophisticated organising principle. To set this process off required an astonishing complex 'innovative principle' and it cannot be seen as the product of chance or accident at all, because there is no way in which hazardous processes could work.

> Just as a bird feathered with the frayed scales of pro-avis could plummet to the ground, so a cell burdened with inefficient proteins, an error-prone code, and choked with junk [redundant forms] would grind instantly to a halt.
>
> p.268

In exploring the origins of the 'code' or translational systems, evolutionary theory has (says Denton) reached a sort of nemesis, for the problem is insoluble in terms of modern biochemical knowledge. So, from our point of view in the humanities, while we have pressed upon us the view that 'science says' life and man are the product of 'accident', the actual

---

3 H.J. Morowitz, 'The Minimum Size of Cells', *Principles of Biomolecular Organisation*, eds. G.E.W. Wostenholme and M. O'Connor, J.A. Churchill, 1966, pp 446-59.

scientific examination of the postulated processes of purposeless hazard (Cuvier talked of 'incoherent combinations') shows that they *would not work* and cannot be explained by the most up to date knowledge of physics and chemistry. So much for the 'facts of science' which are used to daunt us!

Both Monod and Crick admit as much. Monod admits the problem of how the primordial soup 'learned' to mobilize chemical potential and to synthesize cellular components is an 'enigma' : Crick (in *Life Itself*) says that 'the origin of life appears at the moment to be almost a miracle'.

Denton ends his discussion of life; origin by pointing out that the problem is even more complex, because there is more to consider than the protein synthesis apparatus. The cell needs to be held together by a membrane, and this has to be synthesised. Yet the synthesising apparatus only works if it is held together by a membrane! The processes need energy, and this requires proteins, which are themselves manufactured by the synthesizing apparatus. The simplest life form is full of such complex inter-dependent cycles. So, in a conclusion close to those of Polanyi, Denton says

> Complex systems cannot be approached gradually through functional intermediates because of the necessity of perfect coadaptation of their components as a pre-condition of function.

Monod admits that the a priori probability of life was virtually zero : faced with this, scientists like Francis Crick have opted for space fiction solutions like panspermia, the notion that life was placed here by intelligences from outer space. Others have been groping for other influences, 'factors as yet undefined by science'.

But the truth is that 'the new biochemical picture has not had the effect that evolutionary theorists might have hoped'. It has not blessed the distinction between living and non-living objects, while it reveals the whole 'system' of living organisms as being amazingly complex unique to living systems and without parallel in non-living nature - a greater mystery and miracle than ever before, and one which speaks of some 'innovative' principle and power which is as yet beyond the comprehension of man. The bleak effects of mechanism must surely begin to recede from these observations?

As we have seen, Denton points out that the old nineteenth century model in morphology was typological. He recognises the subjective element in all morphology : we cannot quantify the difference between a dog and a cat. But now a new situation has arisen, because in molecular biology it is possible to compare organisms at a biochemical level. The sequences of protein (Gly-Leu-Phe-Gly etc) vary from species to species, as in haemoglobin. This offers a strictly quantitative means of measuring the distances between species.

This could, of course, have produced strong, if not irrefutable evidence for evolutionary theory. However, this has not proved to be so. While the divisions have turned out to be mathematically perfect, *they do not provide any evidence of sequential arrangements* and tend rather to confirm the typological view that nature conforms fundamentally to a highly ordered hierarchic scheme from which all direct evidence for evolution is empirically absent.

Denton's discussion of the consequent classification systems is complex, but he points out that its effect is that 'there is a total absence of partially inclusive or intermediate classes'. Despite the discovery of new, distinct living kingdoms, like the anaerobic bacteria which generate methane, which have been called archaebacteria, these cannot be designated as *ancestral* or *primitive* with respect to other classes (such as eubacteria and eucaryotes).

The molecules, like fossils, have failed to provide the elusive intermediates so long sought by evolutionary biology. Moreover,

> At a molecular level, no organism is 'ancestral', 'primitive' or 'advanced' compared with its relatives.
>
> p.290.

This new era of comparative biology illustrates just *how erroneous is the assumption that advances in biological knowledge are continually confirming the traditional evolutionary story.* If such organisms as the lung fish and ancient amphibia, seen as possible intermediates, were as separate from one another as their present day descendents are by molecular analysis, then *the whole concept of evolution collapses.*

Evolutionists are always, using terms like 'primitive', 'ancestral', 'ideal ancestors', 'basal stock' to imply continuous descent. But in terms of their biochemistry these claims will not bear examination. The divisions of species according to biochemical analysis are incredibly orderly. If we clung to evolutionaty theory, it must seem that every branch of every class has evolved at exactly the same time rate : this is the 'molecular clock' hypothesis. This, Denton declares, is really a tautology, invented to preserve the evolutionary hypothesis : it is really a restatement of the fact that

> at a molecular level the representatives if any are class are equally isolated from the representatives of another class.

Denton's argument here is dense and difficult to follow, but what it adds up to is that having at last found a strict mathematical way of analysing species differentiation, the evidence is that species are different typographically, and show no evidence for continuous descent. For example

> The opossum is another classic living fossil, virtually unchanged morphologically from its ancient ancestors of the late Cretaceous period nearly one hundred million years ago. But when opossum haemoglobin is compared with the haemoglobins of other mammals it is in no way primitive with respect to other mammalian species.
>
> p.309

This finding strains the whole paradigm of evolution. There is no way of explaining how a uniform rate of evolution could have occurred in tiny family of homologous proteins either by chance or selection.

Denton's conclusions here are startling :

> The hold of the evolutionary paradigm is so powerful that an idea which is more like a principle of mediaeval astrology than a serious twentieth-century scientific theory has become a reality for evolutionary biologists.

The sequential comparisons of homologous proteins reveals an order in living nature as emphatic as that of the periodic table. Yet the biologists, clinging to evolution, can offer explanations which are no more than 'apologetic tautologies'.

Yet it is when we challenge evolutionary mechanism that we are accused of being 'unscientific'!

Denton now turns to *chance.* The mutations upon which evolutionary theory depends are entirely random and completely blind as to what effect they may have : 'drawn from the realm of chance' as Monod put it. It is only when an innovation has been disclosed by chance that it can be 'seen' by Natural Selection.

So, all evolution depends upon a purely random *research strategy.* But scientists who have put this kind of strategy to the test know it to be *hopelessly inefficient.*

The question at issue is whether it is feasible that living organisms are capable of undergoing functional transformation by random mechanisms. If we study any analogy between machines and organisms we can see that for one thing the outcome of any random research would be tiny compared with the time and possible combinations tried out. Moreover, if we look at the question of random variants in machines, not only would any workable new entity be surrounded by a vast sea of junk, but random changes would most likely violate the principles of function : while a watchmaker has little trouble in assembling a watch by *following the rules which govern functionality,*

> it is obviously impossible to contemplate using a random search to find combinations which will satisfy the stringent criteria which govern functionality in watches.
>
> p.313

Function is restricted to unique and fantastically improbable combinations of sub-systems, '*tiny islands of meaning lost in an infinite sea of incoherence*' (here Denton's argument parallels those of Michael Polanyi).

Unguided trial and error cannot reach anything but the most trivial of ends : this is the conclusion of computer research into the possibilities of solving problems. Such exploration requires guidance : specific algorithms. Without such guidance trial and error is totally inadequate and, Denton believes, it is because of this that it has proved impossible to simulate Darwinian evolution by computer analogues. One biophysicist has declared

> even some of the simplest artificial adaptive problems and learning games appear practically insolvable by multistage evolutionary strategies.
>
> *H.H. Pattee, Natural Automata and Useful Simulations*, 1966.

Denton concludes :

> The fact that systems in every way analogous to living organisms cannot undergo evolution by pure trial and error and that their functional distribution invariably conforms to an improbable discontinuum comes, in my opinion, very close to a formal disproof of the whole Darwinian paradigm of nature. By what strange capacity do living organisms defy the laws of chance which are apparently obeyed by all analogous complex systems?
>
> p.316.

If we were to have a machine capable of altering its own organisation in an intelligent way, this kind of evolution would be closer to the Lamarckian. Here we glimpse the fundamental objection to Darwinian mechanism : it lacks recognition of the necessary intelligence at work, without which evolution is incomprehensible. Discussing the development of a machine like an aeroplane, Denton assents 'the search for function is intelligently guided'.

Denton talks of 'islands of function' in the space of all organic possibilities. In this, many scientists are naive : the kind of organisation which manifests an organising principle, which involves boundary conditions in Polanyi's sense, is a rarity. To find any such meaningful system by chance would be a miracle. The possibility of a sudden macromutational event on hopeful monster or other lines is vanishingly small.

If we consider the simplest biological requirement, the protein molecule, not only do we have to consider its molecular composition, but its functional form, including its correct way of folding. There are, say Denton, *a priori* rules which govern function in an animo-acid sequence which are relatively stringent. If so, functional proteins could be exceedingly rare. What then are the possibilities of one coming into existence by chance?

> Even short unique sequences such as just ten animo acides long only occur once by chance in 10^{13} average-sized proteins... unique sequences thirty animo acids long once in about 10^{39} proteins.

It can easily be shown that no more than 10^{40} possible proteins could have even existed on earth since its formation. *It becomes increasingly unlikely that any functional proteins could ever have been discovered by chance on earth.*

> To get a cell by chance would require at least one hundred functional proteins to appear simultaneously in one place. That is one hundred events each of an independent probability which could hardly be more than 10^{-20} giving a maximum combined probability of 10^{-2000}.
>
> p.323.

Recently Hoyle and Wickramasinghe in *Evolution from Space* provided a similar estimate of the chance of life originating : 'the chance of obtaining them [i.e. two thousand enzymes] in a random trial is only one part in $(10^{20})^{2000}=10^{40,000}$...' This was an 'outrageously small probability that could not be faced even if the whole universe consisted of organic soup'.

Darwinian claims that the adaptive design of nature has resulted from random search is one of the most daring claims in the history of science : but it is one of the least substantiated.

> No evolutionary biologist has ever produced any quantitative proof that the designs

of nature are in fact within the reach of chance.

Denton ends by quoting Dawkins, in an article on 'The Necessity of Darwinism' from the *New Scientist* (April 1982) :

> Charles Darwin showed how it is possible for blind physical forces to mimic the effects of conscious design, and, by operating as a cumulative filter of chance variations, to lead eventually to organised and adaptive complexity, to mosquitoes and mammoths to humans...

But, as Denton says, we find it hard to determine the probability of the discovery by chance of one single functional protein, let alone the random assembling of mice and men. No biologist has ever calculated the probability of a random search finding in *the finite time available*, the sorts of complex systems which are ubiquitous in nature.

Denton's next point is that so many of the ingenious devices of life - from the kidney to the eye - are 'brilliant'. Intuitively, we feel it impossible that such designs should be the result of chance. Though there are imperfections in nature many biological adaptations seem perfect : as Darwin himself put it, 'a perfection of structure and coadaptations which partly excites our admiration'.

Denton then gives a fascinating picture of a cell, containing about ten million million atoms. If we set out to construct an atomic model, with atoms the size of a tennis ball, it could take us a million years to finish it. Yet of course one cell is nothing compared with an organ like the brain consisting of a thousand million million connections between connecting fibres and nerve cells. Again, we are faced with the question whether any kind of 'chance' process could have assembled such complexities in the time available. As discoveries proceed, as of the adaptive traits in mammalian genomes, Darwinism will be presented with insurmountable 'numbers games' (10^{10} genes each containing 10^{3} significant bits of information) to account for.

But not only is it the mathematical reaches, or the complexity which present problems : it is the ingenuity of life, as in the functions of the eye, or limbs, or the multifarious chemical operations in every organism, and the 'storage of information' in the components of living organisms.

One dimension seems still lacking in Denton's account, when he talks of the 'protein synthetic apparatus' : 'it can also construct any other biochemical machine'. He talks as if this were a matter of the information code operating as in his explanation of DNA theory. As we see from Polanyi's account, the 'code' is not enough : some overall 'intelligence' must be postulated to control the processes.

But while he fails to find 'primary consciousness' Denton does pay tribute to human intelligence. Some scientists talk of 'artifical intelligences' : but even if we knew how, it would require eternity for us to make an intelligent machine like the human brain. '*No machines have yet been constructed which can in any significant way mimic the cognitive capacities of the human brain*'.

And yet this human intelligence is nothing compared with the creations of nature :

> It is the sheer universality of perfection, the fact that everywhere we look, we find an elegance and ingenuity of an absolutely transcending quality, which so mitigates against the idea of chance.
>
> p.342.

Denton admits that Darwin's special theory has been confirmed. In the general theory, however, the idea of a functional continuum in all life and the belief that all the adaptive design of life has resulted from a blind random process have not been validated by one single empirical discovery or scientific advance since 1859.

Yet so entrenched is the paradigm that its priority takes precedence over common sense - and Denton shows himself aware of Thomas Kuhn's discussion of this problem. Everywhere, evolutionists greet evidence that menaces their paradigm with perverse arguments : for instance, if the time between the origins of the earth and the first appearance of life shrinks, rather than be dismayed, they will declare that this just shows that the origin of life had a high probability. The illusion is created that every single fact of biology irrefutably supports the Darwinian thesis.

> Hence, even evidence that is to all common sense hostile to the traditional picture is rendered invisible by unjustified assumptions.
>
> p.353.

Denton's view is that *life might be fundamentally a discontinuous phenomenon.*

Darwinism is the only truly scientific theory of evolution. The Lamarckian model is incompatible with the modern understanding of heredity. Creationism requires supernatural intervention. Science is afraid of rejecting Darwinism, because to do so means living without a scientific explanation.

There is a possibility that there may be properties or phenomena such as science cannot yet understand. But what is obviously needed is the kind of paradigmatic revolution Kuhn discusses. Until that happens, Darwinian concepts will persist, even in the face of 'disproofs'. But this means that Darwinism will continue to dominate '*more by default* than by merit'. What can be no longer doubted is that '*biologists have failed to validate it in any significant sense*'.

But the failure to validate Darwinism will have prodigious effects on the Humanities, as Denton realises. 'The entire scientific ethos and philosophy of modern western man' is based to a large extent on the theory that

> humanity was not born by the creative intentions of a deity but by a *completely mindless trial and error selection of random molecular patterns.*
>
> p.357.

This is the triumph of a naturalistic view of the world. This has become the directing force in modern scepticism, with profound political and philosophical consequences.

But now Darwinism is shown to be not complete, not comprehensive, not a plausible explanation for biological phenomena from the origin of life to man's existence. Darwinism has become a metaphysic, a myth : *but now it is exposed as only that.* The supposed scientific grounds - as a 'fact' are now severely in doubt.

We have a psychological need for an all-embracing explanation for the origin of the world and for 'life'. But now we find, as Denton puts it, 'Nature refuses to be imprisoned'. The mystery of mysteries - the origin of new beings on earth - is still largely as enigmatic as when Darwin set sail on the Beagle.

Denton's is a remarkable book, and it would seem quite incontrovertible. Whatever its effects in the sciences, in the realm of the humanities it is yet one more contribution to our release from the bleak metaphysics of Darwinism, of 'chance' and 'accident'. And though Denton does not say so, it points to the need to rediscover a kind of creative intelligence in life whose existence is far more exciting and challenging than any of those concepts to which religions have given the name of God, while our (unique) existence as knowers in the universe, able to be conscious of our plight, places upon us a terrible responsibility, to seek to understand and to serve the mystery of creation.

I should like to add to my discussion of Denton a note on a paper '*The 'Evolutionary Paradigm' and 'Constructional' Biology* by Goodwin B., Webster G. and Smith J.W. published in *Scientific Explorations.*

This paper discusses the problem of biological form. How are organisms of a specific morphology generated? The implication of Darwinism is that genes create form, and taxonomy is related to this assumption. In Darwinian theory the genotype is take to be sufficient cause of the phenotype, and the genetic information contains a set of 'instructions' for the construction of an organism. Analogies are made with computer software.

Taking information theory, the authors declare that what is absent from an account of morphogenesis based upon the concept of information which 'flows along the germ line' (Maynard Smith 1982) is the origin and nature of the developmental processes among which selection is being made by the genetic information.

Then, if the 'genetic programme' is regarded as something like a computer programme, it has been found by some that 'gene products are not in themselves sufficient to determine the form or morphology of the structures into which these gene products are assembled'. A particular pattern of gene activity is neither necessary or sufficient to generate a particular morphology.

The energy-information or process-instruction dualism embodied in this metaphor seems a modern version of the 'directing spirit' of nineteenth century holists.

The authors turn in seeking an alternative to the work of Driesch, who found that with certain organisms perturbations of certain fields resulted in the restoration of overall order. They argue from this that the cell is not the basic entity underlying the generative process nor is generations to be understood in terms of the properties of the DNA.

> Rather, generation is to be understood as a process which arises from the field properties in the living state, the inherited variety of form observed in organisms arising from the inheritance of particular influences such as specific gene products or nucleation centres, acting as parameters contributing to the specification of critical and boundary values of the field equations.

Their primary influences are 'the generative field' and 'the morphological potential' : these reside in 'diffusion and reaction', and from 'visco-elastic properties of cells, from electrical potentials and ion flows'. They suggest a research programme based on a generative theory of transformations rather on the micro-reductionist model of evolutionary Darwinism.

Once more, Polanyi's concept of an overall directing principle seems relevant. The reductionist paradigm which must explain everything by minimals is again being challenged, and this paper shows scientists once again groping for a more adequate mode of thinking in biology : 'an alternative root image is required'.

13 On new modes of thinking

I have spoken from time to time in this work of the need for new ways of thinking. What can this mean?

The success of chemistry and physics, and the great advances made by Nobel prize-winners, not least in unravelling the structure of life, have generated by degrees the impression that the 'real' is the physico-chemically real. This, in turn, as Marjorie Grene makes clear, has led to a situation in which 'to admit, *au fond*, the reality of living nature seems a betrayal of science itself'. And, I would add, we have now reached a point at which to admit the *reality of our own living nature* may seem an offence to science and so to truth! It seems an outrage (even if no-one actually puts it explicitly like that) to science, to admit and embrace the reality of ourselves as beings. And in the metaphysical sphere, of course, the consequences of only being able to find the non-living real is catastrophic.

There is, thus, a radical conflict between *ontology* and *science*. A mistaken respect for 'objective' science has led, paradoxically, to a situation in which our urgent need to contemplate a due place for man in the natural world - an urgent conceptual reform - is baffled by erroneous concepts of what the scientific truth is. And the *untruth* is that it is the non-living which is real. It is this untruth which the biologists discussed by Marjorie Grene in her *Approaches to a Philosophical Biology* challenge. The central problem, both to science and philosophy, is what is means to be alive.

As Tomlin has made clear, many contemporary 'scientific' beliefs, despite their assumed ruthless materialism and mechanism, rest upon belief in the impossible in miracles – really. For instance, they have to believe in a sudden and magical leap into a new dimension, as inanimate matter becomes animate. Science urgently needs to solve these problems without magic. It cannot, for intance, go on believing that a form or structure as in life can suddenly emerge from an uncoordinated assemblage.

A quite different point of view is put forward on scientific evidence in *The Morphology of Organic Forms* by A. Frey-Wissling, a book to which Tomlin draws our attention. In the retrospect to his book he declares that 'there is no fundamental difference between living and inanimate matter' :

> The complicated process of metabolism is not controlled by some special vital principle, but has its being in the co-ordination of innumerable reactions, each and all being separately accessible to causal investigation, yet no simple mechanistic interpretation can account for their delicately attuned harmony and their purposiveness.
>
> p.371. (my italics)

There are no mysterious life forces and no special entelechy of formative principles, even in the invisible regions of the sub-microscopic world. 'The formative forces in protoplasm and its derivatives are no different from those operating within inanimate Nature'.

There is no evidence of the existence of formative principles beyond the atomic valency and the various molecular cohesive forces in their various patterns.

Substance and form are closely related in the molecular world. Orderly biological processes are unthinkable, says Frey-Wissling, without presupposing structure. It is out of the question that any living constituent of protoplasm could consist of structureless, fluid, independently displaceable particles. The cell is not a pouch filled with ultra-microns suspended in a fluid, whirling about haphazardly and in confusion :

> it is, on the contrary, a wonderful system, the intrinsic structure of which, could it but be seen, would assuredly fill every observer with an enthusiasm equal to that which microscopic cytomorphology inspires.

There are two guiding principles in biomorphology which are recognisable in the conformation of chain molecules. These are first the principle of repetition, which is the foundation of all lattice structures and of every form of banding : and secondly, the principle of specificity. The first principle is represented on the one hand by the ever-recurring members of the chain (intra-molecular spacing) and, on the other, by the assemblage into a lattice pattern of kindred chains (intermolecular spacing), as for example in frame substances, reserve substances and lipid layers. In polypeptide chains the second principle holds sway, the capacity of otherwise similar molecular elementary units to assume a *specific* arrangements which may be repeated for its part in long-range periods.

> We do not yet know how the visible specific forms of cellular organelles, cells, tissues and organisms grow out of this specificity, but doubtless causal relations do exist between molecular morphology and morphogenesis...
>
> p.372.

What Frey-Wissling points to is an approach based on the understanding of highly complex *systems* : and here it is important to note his word 'wonderful' (wonderful system) and his emphasis that 'no simple mechanistic interpretation can account for their *delicately attuned harmony*' and 'their *purposiveness*'. While (as Marjorie Grene emphasises) the growth and development of life are not 'directed' (as by God, Paley's watchmaker) they are evidently directive - a wonderful, delicate, harmonious *purposiveness* is clear even to the scientist who recognises that there are in life no special 'vital forces' or formative principles. There are *forms*, however, which we need to study.

Frey-Wissling notes that metabolic centres (lyoenzymes, mitochrondria, erythrocytes, chloroplasts) are independent of each other, *but their movement does not obey the law of entropy*, they are 'actively directed to the localities where their bio-chemical capacity is needed'.

As Bernard Towers points out, one of the ideas which came across into the Humanities from science was Entropy, the Second Law of Thermodynamics, the theory that everything was 'running down' towards 'equilibrium'. It is important to have it established by Frey-Wissling that at the microbiological level the movement of metabolic centres *does not obey the law of entropy*.

> On the other hand, the special cytological and histological systems which facilitate an appropriate production and distribution of those metabolic centres must have some coherent structures at their disposal. The organisation of these semi-solid structures is responsible for the creation of biological objects of any shape or form and, therefore, is the very foundation of morphogenesis.

The structures of living protoplasm cannot be spontaneously generated like crystals forming an amorphous mass, from unformed solutions. Complicated and delicately inter-adjusted as they are, they can only actualize in contact with already existing structures. The supreme axiom of cytology, declares Frey-Wissling, is that all cells derive from their like, and this applies, too, to invisible, submicroscopic cytogenesis. He ends his book with the Latin motto :

STRUCTURA OMNIS E STRUCTURA.

While Frey-Wissling argues that living forms and structures manifest the coordination of innumerable reactions ('each and all being separately accessible to causal investigation') in which the only formative principles are the atomic valency and the 'various molecular cohesive forces in their various patterns', it is also clear from what he says in this retrospect at the end of his book that 'life' cannot be reduced in a simple mechanistic explanation to *only* 'little causal thingummies'.

Looking through his text, we may pick out various terms which imply that, however it originated and evolved, 'life' has developed forms of *system* which cannot be understood as the mere simple mechanistic operation of the laws of chemistry and physics : that is, life obeys the laws of chemistry and physics, but it obeys other laws, too - as, for instance, by virtually *defying* entropy. These terms, which should be reflected upon, are as follows :

1. Coordination. What is it that coordinates the complex innumerable reactions in living entities? Causal investigation surely needs to ask the question, 'Why should these reactions be coordinated? What is the purpose or directedness of their coordination, and how did it arise?' This coordination masks the difference between such reactions as the growth of crystals in an amorphous mass, and the generation of living protoplasm into structures which can only be actualized in contact with already existing structures : that is, the coordination has a historical perspective.
2. 'Delicately attuned harmony'. Why does this exist?
3. 'purposiveness' or 'directiveness' : a fact in all biology which (as we have seen) is implicitly recognised by the mechanists who try to avoid it by using terms like 'apparent purpose', and deny it in their philosophy.
4. 'formative' : the word, whether used of inanimate or animate things recognises the existence of order throughout the universe. The existence of this order may be set against the implications of such terms as 'chance' and 'mistake' which are the more and more assertively used, the more ordered and harmonious the organisation described. That is, no-one used the word 'chance' about the lovely orderly structure of a crystal : it is only when faced with the much more marvellous and orderly structures of animal or man that some begin to speak of 'mistakes in copying' and 'patterns' (which imply order).
5. 'Wonderful system'. The implications of this are that scientists are interested in living entities *because* of the order in them, *not* because they are haphazard products of chance. It is also an admission of the mystery and the beauty : and so of what Polanyi calls the 'meaning' in life.
6. 'the creation of biological objects' : If we exclude 'creationism' of any kind, the problem remains of explaining why there should at all be systems which 'facilitate' production and distribution of metabolic centres, why there should be an organization of 'semi-solid structures... responsible for the creation of biological objects of any shape or form'. The answers can only be given (as Tomlin emphasises) in terms of forms.

'Creation' as in the poetry of the Bible, is a human poetic attempt to explain the existence on earth of 'wonderful' systems, with delicate harmony and purposiveness and a fantastic ingenuity. The creation, the complex and delicate harmony, the ingenuity and the directiveness *are* the facts! It is no real explanation to say, they 'just growed', out of chaos, which is the implication of much strictly mechanistic reductionist theory, which really depends upon impossibilities.

Nor can it be an explanation to assume, simply, that the marvellous discoveries of the structure and operation of DNA reveal that 'life' is simply shown to be the operation of simple laws of chemistry and physics. The discovery actually implies something quite different.

These questions have, of course, come up in my discussion of Tomlin's work. Here I am trying to indicate the limitations of mechanism. What it does not see is that in the organic world one form of structuration must derive from another form of structuration, of the same or a very similar order : no organic structure can derive from an uncoordinated assemblage. All life, says Tomlin, goes back to the beginning of life : which leaves us, again, of course, with the problem of the mystery of its beginning. And as I have suggested, mechanism cannot solve these problems.

The way out of some of the blind alleys is, Marjorie Grene suggests, throughout her work, is to undercut the conceptions of science from which those problems have developed. Take, for instance, the whole assuption behind Darwinism, of reducibility. Could not more scientists be persuaded just to let reductionism and the impulse to explain everything on the one level *go*, as 'side issues' : as quests no longer worth our devotion or opposition?

Marjorie Grene refers us to some work by Rom Harré (*Principles of Scientific Thinking, 1970). In considering scientific explanation, he suggests, we should distinguish between scientific theories and theoretical models.* While the former have been persistently preferred over the latter, the truth is that science exists in the search for and elaboration of models. Science makes existential choices : is the heart a furnace (as Descartes thought)? Or a pump (as Harvey thought)? We settle for the heart as a pump because this model explains the cause and effect action of the blood being circulated. Scientists are trying to find out how nature works : scientific knowledge is no more than scientists doing this, getting a greater satisfaction (subjectively) from models that 'fit' (like the pump) than from those that don't (the furnace). Science looks for a mechanism which might underlie the phenomena we hope to understand. The discovery of stable mechanisms in nature, not the summary of one flat level of pure phenomena, is what science is after.

Turning back to reductionism, it is assumed that since everything is made of matter, the fundamental laws of physics (quantum mechanics and so on) ought to give us the laws of all systems. But they do not, because the laws of special systems cannot be universalised. The nervous system has laws but these cannot give us universal laws unless we break them down into initial conditions : if we do so, then the laws of nerve action disappear. This means, in practice, problems when, in the analysis of living creatures, we kill them, as under the electron microscope : or when (as some microbiologists point out) we attend to bacteria in the lab, (E. Coli, the white rat of the micro-biology lab.) and not in the open country or naturaal habitat, where their autonomy is more evident. 'Out there' rats obey laws which are not apparent when they are caged in a Skinner Cage - or chopped up or dead, on their way to particulate analysis.

But science, by its traditional paradigms, inclines to drive us towards simplification, towards unification, towards the ideal of one great comprehensive theory. In the pursuit of these that principle has crept in, that only the smallest, invisible part of things only are allowed to be counted as 'real' : minimals. Thus we have an atomistic metaphysic - an attachment to a phenomalism such as implicit the statement 'we are only DNA's way of making more DNA'. As Marjorie Grene puts it :

> Thus when I tell my students, there are quail in my garden in the morning and robins in the afternoon, they say, nonsense, there is one gene pool in your garden in the morning and another in the afternoon. That's what you're really observing. In the name of scientific observation I must admit that my eyes deceive me : the real phenomena are the invisible ones demanded by the most unifying and most economical theory, the phenomena I *see* are only apparent and must be explained away.
>
> *The Understanding of Nature*, p.62.

This state of affairs has arisen from a long tradition, and we may see in it Galileo's distrust of 'secondary qualities', Descartes' concept of 'clear and distinct ideas', Hume's view of the mind as an assemblage of impressions, Hobbes' atomism - and, indeed, a whole British tradition in which what is disastrously missing is the perceptive consciousness and a mind that is constitutive of reality, with an I-being at the centre, looking at living wholes which are themselves experiencing beings. In consequence, our science is not allowed to tell us about the real world, only about *aggregates of phenomena*. In the work of Erwin Straus the serious effects of this have been analysed in great detail, especially in the field of psychology, and especially behaviourist psychology. In these disciplines, because of the need to reduce everything to the psychological or behavioural atom, there needs to be a persistent *misreading of the phenomena - to suit the metaphysics* (See *The Primary World of Senses*, Erwin Straus, and Marjorie Grene's analysis of it in *Approaches to a Philosophical Biology*). On the other hand, of course, scientists watching lions in the bush are not making particulate analyses though they may not yet have become aware of the philosophical implications (as of their recognition of what Adolf Portman called 'centricity'). A great deal in biology cannot be micro-analytic, since (for instance) it is concerned with colonies and populations and the characteristics of living beings.

If we think of science in Rom Harré's terms, declares Marjorie Grene, the demand for reducibility vanishes into thin air. For one thing, it accepts that nature can never be known as a *whole*. Each investigation must accept that it is partial and perspectival. It is the ideal of 'objectivity' that moves into abstractions which must neglect certain aspects of experience in order to establish others sufficiently exact for the scientist to manipulate. Reductionist science takes segments : in its method and content it is partial and plural - a maxim Marjorie Grene finds also in Max Weber.

In seeking understanding by the postulation of theoretical models there is a kind of unity in science, but an *open* unity. The psychologist may, for example, try to understand the brain by analogies from computer science, as Freud tried to explain mental processes on an analogy with physical theories of energy, but such analogies need not mean the acceptance of (say) a metaphysical dogma of 'mechanical man'. Anything may be relevant to anything else but each enterprise is personal and irreducibly pluralistic. Science is one, but also many. But the brilliant postulate of the double helix remains as the application of structural engineering analogy to molecular biology : it does not lead to a universal conclusion that *everything* is to be explained in terms of 'DNA coding'. The workings of the brain, say, or perception, require other models.

The demand to reduce all theories to one level spring from a phenomenalistic programme. No law or theory in a part of science comprehends all the phenomena of every kind that any scientist wants to explain. Explanation works by leading us to see, in the case of a particular set of puzzling phenomena, *how in fact those phenomena are produced.* Thus investigation of DNA etc. was impelled by the desire to ask 'What is the genetic material that produces the effects which we see'? But to show that certain events follow one another is not necessarily to show that they *cause* one another : this is not the problem of phenomenalism. It has never proved possible for this reason to give a reasonable account of scientific method in pure phenomenalist terms.

But if we accept that scientists are trying to look for the hidden mechanisms that produce phenomena, we escape this problem, and the demand for universal reduction. The task of science becomes, as Harré argues, the imaginative contruction of theoretical models which suggest ways in which particular sets of phenomena can be *produced.* Of course, each model will only explain the mysterious complexities of life-stuff, for example the way in which, in some creatures, when they are wounded, cells intended for one thing can be 'pressed into service' to become a different kind of cell, to repair the wound. How is this 'coded' for? In embryology, and in the study of stages of growth, there are many such mysteries which certainly could never be the product of the laws of physics and chemistry operating through the DNA molecules, since the autonomy in them, and especially the variability and flexibility of time-dynamics in them, defy any explanation of cause and effect in such a one-level mode. The concept of a 'blueprint' for example is inadequate : who is 'using' the plan? There is a second level, of hierarchical principles here. But drop the ideal of one unified system, and the demand that all sciences be reduced to physics and the scientist can investigate such phenomena with an open mind by a variety of models. Moreover, from the anti-reductionist point of view it becomes unnecessary to combat reductionism : it becomes a side-issue.

If it is a side-issue in science, then it becomes exposed in the Humanities as a nonsense. To say that man is 'nothing but a complex biological mechanism powered by a combustion system which energises computers with prodigious storage facilities for retaining encoded information',[1] is an absurd extrapolation from a number of legitimate applications of theoretical models. Yes, the stomach is a kind of combustion system : the brain is a kind of computer. But the fault is in the phrase 'nothing but' which implies that every complexity of the organism, of the *man*, can be reduced to mechanics, with (in the background) the implication that everything can be reduced to chemistry and physics or at least to mechanism. By Rom Harré's approach the analogies may be valid, but the gesture at overall reducibility may simply be no explanation at all.

In molecular biology micro-reduction has been a powerful tool, it is true. It has provided a valuable model which offered a new way of thinking : it must be accepted as 'real'. Because of this, some scientists even argue that all advances in science are advanced to, and in, particulate thinking : all moves away from such thinking is retrogressive. (*The Edge of Objectivity*, C.C. Gillespie, Princeton, 1960). It seems '*progressive*' to talk of man being 'nothing but DNA', and to opt for the 'selfish gene'.

But it is a serious error, Marjorie Grene suggests, to believe that the only genuine science is this kind of science which reduces everything to the smallest particles. This is no longer assumed, for example, in physics (*Experience and Conceptional Activity, J.M. Burgess*, MIT, 1965 ; and *Philosophical Impact of Contemporary Physics*, M. Capek, Van Nostrand, 1961). In biology, Marjorie Grene argues, it is an absurd oversimplification, to postulate microreduction, as a way of explaining communities in terms of the aggregation of cells. The interaction between organisms and their environment, whether we take bees, clovers, cats, mice or spinsters, if studied at its own level, makes it clear that here the realities can neither be predicted not adequately understood by a study of the elements which make them up. Evolutionary theory itself depends upon thinking about *populations, not* individuals and so is not macro-reductive or micro-reductive and is even anti-reductive. While the study of populations (as of 'gene pools') looks particulate, because it works in terms of collections of bits, yet because there is so much thinking about laws of the collection, it isn't really the kind of thinking that depends upon reducing everything to its elements.

---

[1] Quoted by Viktor Frankl in *Against Reductionism* from a book he came across.

If we turn to DNA, we are not saying there is a special 'organic matter'. To operate, however, DNA molecules must be programmed : and a programmed system is hierarchical. It can be made, predicted or understood only in terms of an engineering principle, as well as in the terms of physics and chemistry. It is a system (as Polanyi has argued) with *dual control* (see also Richard Gregory on 'The Brain as an Engineering Problem' in *Current Problems in Animal Behaviour*, ed. Oliver Zangwill and W.M. Thorpe, CUP, 1961 and in *Towards a Theoretical Biology*, ed. C.H. Waddington, Languet Higgins, on 'What Biology is About').

We are at liberty, in Rom Harré's view, to draw our models from wherever we wish. They will help our understanding if it seems that they 'produced the effects we see'. If we look at machines, and try to find one capable of improving its own programme, we may illuminate the fantastic capacities of living creatures. We can also look at atomic physics and plant-breeding, to illuminate genetic selection. But there is no need to draw models all from one source, no need to attempt to think everything through in terms of 'systematic physico-chemical fraction of organisms, and the intensive study of the resulting fraction by means of the standard techniques of physics and chemistry' (Robert Rosen, *Journal of Theoretical Biology*, 1968). Such techniques of fragmentation have limits as well as strengths : for example, sometimes minute attention to the fragments provide no illuminations of the overall system. The models invoked don't help us understand. Microreductive models must be supplemented by models drawn from other sources which may shed new light : perhaps models which can illustrate the whole organism in Kurt Goldstein's sense.

If we look at scientific enquiry in this new way, as Rom Harré indicates, as an effort to solve particular problems by seeking from whatever source hypothetical mechanisms that might explain them, there is no need for the reductionist ambition, no need to seek the one source, of fragmentation - a source which too often makes for over-abstractness and for artificial analogues.

While fragmenting analysis has achieved much, it has also created over-confidence (as evident in Crick). Many aspects of biological activity need holistic and relational approaches (such as Kurt Goldstein argues for). Systematic analysis is as valuable as reductivist analysis and it is a serious distortion of biological science to prefer the fragmentation approach to the *exclusion* of the relational or holistic, which is what has happened. Dr. Leakey's approach in *Origins* to 'co-operation man' is a holistic mode of approach to our origins. Ten million years of evolution suggest a hierarchical dimension to be taken into account as well as the DNA : that is, something was striving to come into being. Maybe it was the opposable thumb, or maybe our kind of brain. Its emergence is not explained by evolutionary theory or Natural Selection which attribute development to minute differences occurring by change mutation. There were clearly major leaps forward which are inexplicable by 'chance and necessity' but require quite new models to explain.

Marjorie Grene ends her essay by referring to subject directly relevant to poetry : perception in relation to reductivism. Under the influence of Descartes' insistence on 'clear and distinct ideas' philosophers have tried to create a theory of perception based on the aggregation of minima : the assumption was that only by an aggregation of sensations do we perceive. This theory dominated psychology until the Gestalt school, and still persists in many quarters.

But in a recent book J.J. Gibson declares that such elementary sensations don't exist or hardly exist. The senses are detection systems by which organisms significantly act to obtain information about their environment. This more active view of perception is, of course, also put forward by Erwin Straus : we do not passively receive - we *sense*. The scientific question should be how we use our senses to find out about the world. The old reductionist approach used a building brick model, derived from atomistic physics, with a Humean philosophy in the background. We passively receive sense data : physical optics have to do with the mechanical interpretation of the image on the back of the retina, and all this seemed compelling by its analogies with technology. But where is the experiencing 'I'? It is excluded from the picture, and so too is the capacity *to sense, to experience*, in the positive way which Erwin Straus emphasises : so is our creative relationship with the world - what F.R. Leavis called 'the living principle'.

Other approaches imply other models and existential choices. Not only the eyes but their muscles, and the relationship of eyes to other organs, the head and whole body provide information that can be specified by the variables of optical structure (information about objects, animals, motions, events and places). That is, we *see*, and (as Polanyi and Merleau-Ponty suggest) the *whole being* is involved. The models used by Gibson derive from evolutionary theory, ecology, ('ecological optics') illumination engineering and 'perspective geometry'. What this all points to is *new forms of thinking* quite different from traditional reductionist accounts of perception. (The thinking of Straus is again clearly relevant here : he said 'Man thinks - not the brain'). We must recognise ourselves as beings looking (as in biology) at whole beings, with a subjective life in each.

These new approaches are revolutionary and require suspicion of 'sensations' as universally available (as Straus is suspicious of the very concept of 'stimuli'). The new view requires acceptance of immediate contact with the external world and, incidentally, the investigation of perception by these models comes closer to the way in which the poet explores the nature of what he sees, of the world he sees - as a being-in-the-world. Both poet and scientist are seen as human beings trying to make sense of their world : as men living, in a world of life : the *Lebenswelt* is the true focus of attention, as Tomlin says

> The Philosophy of Organism, of which Whitehead laid the foundations, has been called the last metaphysic. There is a sense in which this description is just. There is another sense in which it may prove misleading ; for in constituting the last metaphysic of the old order, it maybe from another angle the first metaphysic of the new. Nor is the philosophy of organism to be considered a special branch of enquiry. It is the metaphysics of the *Lebenswelt* (Husserl's term; term), the living world.

The conception of a *Lebenswelt* provides a way of liberating us from the bands of scientific materialism, and develops naturally from biology which, as I have said, will save us. - its proper concern being with organisms, individuals - as Tomlin puts it, *beings*.[2]

---

[2] Tomlin's article referred to here is again *Books and Issues*, Vol 1. No. 2, 1979 p.4.

14 The mind on its blind origins: Jacques Monod

Next, let me bring the work of those who have examined and rejected Darwinism to bear on a significant work in which we find extrapolations from mechanistic materialism, into the spheres of ethics, metaphysics, and the philosophy of being - Jacques Monod's *Chance and Necessity*. Can mind be the product of 'blind chance'?

In fact, such books belong to a tradition in which scientists are calling for a new 'realism' in the realm of morality and philosophy, on the basis of mechanistic materialism. T.H. Huxley, in his Romanes lecture, concluded : 'Let us understand, once and for all, that the ethical progress of society depends, not on imitating the cosmic process, still less running away from it, but in combatting it'. P.B. Medawar, in his Reith lectures on 'The Future of Man' in 1959, said : 'It is a profound truth - realised in the nineteenth century by only a handful of astute biologists and by philosophers hardly at all, (indeed most of those who held my views on the matter held a contrary opinion) - a profound truth that nature does *not* know best; that genetical evolution, if we choose to look at it liverishly instead of with fatuous good humour, is a study of waste, makeshift compromise, and blunder'. (*The Future of Man*, 1960). Books like *The Naked Ape* and *The Territorial Imperative* belong to the same tradition : with these it has to be added that some scientists accuse the authors of those works of 'serious (and I mean *really* serious) misunderstanding, and interpretation of the facts of biological evolution'.[1]

What must certainly strike the individual in the Arts and Humanities about this tradition is its pessimism. Here, indeed, we encounter that cold, bleak and dark view of existence which has come over from science in the last hundred years, and which daunted Tolstoy, Conrad, Hardy and Yeats. We find it, for example, in Bertrand Russell, who believed of man that 'his origin, his growth, his hopes and fears, his loves and his beliefs, are but the outcome of accidental collocations of atoms'.[2]

In 1935, Russell said :

> The same laws which produce growth also produce decay. Some day the sun will grow cold, and life on earth will cease. The whole epoch of animals and plants is only an interlude between ages that were too hot and ages that will be too cold. There is no law of cosmic progress, but only an oscillation upward and downward, with a slow trend on balance owing to the diffusion of energy. This at least is what science at present regards as most probable, and in our disillusioned generation it is easy to believe. From evolution, so far as our present knowledge shows, no ultimately optimistic philosophy can be validly inferred.

1 Professor Bernard Towers, in the paper : 'Ethics in Evolution' given at the Third Trans-disciplinary symposium on Philosophy and Medicine, Connecticut, 1975. See also : *Naked Ape – or Homo Sapiens*? by Bernard Towers and John Lewis.

2 Quoted by J.W.N. Sullivan renewing E.A. Burtt's *Metaphysical Foundations of Modern Science* in *The Calendar of Modern Letters*, Vol.I., 1925, p.

Religion and Science, 1935, p.81.

It is thus hardly surprising that today we even find artists believing that there is no kind of meaning, pattern, order or source of right and wrong in nature : Francis Bacon, the painter, for instance, has said, 'I think that man now realises that he is an accident, that he is a completely futile being, that he has to play out the game without reason'.[3] Evolutionary theory and especially evolutionary genetic theory are now at the heart of this pessimism, as we have seen. As Jacques Monod declared, the theme is that 'Man has to understand he is a mere accident...' Bertrand Russell declared :

> The most fundamental of my intellectual beliefs is that the idea that the world is a unity is rubbish. I think the universe is all spots and lumps, without any unity and without continuity, without coherence or orderliness, or any of the other properties that governesses love.
>
> Bertrand Russell in *The Philosophy of Bertrand Russell*, ed. Paul A. Schlipp.

However, again from the Humanities point of view, there is a phenomenological aspect to all this. As F.R. Leavis pointed out, Bertrand Russell's attitude to man in the universe comes within the scope of our consideration of tragedy : and if we examine some of his prose around this theme, it is emotionally inept, full of self-pity and self-dramatisation. Whatever the scientific truths upon which these essays into metaphysics are based, the scientists writing them find themselves on the same ground as the authors of *The Book of Job* or *King Lear* : and from our point of view, when certain scientists and philosophers move into this ground, their engagements with problems which belong to the philosophy of being are not very good. Examined in the light of our subjective disciplines they are often immature and emotionally barren, without any metaphysical strength.

From our point of view, there is another strange enigma about pessimism : although it pronounces about life in a bleak, nihilistic way, it is extremely popular. I have referred to the strange paradox of the success of such books as *The Naked Ape*, which has sold 8,000,000 copies in many languages. Why is such a reductionist approach such a commercial success? Why does it have such a subjective appeal? We may note that Russell admits the subjective element, by referring to 'our disillusioned generation'.

Professor Monod's *Chance and Necessity* was no exception. His jacket showed figures of men, each exactly the same, reduced to a homunculus, marching around aimlessly, from infinity to infinity of darkness - black, primitive, anti-homunculi. The march of man - this illustration cenveys by its use of symbols - is meaningless, and leads into nothingness. Our life (the drawing implies) is a mechanical, aimless procession of organisms - winding into the darkness of nothingness, under the shadow of universal entropy. The message of the jacket is one of spiritual aboulia and existential nothingness.

Of course I should hold his publisher to blame for this, rather than the author. This may not be Professor Monod's message : but it is the message of the book wrapper, which he must have approved, and it was no doubt taken to be the message of his book by those who are impressed by his scientific achievements. Thus, despite his essential humanity, Monod's book must have contributed to the widespread feeling at large that the unlocking of the DNA code and the rest - the most recent findings of science, with all the mystical authority of science behind them - inevitably endorse the prevalent fashionable mood of doom and nihilistic dehumanisation in our culture.

Monod's distinction and his motives were, of course, unquestionable. He was a Nobel Prize winner, and Director of the Pasteur Institute in Paris. He was one of the world's most able molecular biologists. No one could question his authority in his own field, or his complete integrity in developing his views. He spoke on the Radio once of the 'urgent need' for a 'spiritual change' in man, towards a better sense of meaning in life, if man were to solve his ecological problems.

However, one cannot help feeling that the effect of his pessimistic philosophy, in such a pessimistic book-jacket, must have been to contribute to the deepening of the current spiritual crisis itself.

---

[3] *Interviews with Francis Bacon* by David Sylvester, 1975, quoted by Bernard Towers, in his paper : *Ethics in Evolution.*

How can we defend ourselves against a brilliant man like Jacques Monod?

The answer, again, is by philosophy. Monod himself said of his colleagues, 'a number of very able scientists... have not applied in other fields the same competence that they would exert in their own' (Radio broadcast). This, as John Lewis pointed out (in *Beyond Chance and Necessity*, Preface), Monod did not sufficiently apply to himself. When he stepped into the realms of metaphysics, he left the realm of his own competence as a scientist and must come in for criticism.

First, he must be criticised for his Cartesian reductionism :

> anything can be reduced to simple, obvious, mechanical interactions. The cell is a machine. The animal is a machine. Man is a machine.

Secondly, Monod's emphasis on 'chance' must be examined. In his inaugural lecture, 1967, Monod declared that the rise of modern biology was associated with the working out of the theory of evolution in physical terms. He attacked vitalism, and quoted Democritus : 'Everything in the universe is the creation of chance and necessity'.

Thirdly, Monod's view culminates in the 'ethic of knowledge' :

> The ethic of knowledge is radically different from religious or utilitarian systems, which find in knowledge not the end in itself but merely a means to an end. The only end, the supreme value, the 'sovereign good', in the ethic of knowledge is not, let us admit, the happiness of mankind still less its temporal power or comfort, nor even the Socratic 'know thyself' - it is the objective knowledge itself.
>
> Quoted by Horace Freeland Judson in *The Eighth Day of Creation*, p.592.

As Polanyi has argued, 'objective knowledge' is a false ideal : knowledge is never anything more than men trying to make sense of their experience, and this must always be contingent : knowledge is something that men do, and since it is directed towards their question of the meaning in life, it is always value-bound, and inseperable from their pursuit of inner truth. If Monod had said 'truth' we might have agreed with him, if that truth was allowed to include the endless pursuit of the Dasein : for 'objective truth' has no value simply in itself as a goal : it is part of that ideal towards which men set out in the original *telos* of Greek civilisation, which includes both subjective and objective truth. But there is no body of objective truth 'out there', once and for all, to which we must defer.

The goal is total truth : and (as Edmund Husserl pointed out) it was the Cartesian division of *res mensa* from *res extensa* which allowed the pursuit of 'objective truth' to strip the world of meaning, because the pursuit of inner truth was neglected, in favour of the objective. Monod's kind of approach and his idea of knowledge, contributes to that stripping – since, in a sense, his goal is that kind of science whose lowest form is mathematical measurement : making a topography of 'where the atoms are'. And, as both Husserl and Polanyi have pointed out, this answers none of the real questions at all.

And then, finally, we have to say of Professor Monod that the way he often writes reveals that he implicitly acknowledges truths which he attempts to deny by his main mechanistic emphasis, and thereby demonstrates the inadequacy of his concept of what knowledge is, since these truths concern aspects of living beings and processes in life which are not susceptible of 'objective' empirical approaches.

Elsewhere Professor Monod speaks of 'roulette' and 'a vast lottery in which natural selection has blindly picked the rare winners from among numbers drawn at utter random'. 'This conception alone is compatible with the facts', - though 'it does strike us as any less miraculous' (p.131). This way of talking is like that used by evolutionists when they speak of 'apparent purpose' : admitting a teleological dimension while denying it at the same time. So, Monod sees the origin and development of life as a *miraculous* game played by 'Nature' in which the multiplicities and mysteries happen, and even their directiveness is acknowledged, but all positive yearning, fulfilment or achievement is vigorously denied. But then, how can he use language such as 'From bacteria to man?' or speak of the '*desire*' of the cell to divide (p.122)? - the '*dream*' of the ancestor inherent in some organisms? The '*perfecting*' of species? The '*efficiency*' of the chemical machinery of living beings? '*The secret of life* (if there is one)'? In one poetic passage Professor Monod speaks of :

> the deep significance of the mysterious message made up by the sequence of residues in a polypeptide fibre. A message which, by every possible criterion, seems to have been composed completely haphazardly; a message nevertheless charged

> with meaning which comes out in the discriminative, functional, directly teleonomic interactions of the globular structure. Globular protein is already at the molecular level, a veritable machine – a machine in its functional properties, but not, as we now see, in its fundamental structure, where only a play of blind combinations can be discerned : random chance caught on the wing, reserved, reproduced by the machinery of invariance and thus converted into order, rule, necessity. A *totally* blind process can by definition lead to anything; it can even lead to vision itself.

But if 'the origin and lineage of the whole biosphere are reflected in the ontogenesis of a functional protein', what is the word 'blind' doing, repeated so often in Professor Monod's prose? The profounder meaning 'of the message which comes to us from the most distant reaches of time' is that the 'ultimate source of the *project* that living beings represent, *pursue*, and *accomplish*' is revealed in a 'neat, exact but essentially indecipherable text formed by primary structure'. :

> indecipherable, since before expressing the psysiologically necessary function which it performs spontaneously, it discloses nothing in its structure other than the pure chance of its origin.
>
> p.97.

We may grant that everything we are looking for is 'matter', and we need not read into the things we observe any ghosts or God. But every other word that Professor Monod uses is a word that indicates that he is finding aspects in what he observes that point to a quite different interpretation from the one on which he dogmatically insists : 'significance', 'message', 'charged with meaning', 'caught on the wing', 'order', 'rule', and 'vision'. These imply meaning and directiveness.

Somehow, this whole process of life was set in motion. Professor Monod does not tackle this problem. Somehow the possibilities of chance being (as it were) 'employed' to interact with 'necessity' arose. The verbs Monod uses all demand subjects; they are active verbs, full of 'intentionality', and they point towards Merleau-Ponty's sense of consciousness in creative time. That is, what Monod exhibits is mind recognising the directive processes of which mind is a culmination. Here we are, conscious beings, inquiring into the meanings of these processes and marvelling with Professor Monod at them. Machines don't and couldn't do that, and neither is this a manifestation of 'objectivity'.

Yet he declares, we are to be overcome at the definite and unquestionable mechanistic blindness of these processes, - that we must expunge ourselves, and our sense of meaning, from the scene :

> There is no scientific position, in any of the sciences, more destructive of anthropocentrism than this one, and no other more unacceptable to the intensely telenomic creatures that we are. So for every vitalist or animist idealogy, this is the concept or rather the spectre to be exorcised at all costs.
>
> p.110.

I am, I believe, no vitalist or animist. But everything Professor Monod says, in his interpretation of the more recent findings of molecular biology makes it clear that there is no more complex, conscious, marvellous or meaningful manifestation of the biosphere than man - and his consciousness, which is no 'spectre' but gives all the rest meaning. And while life and man *can* be seen as unique occurrences, the fact is that he *exists*, and *sees his world*, and makes it thereby *meaningful*. Moreover, as Straus insists, he must be seen in a continous stream of life with the animals, which have a kind of consciousness.

It can only be the most arrogant perversity that impels a man to suppose that to find that mere chance and necessity 'produced' him, means that his 'anthropocentrism' must be abandoned, or that this disposes of all philosophical and religious interpretations of the universe, except reductionist objectivity, which must see only (dead) 'machines'. The universe which Professor Monod confronts with his consciousness is not one which has been content to produce machines, because it also produced Professor Monod. Machines do not have his vision, or his kind of consciousness, and could never have initiated science, conducted experiments, devised philosophies, or either conceived of anthropocentricism or rejected it. Man is anthropocentrism embodied - because he sees the universe and investigates its meaning as no form of life does (and is matter investigating its own nature through science). In seeing it, he makes it and gives it the highest meaning it can have.

At one point it seems Professor Monod decided to restrain himself from metaphysics.

> For science the only *a priori* is the postulate of objectivity which spares it from taking part in the debate.

Professor Monod, however, would not be spared. He sought in his book to relate the findings of molecular biology to thought as a whole. Life, he believed, had been shown to be the result of chance mutations which could not have been predicted. Monod bases on this assumption that there can be no general purpose in the existence of life : it need never have existed. Man need never have existed and is the product of an accident. So, no system of religion or ethics which presupposes any reason or purpose in life can be viable.[4]

So, the only basis for an ethical system must be the pursuit of 'objective truth' : that is, a non-emotive approach which is not prejudiced in favour of human affairs.

Up to now, all systems of thought have been vitalistic or animistic. Both kinds of system are incompatible with true objectivity, though both may accommodate the existence of necessity in nature, as in evolutionary theory. Vitalism makes an absolute distinction between animate and inanimate, and holds that there is a purpose or direction in living things alone. Animism holds that a purpose is being worked out throughout the universe, in evolution, in history or in some other way. Monod saw animism both in Marx's belief in historical processes, and he sees Teilhard de Chardin's law of increasing complexity and consciousness as a form of animism, too.

For Monod, the whole idea of purpose must be expurgated, both from the whole universe and from the human.

The naturalistic fallacy is to suppose that values may be based on such a concept of purpose in nature. But now there is no excuse for perpetuating this fallacy, since objective science has shown there is no purpose there.

There is of course the problem of the human spirit, but from that, according to Monod, we cannot derive any values whatever. And here Monod takes an 'objective' developmental view of man. He has simply developed by social and cultural evolution to be what he is. The factors which make for individual success are not necessarily those which make for human survival. And the most contemporary problem is man's sickness of spirit.

Various ideas have appeared in human history, and those which have survived are those which deserved to have survived (this is both an application of evolutionary theory to culture, and tautologous). Now the idea most likely to lead to survival is objective science. While objective science has success and authority in the field of its application, as in technology, it has not yet been accepted with respect to its ethical and religious consequences. Men still insist on pretending to a source of values that doesn't exist. Man's sickness lies in his authenticity. Many blame the products of science, like the hydrogen bomb, or chemicals in agriculture : but these are but disguised antipathies to science itself.

It is time for man to wake up out of his dreams, and to accept the atheistic implications of science. Then he will recognise that there is a huge gulf between facts and values. To confuse knowledge with values is to indulge in intellectual inauthenticity.

We must develop an ethics of knowledge. Ethics must be founded on the scientific attitude and on nothing else. Other ethical values must be seen to be derived from the basic value which is understanding. 'We must choose knowledge', writes Mary Warnock, at the end of her summary. 'Choosing this, the highest value, is choosing the kingdom of ideas, and rejecting the darkness of ignorance, deception and superstition'.

We have seen a similar impulse in E.O. Wilson and in Richard Dawkins : 'science' must now take over ethics, philosophy - and all the other Humanities. but, as Dr. Warnock points out, the impulse to base ethics on the hard facts of molecular biology, is itself a monstrous example of the naturalistic fallacy. Oughts are to be based on *what is*, despite the fact that ethical principles cannot be based on anything of the kind. Moreover, as she also points out, his theory offers satisfaction to those who look always for one solution, an answer to everything.

The place to begin to resist Monod is within science itself. The biologist must not only study minimals, as we have seen : he must study the behaviour of whole living organisms : the whole nervous system, rather than its component bits : animal signs and behaviour, as

---

[4] I am here paraphrasing Dr. Mary Warnock's useful chapter in *Beyond Chance and Necessity.*

manifestations of the life of a whole being. Monod demonstrates himself in his book to be essentially reductionist in this matter. Beginning from the confused statement 'Nature is objective', he declares repeatedly that :

> 'the cell *is* indeed a machine...'
>
> p.108.
>
> 'living beings *are* chemical machines...
>
> p.51. (my italics)

Such statements would be acceptable if they were framed in some such terms as this :

> For the purposes of molecular biology, I intend, for the time being (and while allowing always for possible evidence which might prove otherwise) to look at living creatures in the world and their structure, as if these were machines, as if their workings were like the workings of an alarm clock, computer or robot. I will ignore for the purposes of my research the question of what started this machine going, or how it came to be created or how incredible it is (compared with man-made machines) that these machines repeat and renew themselves, reproduce and project themselves (which does not account for why they do) : that they are living beings. And I will for the time being ignore the possibility that there are other aspects of the matter and dynamics of these organisms of which I am at present ignorant, and other dimensions by which these might be seen. I will ignore their 'centricity', that is their subjective life, their autonomy, and their primary consciousness. I will ignore all those characteristics which are not found by the space-time concepts of physics or explained by the laws of chemistry. But I make these reductions for my special purposes. I realise there are other perspectives on the truths of the world I am trying to understand. I recognise I am only looking at *one* perspective or aspect of living creatures, and that there is always a need to go 'back to the things themselves', and look at them holistically. I hope I can still be rational as I do this.

Monod, however, did not have such humility, despite his evident eminence and courage as a scientist and citizen. He asserts that 'the cell is *nothing but* a machine' which is patently not true since no machine can divide itself in two to reproduce : he says : 'living beings are chemical machines', implying that they can be explained by chemical laws, when they cannot be. His 'is' and 'are' have the force of an equals sign, and he does not see how naively dogmatic he is being. As a scientist he should see that ' = ' here means 'for the purposes of the physical sciences may be seen in part as'. But he does not, and the implications of his failure to be meticulous here are seriously falsifying, even of the nature of knowledge.

So Professor Monod's 'machine' way of thinking indeed skates without being plagued by doubt over some of the most complex philosophical problems.

As so often, a phrase will slip into the most mechanistic piece of writing, to admit by implication dimensions which on the basis of the account could not be there :

> A structure which already possesses a property of invariance - and hence is capable of preserving the effects of chance....

A structure which is 'capable of preserving' reveals a dynamic which is known to us in life : but in what could it reside on the mechanistic account? Monod manages to have it both ways . For instance he says :

> Objectivity nevertheless obliges us to recognise the telenomic character of living organism, to admit that in their structure and performance they decide on and pursue a purpose.
>
> p.31.

An organism which decides on and pursues a 'purpose' cannot satisfactorily be described as merely a machine or a chemical machine.

And this brings us to the fundamental question of 'chance'. This central dogma in Monod is itself questionable. Perhaps the most interesting chapter in *Beyond Chance and Necessity* is that by Professor C.H. Waddington : 'How much is evolution affected by Chance and Necessity'? Waddington discusses some of the more recent theories of evolutionary genetics, and how the 'chance variations' of Darwinian theory became translated into the random mutation of genes.

Gene mutation, however, he says, is 'certainly not random in any complete sense'. The gene consists of a string of nucleotides, of which there are only four kinds. '*Only certain types of change are possible*' :

> But although we know a great deal about the kind of changes that may occur, and although the study of the causal mechanisms involved is a very active research field, this does not completely remove the grounds for referring to mutation as 'random in the context of evolution'. What is meant by that expression is that the alterations produced in a gene, and the effects which this alteration will have on the phenotype of the individual which develops under its influence, are not causally connected with the natural selective forces which will determine its success or failure in producing offspring in the next generation.
>
> *Beyond Chance and Necessity*, p.98.

Those environmental influences which affect natural selection have nothing to do with induction of mutations appropriate to meet the pressures from environment.

Waddington then traces genetic theory, from those who believed that populations of organisms must be studied in relation to single indentifiable genes, to those after Dobzhansky, who turned attention to populations of genes. Dobzhansky demonstrated that any individual in a population differs from every other in many genes. There is no uniform 'wild type'. We have to think of the population as containing a pool of genes, from which each individual draws its particular complement, and returns them to the pool again when mating with another individual and producing offspring (Dawkins' book is based on 'gene pool' theory). These are successive developments in neo-Darwinism.

Waddington himself put forward an epigenetic theory, which brought him the appellation of being a post-neo-Darwinian. The point here is that natural selection does not act on geneotypes, but on phenotypes which are produced by epigenetic processes in which the environment as well as the genotype plays a part.

This is based on what is *known* : and what is known is that when 'unusual environmental circumstances succeed in modifying the phenotype of an organism', the resulting modification is, '*far more frequently than would be expected on a random basis, of such a kind as to improve natural selection fitness*." (p.96).

> biological organisms have a strong tendency to *adapt* epigenetically and phenotypically to the environments they meet. Even bacteria, as Monod knows much more thoroughly than I do, can often react epigenetically to the presence of an unusual substrate in their environment by producing a phenotype which is better able to survive and multiply in that particular medium.
>
> p.97.

So, while environmental changes do not produce appropriate changes in the genotype, they do very often tend to produce appropriate alterations of the phenotype :

> and it is on the phenotype, not on the genotype, that natural selection operates.

Then Waddington speaks in a way that seems more appropriate to continental biology (the kind Marjorie Grene discusses in *Approaches*) :

> there is always the possibility that at least some members of the population may decide to opt out and try something else. Some mammals, at some point in evolutionary history, must have said : 'to hell with all this dry land stuff, let's take to the sea', and became whales, dolphins and so on...
>
> p.97.

Waddington offers this as contrary to the cherished neo-Darwinian myths : 'but, if we accept it, it has profound implications about life and the general attitude to life we draw from it, for it recognises that, on good biological grounds, we are *not* at the mercy of chance' : that there is in life an impetus, an intelligence, a primary consciousness that 'tries'. *A machine does not try*.

> In the neo-Darwinian paradigms, we have genotypes or gene pools subject to mutational alterations 'at random', and to natural selection by external forces unrelated to the character of the things they have to select. In the epigenetic paradigm we have to contemplate a subtle system of feedbacks, in which the environmental factors which exert natural selection operate on phenotypes which they themselves have influenced in relevant ways (usually to improve their

efficiency), and in which the phenotypes in their turn have an influence on the nature of the natural selective pressures to which they will be subject.

pp.97-8.

In the theories of Dobzhansky, 'chance' is still important, because 'If the antelope needs to run faster, if the whale to swim better, success must very rarely be dependent on a precise amino-acid substitution in one particular protein'.

But the gene pool of a population can offer an enormous number of variables, and 'the 'random' nature of the processes .... comes to be unimportant when the building blocks are employed in statistically large aggregates'. As evolutionary theory moved into the orbit of Dobzhansky, *chance lost its dominant position.*

> *within the epigenetic paradigm it plays an even more subordinate role....* The basic biochemical variants which genuinely do depend on chance mutations, such as the position of particular amino acid residues at specific points in a primary protein chain, appear *as of very little consequence indeed...* Le Hazard has to be content with a very back seat role....[5]

p.99.

Having rejected Monod's emphasis on 'Chance', Waddington turns to 'Necessity', : here again, it becomes an inadequate term because 'the phenotypes resulting from previous natural selection have a strong influence on the character of the natural selection to which they themselves will be subject'.

Waddington even goes so far as to talk of acquired characteristics (Monod called him a Lysenkoist) : acquired not through the genes, but through epigenetic processes :

> 'Characters acquired' by individuals are not inherited by their individual offspring but characters acquired by populations are inherited by their offspring populations if they are adaptive'.

pp.101-2.

In these later neo-Darwinian paradigms, evolution is not concerned with individuals, but with populations. Waddington gives three things which are known about populations :

> One, organisms usually react to abnormal environments in ways which are adaptive, that is to say which increase their probability of leaving offspring. Two, there will be natural selection for those individuals which do adapt most successfully. Three, there will be some genetic component contributing to the appearance of the adapted phenotypes.

p.101.

It is perhaps worth pointing out that in Waddington here natural selection is *only* a negative force which selects out inefficient creatures, : the positive, formative force is in the creatures themselves which adapt and *try*. So, Waddington suggests that instead of the slogan 'chance and necessity' based on a theory whose foundation was accidental mutations, one might arrive at one more like '*learning and innovation*' : making a radical reversal of all biology and thinking about life.[6]

In a footnote, Waddington, expressing himself hostile to dogma, suggests other ways in which the 'coding' processes of DNA and RNA may be affected by other factors. Here, the complexities of molecular biology tend to be beyond most of us – but it is fascinating to see a distinguished mind like Waddington's at work resisting orthodoxy :

> It seems to be not inconceivable that the imposition of certain metabolic conditions on an organism might change the proportion of mutant forms of gene within the population to be passed on to the next generation...

What might be happening – indeed, what *is* happening, is that, as the secrets of life are explored, the questions became more complicated, not less. The idea of a simple mechanical principle to explain everything should recede.

---

5 In the volume edited by Lewis, Theodore Dobzhansky also finds natural selection an 'anti-chance agent', and says that the question of man's emergence is too subtle and complex for it to be said that man emerged by chance. p.135.

6 In his *Darwinian Impacts* D.R. Oldroyd points out that Lamarckian thinking persisted in Darwin himself : see his p.175.

That actually is what has happened – and Monod has contributed here. As Professor W.H. Thorpe points out in *Animal Nature and Human Nature*, Monod has suggested that the combination of processes which must have occured to produce life from inanimate matter is so extremely improbable that its occurence may have been a unique event (an event of zero probability). Monod also rightly points out that the uniqueness of the genetic code could be the result of natural selection. But the enigma is that the 'code' doesn't mean anything unless there is 'life' to interpet it – a problem the mechanist cannot explain :

> The extraordinary problem remains that if the genetic code is without any biological foundation unless and until it is translated, that is unless it leads to the synthesis of the proteins whose structure is laid down by the code.

Monod himself showed that the machinery by which the cell (or at least the non-primitive cell which is the only one we know) translates the codes 'consists of at least fifty macro-molecular components which are themselves coded in DNA'. Thus the code cannot be translated except by using certain products of its translation.

> This constitutes a really baffling circle : a vicious circle, it seems for any attempt to form a model or a theory, of the genesis of the genetic code.

The break-through in molecular theory, far from solving the problem of the origin of life, has made it a greater problem than ever.

To both Waddington's and Thorpe's investigations, of the nature of DNA coding we perhaps need to bring Polanyi's emphasis – that the way 'information' and form are conveyed through these molecules requires, if we are to understand it, some 'systems' thinking which does not simply reduce it to the laws of physics and chemistry (see above p52).

There is a problem of time which Monod evades. We know that time, as lived time, is 'telic in structure' (Marjorie Grene, *The Knower and the Known*, p.245). Only bring in such concepts as intentionality and lived time, and we are faced with the need for teleological thinking. She lists a number of teleological thinkers. (See above page 65 and *The Knower and the Known*, p.239). But this new teleology must be carefully distinguished from that which seems to demand a reference to a supernatural 'maker' of organic beings, or a new kind of Paley's watchmaker :

> It is an inner teleology, where each part of an organ possessed by the subject is a means to the end – the subject itself...
>
> *Approaches*, p.94.

This new approach seeks to rescue us from the twin intelligibilities of mechanism and the will of God.

The whole philosophical issue may be studied in Marjorie Grene's two books. But even if we do not understand the full argument, it is clear that there are ways out, even in science, from the conflict between rigorous materialism (as in Monod) and a belief that brings in 'God's purpose'. It is also clear that what the modern mind is faced with is the problem of the *nature of things*, which it is too often assumed that we have solved. We have not begun to solve it, and Monod's aggressive mechanism is no help here. Amoeba are different from molecules : but how did the evident capacity for *behaviour* in the former emerge from the latter, as a consequence of mere chance and necessity? Even if we attribute the change to 'structure and performance' we are still left with questions that demand some kind of teleological thinking.

Monod himself implicitly recognises that there are levels of being and development, 'gradients', though Darwinism forbids this. He used the phrase '*from* bacteria *to* man'. He does not say 'alongside the primitive simplicity of life-forms there is by accident the muddle of life-stuff we call man'. Implicit in his way of talking are levels of being, and a view which recognises man at the end of a long creative process : and so there is a recognition of *ends*. When dealing with the origin of life he admits that 'science cannot say anything about a unique occurence' and thus both recognises the momentousness of the event, and the fact that his kind of science cannot pronounce on it – it cannot thus be the complete and secure basis of the metaphysic implicit in his position.

It is true that Jacques Monod saw that there is an 'epistemological contradiction'. The 'prodigious development' of science over three centuries has been guided by the 'postulate of objectivity' which is 'consubstantial' with it :

It is impossible to escape it, even provisionally or in a limited area, without departing from the domain of science itself.

p.31.

Yet there are certain things, as we shall see, which cannot be found within his paradigm, such as animal dreams and the tendency towards ends : yet he must try to make his 'objective' paradigm into a metaphysic. To do this, he must not only equate science with 'objectivity', but try to show that there is nothing else other than what an objective science can find : so, 'The corner-stone of the scientific method is the postulate that nature is objective'. Indeed, he goes further and declares that 'nature is objective and not projective'. To say that 'Nature is objective' is to equate 'what is' with 'my way of seeing it', and leads to the confusions which Polanyi has examined. It is an assertion that there is (or could be) a body of objective fact (in science) that is never to be questioned, but is Really Real, once and for all, and *is* the universe. This is a Cartesian view of science, in which there is a long chain of acts of a pure and attentive mind – all equal in purity, clarity and certainty. There is a body of unquestionable fact which is to be equated with the universe, conceived of as matter in motion. The problem is that in such a universe there is no-one to see it : where does Monod himself come in?

The claim is also question-begging, since implicit in the statement 'nature is objective' is the assumption 'nature is only matter'. Such a categorical declaration is absurd as soon as uttered, since the sentence itself could only be conceived and spoken by a consciousness which must be understood in quite another dimension from that implied. Monod's statement is an attempt to keep nature dead – or, certainly, to hold off the recognition of the 'category of life' because that is philosophically awkward. It is awkward because as a Cartesian *res extensa* must be rigorously kept apart from *res mensa.* But there is knowing mind, and where does it connect?

So from this idolization of 'objective knowledge' in his concept of the ethic of knowledge, Jacque Monod turns on us, and declares, as over DNA theory :

> with (the defeat of Lamarckism) and the understanding of the random physical basis of mutation that molecular biology has provided, the mechanism of Darwinism is at last securely founded. *And man has to understand that he is a mere accident.*
>
> Quoted in *The Eighth Day of Creation*, p.217.

Here Monod has moved on from his elevation of objective knowledge to the status of a value higher than man to a metaphysical dogma. He is saying that the latest discoveries of our objective knowledge of the structure of life (and in principle, he says, we 'already know the secret of life') tell us that man is a 'mere' accident. An objective body of truth seems to impose upon us a necessity to accept a certain metaphysical view on 'the basis of unquestionable 'fact'. This is absurd.

It is worth pondering the force of Monod's word 'mere' : it has a certain energy which thrusts that aggressive metaphysic upon us. It seems, from Monod's way of putting things, that 'science' establishes a body of unquestionable truth which is superior to our welfare or comfort, and that this objective truth now tells us that we are a chance by-product of haphazard material events. We have only a fortuitous place in the natural world and as far as our relationships with other creatures, we are only one accident among many.

But maybe we may glimpse something ambivalent in the heart of all this. Monod himself for example, in *Chance and Necessity*, not only speaks of 'nature's roulette', but 'nature's purpose'. As Polanyi might say, the scientist is not interested in random scatters, chaos or 'noise' : he is interested in order, and in speaking like this Monod admits as much.

Despite the emphasis on 'chance' thus, it is evident that as with their excitement about the elegance and beauty of DNA, combining autocatalytic and heterocatalytic powers or its capacity to carry 'information' : that is, they find in their objects of study an intrinsic interest.

Monod however must implicitly deny any such intrinsic interest : he reports that modern biological research has made it possible to analyse the different types of discrete accidental alternations of DNA sequence may suffer.

Various mutations have been identified as due to :

1. The substitution of a single pair of nucleotides for another pair;
2. The deletion or addition of one or
3. Various kinds of 'scrambling' of the genetic text by inversion, duplication, displacement

or fusion or more or less extended segments.

His appendices support these scientific facts and they are no doubt unquestionable deductions from the laboratory. But he goes on :

> We say that these events are accidental, due to chance. And since they constitute the *only* possible source of modifications in the genetic text, itself the *sole* repository of the organisms herediary structures, it necessarily follows that chance *alone* is at the source of every innovation, of all creation in the biosphere. Pure chance, absolutely free but blind, at the very root of the stupendous edifice of evolution : this central concept of modern biology is no longer one incompatible with observed and tested *fact.* And nothing warrants the supposition (or the hope) that conceptions about this should, or ever could, be revised.
>
> *Chance and Necessity*, p.110 (my italics)

By now this paragraph should appear not only dogmatic but bigoted. It will not do, in order to avoid the teleological issues, to opt only for what Marjorie Grene calls a 'negative and external agency', that is, to imply that we and nature are the product of impersonal, indifferent, dead forces, whose 'blindness' removes from the whole universe any element of meaning, development of 'striving' towards 'a vast variety of forms, energies and events, living and non-living, sentient and insentient, or processes alive, active and striving...' (Marjorie Grene, p.14).

Professor Monod offers to supply an 'explanation' on the neo-Darwinian account. But all he does provide is the molecular conditions, and (as Polanyi has pointed out) these do not solve the problem. The theory remains (as Waddington has said - quoted by Marjorie Grene) - 'inadequate'.

To apply the concepts and methodologies of physics, strictly, and without there being a need for revision, to animals and especially to man, can lead to a situation in which complex realities in 'the category of life' cannot be found. If the whole complex of life is not found, it is obviously absurd and dangerous to draw wide conclusions, especially when these impinge on moral values. It is even more absurd and dangerous to extrapolate such science into a general philosophy and draw from it conclusions about the ultimate nature, origins and teleological aspects of existence. There is a sense in which Professor Monod begins to see such problems when he turns to symbolism and language. But even with these, his belief seems to be that the processes of thought and speech, too, are eventually to be explained again in terms of minimals and the laws of physical entities. But this is absurd, since when it comes to signs and cultural meanings we are in a different dimension. To try to account for the phenomena of consciousness by 'objectivity' in terms of mere functional processes is to pursue an absurdity. There is no possibility whatever of explaining the processes of consciousness in terms of the 'workings' of 'parts' of the brain, synapses and loops, electrical currents, molecules and their interaction. Not only do we not know enough about their connections between levels : molecule, cell, nerve, lobe : the operation of the whole could never be understood by being broken down into 'bits'. It would be utterly impossible to make connections between operations and meanings, and in any case the ambition confuses dimensions.

Take, for example, the nervous system. It operates as a whole and has its own laws. Reduce it, and the principles of the whole tend to disappear. As Erwin Straus declares : 'Man thinks, not the brain'; and once again we must insist that the operations of this sphere of life (mind) be explored in terms of its whole higher dynamics. Such realities can be studied *only* in terms of the life of whole persons - as Goldstein argued from work with brain-damaged patients, and as Erwin Straus argues - by demolishing the philosophical assumptions of behaviourism. We need to study the *Lebenswelt*, by phenomenological disciplines, the life-world of which science is a part.

Knowing and consciousness are themselves to be seen as products of certain dynamics in matter itself. Many scientists find an urge towards potentialities as yet unrealised a fundamental property of matter itself. David Layzer, for example, argues that matter, when seen in 'duration', inevitably tends to arrange itself in groupings and hierarchies of increasing order, of increasing complexity, of increasing information. In an article in the *Scientific*

American[7] he writes; 'The present moment always contains an element of genuine novelty and future is never wholly predictable. Because biological processes also generate information and because consciousness enables us to experience those processes directly, the intuitive perception of the world as unfolding in time captures one of the most deep-seated properties of the universe'. Other philosophers have drawn attention to the problem of the dimension of time, as it is missing from much mechanistic thinking as have Professor Marjorie Grene and Maurice Merleau-Ponty, the French phenomenological thinker. (see *The Phenomenology of Perception* 1962, pp.410 ff, also Marjorie Grene on Merleau-Ponty, *The Knower and the Known*, chapter 9). Majorie Grene, as we have seen, quotes Merleau-Ponty's phrase, 'we are the upsurge of time'. so, there are biological realities which cannot be found by the strict conceptual and methodological discipline of 'objectivity' in which Monod was trained. It is quite unscientific for Monod to declare that his methodology is the *only* possible one :

> for science the only *a priori* is the postulate of objectivity.
>
> p.98.

If the strictly 'objective' approach is inadequate in the biology of the higher categories (or 'more complex categories', if we defer to Monod's kind of approach), it is obviously totally inadequate as the basis for a theory which may be elevated into a general philosophy.

In an article on *Why the Brain is not a Mechanism* (*The Times*, 25th January, 1969) Professor W.H. Thorpe said :

> There is a widespread tendency among ordinary people, and indeed among some scientists, to assume that the method of science based on the assumption that events can be reached primarily - if not entirely - by regarding the world as an interlocking series of mechanisms can be a basis of a general philosophy. Scientists make these assumptions as a working hypothesis : deliberately for practical purposes leaving out of the picture the mind, the perception, and the personal devotion of the scientist making such a study. This approach may be called 'mechanistic monism'.

He goes on to say : 'As a general philosophy it can have no validity whatever'.

To extrapolate reductionism into a general philosophy is Monod's fundamental fallacy, and his pursuit of this fallacious extrapolation is both absurd and menacing. He tell us that 'our societies' are still trying to live by and to teach systems of values already *blasted at the root* by science itself. He tells us :

> It is perfectly trut that science attacks values. Not directly, since science is no judge of them and *must* ignore them : but it subverts every one of the mythical or philosophical ontogenics, upon which the animist tradition... has based morality : values, duties, rights, prohibitions. ...
>
> p.160.

It is true that science has had this effect, but the question remains whether this effect is a valid one. Polanyi and Prosch quite clearly declare it is not valid in their book *Meaning* : scientists are themselves interested in phenomena because of the intrinsic interest of living organisms, and the elegance, order and significance that they find even in life-molecules. Admit this subjective participation and you admit another dimension of truth.

Whitehead declared that there was too much order in the universe except to be explained by the recognition of some 'ordering principle', and this seems hinted at in some of the language Monod uses. Moreover, clearly Monod recognises that man has a culture, and if this is part of his nature, it surely emerges – values emerge – out of human natural and cultural evolution? Not all values have been based on the naturalistic fallacy : there are such things as humanist values, whilst values also exist in subjective disciplines (as in the works of, say, Buytendijk or Winnicott) which are neither mysticist nor metaphysical. As John Wisdom has shown, there are collocations of naturalistic description by which values can exist, without animism, religion, or reference to a transcendental realm or a categorical imperative. In *New Knowledge in Human Values*, scientists and thinkers in the humanistic disciplines seek to base new values on truths from philosophical anthropology. (See my *Education and Philosophical Anthropology*.)

Science may not have it both ways. If it is not entitled to enter into the discussion of values, then it should declare (or rather every scientist should declare) that any use of science

---

[7] 'The Arrow of Time', *Scientific American*, 1975, p.233,6, 50-59

to subvert values is to make an obscurantism out of science. It should simply present 'data'. But of course, science is not content to do this, because it rests itself upon passionate conviction and so is bound up with values. It cannot provide the principles for interpretation of its data from mere data, but is necessarily involved in making choices, and so is value-bound. The only thing for science to do in this situation is to be open-minded. As we have seen, Monod is not : he is dogmatic, on the basis of his maxim 'Nature is objective'. As John MacMurray says :

> a faith in science does not involve an uncritical acceptance of the conclusions of particular sciences as eternal truth. Such as acceptance is entirely unscientific. All scientific conclusions are liable to revision and it is in true revision that science lives. To treat any scientific conclusion as final is to turn it into a dogma, and at once it ceases to be scientific and becomes part of a system of dogmatic belief. Science is not a body of ascertained truth, even where its basis is a formidable body of ascertained fact. A scientific theory is never itself a fact, but the interpretation of facts in the light of an hypothesis. Science itself is a never-ending process of the revision of beliefs which are not considered liable to revision. To turn a scientific theory into a dogma is to remove it from the field of scientific investigation. When we do this we are offending against science itself, and becoming obscurantist.

Monod's adherence to chance as a principle is obscurantist, within the general dogma of Darwinian mechanism. And the answer must lie in the question of mind. Of course, when it comes to thought, Monod becomes reductionist and mechanistic : he believes that 'the cognitive function' depend upon 'the cortex', rather than the whole development of a creature in its world. Again, we are at the question of the place 'consciousness' has in 'evolution'. The cortex is such a subtle and complex part of the creature. How could such a complexity come into being through mere necessity? Could it come into being through small changes brought about by accidental mutation? Thinking is one of the great leaps forward in life, about which Darwin and strict Evolutionary Theory has nothing to say. As Tomlin says : 'It is inconceivable that consciousness could have come into existence by accident'.

Monod's thinking here shows a profound inability to conceive of the problems implicity in the study of life. He cannot help recognising realities which exist in another dimension - and then struggles to reduce them to mechanism. He says :

> In animals, as in young children too, subjective simulation appears to be only partially dissociated from *neuromotor activity.*
>
> (my italics)

This observation arises from a recognition of the 'simulative functions' in man. At least there is a *recognition* of 'subjective simulation' : but Monod is uneasy at letting it exist in any dimension which eludes reductionism. He recognises that the puppy 'imagines' – but then hastily defines this in behaviourist terms – '*that is,* simulates through anticipation'. It imagines 'the discoveries it is about to make, the adventures ... it will face' - and that 'later on it will simulate the whole thing again ... in a dog's dream'.

So, Professor Monod encounters the kind of 'consciousness' in animals, and their capacity to dream (dreaming seems to be the basis of their sustaining of viability in higher animals and man). But he must struggle to reduce it to functions, though 'dreaming of an adventure' eludes such an explanation.

To say that 'to imagine' *is* to 'simulate through anticipation' is nonsense. When Professor Monod said : 'I am going to write a book called *Chance and Necessity*' he was using his imagination in a different way from what is implied by 'simulate through anticipation'. He *intended* something. There are, of course, differences between puppies and human beings. But the aspects of *experience* which Professor Monod examined here as 'dissociated from neuro-motor activity', can never be 'accounted for' by science, until it finds its way beyond reductionist approaches, and explores 'sensory perception, memory, language' - and consciousness and choice - which it cannot do by the methods of strict mechanism.

As Straus's analysis shows, physics, when it approaches creatures, fails to find certain creative relationships between one creature and another and between creatures, space and time. Yet Monod glimpses these in the puppy, even though he fails to see that here are manifestations which require more adequate concepts and methods to investigate than those of the Cartesian tradition. As Marjorie Grene says :

Descartes' world, of spatial relations on the one hand, and the geometer's understanding of them on the other, is a temporal one ... his failure to understand the unity of our psycho-physical nature can be traced to his failure to acknowledge the ultimate reality of time ... Perception is a pervasive power of animal life, and perception ... has been found to be at bottom the grasp of spatial and temporal relations. Even the perception of spatial relations, however, is temporal in its occurence. It takes the time of stimulation and response, it takes the ongoing apprehension of the organism, to produce it. All living beings beyond the simplest level, certainly all bisexually reproducing organisms, are four-dimensional entities, stretching within a certain niche in nature from their birth to their death, and it is that stretch, taken in itself in its intensity, which, though inseparable from its extension, from its embodiment, is the organism's mind.

The Knower and the Known p.89.

By his reference to the puppy, Professor Monod shows that he dimly recognized these dimensions of existence and perception in time. But yet he implicitly denies this in man by his reductionist way of talking.

As Marjorie Grene says, 'we are not pure cognitions absurdly attached to a machine' - and she quotes Merleau-Ponty :

We are not in some incomprehensible way an activity joined to a passivity, an automaton surrounded by a will, a perception surmounted by a judgement, but wholly active and wholly passive, because *we are the upsurge of time.*

Phenomenology of Perception, p.428.

Monod's is the kind of science which, by ignoring such mysteries and intangibles, turns scientific procedure into a 'mystic chant over an unintelligible universe' (A.N. Whitehead), and blinds us to the multi-farious scene of living processes, reducing everything to an 'invisible billiards game played by chance against necessity'. (Marjorie Grene).

By adhering to an exclusively reductionist view of the biosphere he reduces himself to an organism whose level of being and meaning is not seen : he denies his own creative, learning, exploring, purposeful manhood - his capacity for personal knowledge, his humanness. *He* is 'projective' even if Nature is not : is he then not in Nature? This is the trouble with his kind of science : mind-created, it denies mind.

Professor Monod, endorsing this view, his nature being 'objective' and his organisms 'machines' 'knowing' only by a mechanistic axiom system, obviously contributes to the profound spiritual crisis in our world. He obliterates the knower, and he obliterates life with its subjective centricity. He forgets that science has a 'tacit core' and

because it always has a tacit core, (it) is always rooted in my reliance on all possible clues available at all levels of awareness, even of non-conscious excitation, through which I strive to make sense of my world ...

Grene, p.87.

We feel depressed by a Professor Monod, and he contributes to our spiritual crisis, because we know that everyone is trying to make sense of their experience, by many skills, disciplines, intuitions, and modes of living, thought and creative insight : indeed, this *is* life, and what makes it worth living. It is a manifestation of the projective elements of Nature, and in man a higher version of the animals' positionality. But then, along comes the scientist, with his one narrow specialisation - albeit a very convincing and effective one - who works out a number of 'axiomatic systems' which enable him to postulate a certain description of the workings and the structures of life-stuff, in one dimension. He then turns round to us, to declare every other mode of trying to make sense of our experience invalid. Moreover, he declares that his incantatory axiom systems are the one *metaphysical* truth, and that the implications of this pseudo-metaphysic is that the universe we live in is meaningless and dead, not least in life-stuff (and in total contradiction to experience), running down. Yet, he declares, every other discipline except his own is full of subjective irrationalities, and is to be distrusted. There is 'nothing but' the dead objective matter, he, objectively, deadens by insisting on seeing it only in his specialist test-bench way. There is nothing but this kind of not-seeing 'and all else is vitalism and animism' mere witchcraft, to Monod (Polanyi is 'stupid' while Teilhard de Chardin's style is bad). In such a world, there is, and can be, no meaning - no mystery, no immediate existence, *Geworfenheit.* His world is one in which it is not worth living. Yet this 'chance and necessity' dogma came from a scientist who urged on the world the need for a new sense of values!

15 Rescuing the humanities from Darwinism

I began of course, from the question of the impact of Darwinian theory on the Humanities - especially on my own subject English, on poetry and the problem of belief in our time. Undoubtly, Darwin's influence has been enormous, throughout many spheres. Here a useful survey has been made by D.R. Oldroyd in his book *Darwinism Impacts : an Introduction to the Darwinian Revolution* (Open University Press, 1980).

However, what I believe to be true now is that Darwinian theory no longer provides a satisfactory answer to the problems of the origins of species or the evolution of life. Although Oldroyd, in the book just mentioned, discusses questions of the philosophy of science in relation to Darwinism, he does not really tackle this question, though he does mention Norman Macbeth's book as being an effort by a non-scientist to show that evolutionary theory lies in ruins.

Oldroyd himself seems to believe that Darwinism is satisfactory. It was a logical possibility, he declares, that the world had caused itself:

> ...this is just what Darwin later showed to have been the case. For he demonstrated that by a process of natural selection the apparent design could in fact be produced without any external designer.
>
> p.63.

As we have seen, the fundamental principle in Darwinism allows for minute changes in creatures, however produced, to be selected out if they are successful in assisting survival in the conditions of this or that environment. By such means, some emerging features would seem to have been shaped by the controlling influence of the 'sieve' of natural selection.

But the changes only come by chance - that is, if one accepts the theories of the synthesists. This, however, not only leaves unexplained how highly complex and coordinated changes could come about, but leaves unexplained the very impulse of 'life' to become anything - any form, any 'solution' or 'achievement' of the kind we marvel at as we look round the world. It may be a logical possibility that 'the world had caused itself', but the question as to why it should 'cause' all the multifarious forms of being (against such laws as the second law of thermodynamics) remains unexplained and is not explained by Darwinism. To quote Marjorie Grene again:

> how from single-celled (and for that matter from inanimate) ancestors there came to be castor beans and moths and snails, and how from these emerged Llamas and hedgehogs and lions and apes - and men - that is a question which neo - Darwinian theory simply leaves unasked.
>
> *The Knower and the Known*, p.197.

Neo-Darwinism is a theory deeply embedded in a metaphysical faith ; a faith which believes that science can and must explain all the phenomena of nature in forms of one hypothesis of maximum simplicity. But, as we have seen, there are two important categories which need to be invoked for us to understand the nature of living things : structure or *form* ; and goal-directed process or *end*. However much biologists passionately deny 'teleology', they acknowledge, implicitly, many and diverse shapes, and they acknowledge temporal patterns, life-histories ordered through time - and so they do recognise the existence in nature of ordered processes and therefore of ends.

If we accept this, and reject Darwinian mechanism, with its emphasis on 'chance' and 'necessity' what happens to the modes of thought on which Darwinism has made such an 'impact'? Mr. Oldroyd's book, mentioned above, offers to introduce us to the Darwinian 'Revolution'. But perhaps that revolution has hardened into dogma - and now we need another revolution, of teleological approaches, perhaps : or, certainly, of modes of thinking which can find 'the category of life'?

The mechanist model is not true. It may explain some structures, but it does not explain forms in time : large parts of the secret of life are still missing, and it is clear that we need new modes of thinking to begin to comprehend how to understand the way in which systems are organised by the formative principles in life. There is no vitalism or mysticism required here : only attention to evident facts, as of directiveness and striving : the telic nature of life. There is also a need to take into account the nature of living things as *beings*, and not to fall into the fallacy of believing that only the non-living is real.

This is to restore the 'vital'. And this is also what may be called an existential shift in our thinking. For all this is also at one with the need to invoke phenomenological disciplines – that is to restore mind, or consciousness, to the picture, both in what reality is, and in our concept of knowing. It is absurd for any mind to look at the world and pronounce it mindless.

Anyone who has read Polanyi, Grene, Tomlin, Collingwood and others will recognise here an enormous task.

There is 'something else', other dynamics, and other dimensions in which life needs to be understood – not in animistic or vitalistic terms, postulating some entelechy or supernatural force. This 'other' dynamic is variously hinted at in terms of an 'organising principle' or 'intelligence' or 'primary consciousness' such as is gestured at in Sheldrake's 'morphological resonance'. Life cannot be properly understood unless and until this dimension of intelligence or primary consciousness is recognised, and its relationship with the mind and being who is studying it. So, besides developing a new sense of *what there is to know*, we also need a new sense of *what knowing is*.

We have to find a new and adequate way of trying to study and understand life. We have looked at the superficial way in which a scientist like Richard Dawkins discusses this question, and we have seen Rupert Sheldrake trying to answer it. I have glanced at some of the writing of a philosopher, E.W.F. Tomlin, who has written on the 'concept of life'. He was himself taught by R.S. Collingwood, and has been following the injunction as set out by Marjorie Grene, who urged that we need to find, in a new way, 'what it means to be alive'.

This cannot be achieved without a considerable philosophical change of perspective. Again, most scientists prefer simply to get on with their work, and to enable themselves to do this they make philosophical assumptions to evade the perplexities. Yet they now have a frightening power : it is frequently reported in the press for instance that, through their knowledge of life-molecule structures, they are going to manipulate new tissues, protein for food and new species – much of which work is sponsored by industrial enterprises.

It is always possible, given the combination of immense technological power with philosophical naivety – to say nothing of scientific hubris – that there could develop some terrible threat to life, from scientific over-confident ignorance in a more silent and sinister way than nuclear explosion.

The philosophical revolution that is necessary surely requires a reconsideration of the original Cartesian division between *res mensa* and *res extensa*. There is a pathological quality about the mode of approaching the world which Descartes bequeathed to western civilisation, and it has led to our alienation from nature. This split needs to be healed, and it concerns *how we know*, and so our capacity to understand the world : we need to heal the gap between objective and subjective disciplines, and develop a holistic way of understanding the world and ourselves.

This healing is of profound importance to the Humanities and to poetry – indeed to all literature and the Arts, for it requires a new investigation of man's place in Nature.

As we have seen, one of the main problems which arise from science is that of the metaphysical impingement upon us of extrapolations from mechanistic theories into a general philosophy of life. Here again, we encounter a naivety among scientists who have strayed – nay, who gallop enthusiastically – into territory where they have no authority whatsoever, yet they have little or no sense of their own limitations.

As we have seen, looking at Dawkins' confused arguments, scientists make breathtaking assertions about the 'why' of human existence, when they only have authority in dealing with the 'how' : and then launch into declarations on what 'ought' to be, rather than on 'what is', while insisting that the humanities must now be handed over to genetics and other particulate sciences.

It is absurd to say that Darwinism, for instance, explains 'why' we are here, and solves the problems of human existence : or to declare that approaches to such questions written before Darwin are anachronistic, or to be disregarded.

Such questions may be approached in existentialist art, and are perpetual problems of the kind examined in tragedy. Here, human psychology, man's inner experience, his subjective life, remain more or less constant. Here the philosophical problems remains – man knows that he becomes nothing : what meaning can be found in his life to set against that nothingness? How shall he face his suffering? On what grounds shall be made his choices?

Here the problem is no different essentially from what it was behind Sophocles' *Oedipus Rex*, *The Book of Ecclesiastes*, *The Book of Job*, *Hamlet*, *King Lear*, *Anna Karenina*, Mahler's Ninth Symphony and *Four Quartets*. The existential and metaphysical questions require that kind of whole, symbolic engagement with the human problematic. And while scientific discovery contributes to our awareness of the reality of man and the world, it is so predominantly devoted to 'objective' approaches that it finds it difficult even to find the 'subjective' (or 'intrinsic') realm in which the existential anguish is encountered : so it is absurd for anyone to suppose that 'science' has settled, or can settle, such problems.

As scientific thought develops under the grip of Darwinian mechanism, the metaphysical situation gets no better. We are caught in a terrible dilemma, because the very success of science (as with DNA, yielding the use of drugs to treat cancer) seems to confirm that the mechanistic model is true. Yet at the same time this seems to exclude all meaning from existence, as Gabriel Marcel and other existentialists have pointed out (See *The Philosophy of Existence*).

Here we often see the scientist slipping over into 'our' area of the philosophy of being. An example of the enigma may be detected in the statement, 'nature does *not* know best ...genetical evolution ... is a story of waste, makeshift compromise and blunder' (Peter B. Medawar, *The Future of Man*). If man is a waste and a blunder, how is it this scientist can give a Reith lecture over the marvellous system of wireless communication, knowing that he will be understood and argued with freely by thousands of intelligent beings? What, if not 'blundering' nature, produced Medawar's mind?

Under the Darwinian impact, man forgets himself. Sophisticated Darwinians like Konrad Lorenz assume without question that the origins and formation of species can be explained as a succession of fortuitous variations and mutations passing through the mesh of selection, as Tomlin points out in his essay on 'Fallacies of Evolutionary Theory' in the *Encyclopedia of Ignorance*.[1] But, as Tomlin goes on to show, there are many well-concealed assumptions in the theory, one being that the negative elimination of unfavourable mutations is enough to explain the development of new forms, when it is not. In this there are contradictions. Evolution offers a single unified process : yet we are expected to believe that while the growth of each individual animal is an ordered process, the growth of a species can happen only by accident. While science everywhere displays order and coherence, and ontogenesis may be ordered, according to evolutionary theory, phylogenesis must remain a chance matter of trial and error, yet with the negative dynamic assuming a positive role. As we have seen, negative forces are quite illogically turned by scientists into magic positive ones : 'selection is one of the great constructors of evolution. The other constructor is mutation' (Konrad Lorenz, *On*

1

See also his essay, 'The dialogue of evolution by natural selection' in *Universities Quarters* Winter 1983/4, volume 38, no.1.

Aggression, 1961).

This kind of manoevre is forced upon scientists by the evident facts of life which require a *driving force*. Polanyi writes 'the process must have been directed by an *orderly innovating principle*, the action of which could only have been *released* by the random effects of molecular agitations and photons coming from outside, and the operation of which could only have been *sustained* by a favourable environment'. (*Personal Knowledge*, 1953, p.386, the italics are those of E.W.F. Tomlin), There is also the existence of mind as life. Teilhard de Chardin argued that if there were not built into matter (even in its simplest form), the possibility of combination and co-inherence with other matter : and further, that if there were not built into matter-in-time an inherent tendency to form, by such advancing integration, increasingly complex systems, then self-reflective consciousness as we know it in man would never have been possible.

In many spheres of scientific and philosophical thought, men are thus striving towards a more adequate view of the evolutionary process, and thus of man's place in the universe. Tomlin argues that the entire micro-physical world is now seen to come within the organic sphere, revealing a 'line of continuity between all individual or beings'. All evolution in an 'interior' process, characterised by thematism (that is, 'the recognisable similarities in bone and other structures').

The present impass in evolutionary thinking is due to the intpretation of biological facts in terms of out-of-date physical theory, in attempts to find one unifying principle to explain the mechanistic terms.

Scientists need to be urgently obliged to examine the illogicalities in their traditional ways of discussing evolution.

For instance, some approaches seem simply tautologous. We find this kind of weakness even in the best books : I found an example in Dr. Leakey's *Origins*: 'The keen selection pressures of this new ecological niche must have produced a basic stock that was superbly adapted to the new environment, for we have the highly successful primate order to prove it' (p.41). Evolution theory may be an attempt to show how the 'highly successful primate order' came into being : but it does not demonstrate the accuracy of the hypothesis (of 'survival of the fittest') to point to that successful order's existence, because it is from this fact that one begins, in trying to explain *how* it came into being : that *is* the question.

Today, many of the opinions of evolutionists depend upon blind faith, as Professor otto Frisch demonstrates, in his essay in *The Encyclopaedia of Ignorance* Discussing the development of feathers, he says, 'Even if a very unlikely mutation caused a reptile to have offspring with feathers instead of scales, what good would that do, without muscles to move them and a brain rebuilt to control those muscles? *We can only guess*... Much about the process of evolution is still unknown' - but then he goes on, amazingly... 'but I have *no doubt that natural selection provides the justification for teleological answers*' (p.3. my italics). Having admitted ignorance, then admitted the partial nature of knowledge, he thus lapses back into an insistance that the dogma must be correct.

As Tomlin says, Darwinism hardened into dogma, before it had been thoroughly analysed. The reconsideration of beliefs here is urgently overdue.

Here an important chapter in writing about the philosophy of science is that on 'Order' in Michael Polanyi's last book, *Meaning*. In this he points out that although there have been teleological arguments about the nature of life, the modern mind has turned against these, in favour of mechanical explanations, combined with the metaphysical view that life is pointless and absurd. From this it is only a step towards a completely behaviouristic approach to psychology, in which it is possible to abandon all 'mentalistic' talk of purposes or aims or goals in the discussion of animal or human behaviour.

A great deal of the secret of life still evades us, and so it is absurd to base a nihilistic stance in the Humanities on the implications of attempts to reduce life to the chemical and physical operations of molecules - attempts which a distinguished chemist like Polanyi declares must fail for philosophical reasons. An admission of scientific ignorance here would offer us in the Humanities a reprieve from despair. Even Crick himself admits, 'At the molecular level we understand some of the most fundamental developmental processes... but, how an organism constructs a hand, with its thumb and forefingers, with all the bones, and muscles and the nerves, all assembled and correctly connected together, that we cannot explain...' (Pergamon *Encyclopedia of Ignorance*, p.303). In his work Michael Polanyi suggests a totally new approach to the nature of life and being is necessary before we can make progress, and so we may hope that a new sense of man's place in the scheme of things may emerge, slowly and painfully. This will require a new 'model' of man and his nature, a new way of thinking about him.

Moreover, this new thinking about what it is to be human, and what man's place in nature may really be, requires of us no less than a reconsideration of what knowledge is, since the reason why we have not made so much progress as we should have done here is that we are still too much influenced by the scientific ideal of 'objectivity', and the exclusion of all that belongs to poetry and inwardness. We need to pick up the original *telos* of Greek thought, which required an equal attention to the subjective realm as to the objective, and sought the whole of truth.

The conclusions towards which we are drawn are that there *is* order and meaning in the world : science does *not* tell us that existence is meaningless and absurd. Polanyi quotes A.N. Whitehead - there is too much order in the world to be explained without recourse to some kind of principle operative in the direction of the achievement of more meaningful or orderly or regular relations. These orderly or more regular relations are clearly visible in the world. The human mind and human consciousness may be seen as part of this striving towards order, and they are a product of interrelationship, of the 'I-Thou',and of certain benign and creative dynamics in the universe. These all in turn are the product of qualities found in all animals, which enable them, to one degree or another to be autonomous, to experience 'centricity', and to operate from a subjective centre. Living things cannot be understood unless this subjective capacity as 'beings' is grasped : biology which fails to find its subjects as beings is inadequate, as Tomlin points out.

Man's capacity to be free, and to choose, is a product of this centricity, with his extra dimension as the *animal symbolicum* with his culture. It is consequently his nature *to be moral*, and this is bound up with his *meanings*. So far, then, we can see here glimmerings, of what we can believe - and I think this opening up of a new view of man's place in nature, with a sense of the place of his thought, culture and ethical living in nature, promises something far more true and insightful than religion, which asks us to believe and have faith and obey moral precepts because they are strongly expressed injunctions ('the voice of God'). As Maslow said, the new sources of an understanding of our being, and the values we may derive from it are to be accepted because *they are true*, and not because we are simply conjoined to 'believe and have faith'. The kind of philosophical anthropology to which I try to point here finds 'ethical living' natural to us, bound up with our search for authenticity, our life-tasks. In each human being there is a formative principle urging us towards fulfilment. This view denies neither moral values nor cultural resources - since these are natural products of human engagements with reality.

*

Gillian Beer has written an interesting book on Darwinian evolutionary theory in relation to the literature of his time and since.[2] She properly sees Darwin's achievement as an imaginative one – as 'a determining fiction by which to read the world'. She notes the radical change in thought brought about by Darwin, and sees it running parallel with Freud's influence: Freud was glad that man's 'self-love' had been dealt a blow, and believed that we had to recognise that 'man is not a being different from animals or superior to them'. From Darwin's work emerged the awareness of the possibility that the evolution and development of life had its own characteristics and dynamics of which man was not necessarily the supreme outcome. (D.H. Lawrence occasionally expresses this attitude, not least when he wishes all human beings to be swept from the earth).

Gillian Beer obviously admires the Darwin who is aware of the fecundity and multiplicity of life, and this she suggests was influenced by his reading of Milton's *Comus*, though she

---

2 *Darwinian Plots: Evolutionary Narrative in Darwin, George Eliot and Nineteenth Century fiction*, Gillian Beer, RKP, 1983.

does not quite demonstrate that Darwin had pondered Comus's famous speech on fertility. Her literary approach enables her to be aware of Darwin's language, and the problems this poses for his thought. She points out usefully certain proccupations of his which are ancillary to evolutionary theory, such as his sense of the 'entangled bank' and the 'inextricable web of affinities'. To take the latter in particular, Darwin was aware of the similarities in the structure of many living things, and in recent times the discovery of the structure of DNA has given an added meaning to the phrase. She also notes that Darwin could never fully succeed in extinguishing the language of intention from evolutionary theory, and quotes him from various editions – talking at times of the origins of life in terms of how life 'was breathed' into them, then of how life 'was breathed into them by the Creator' and then of how all the organic beings which have ever lived on this earth may have descended 'from some one primordial form'. Gillian Beer also shows how Darwin made 'natural selection' a positive, creative force, and yet was uncomfortable about this, since his fundamental evolutionary theory did not allow for such a shaping dynamic. He failed to reconcile this matter, and, as Gillian Beer says, his language was expressive rather than rigorous. Another unsolved problem arises over the relationship of ontogeny to phylogeny: there is in Darwinism no evolution in the the independent life, so the individual is both vehicle and dead end. As we have seen, critics like Tomlin ask why we cannot accept that species can develop as we accept individuals develop, by a natural inherent dynamic?

However, over such questions severe difficulties arise which Gillian Beer does not tackle. How could such a delighted awareness in Darwin as an observer, become such a negative philosophy of existence as Darwinian mechanism? At times Gillian Beer glimpses the dangers of nihilism in Darwinism, and quotes Condrad's Heyst: 'Man on this earth is an unforeseen accident'. Lamarck, she says, 'give primacy to mind'. Why, then, has science settled for an explanation of life which so rigorously excludes mind, or what Tomlin calls 'primary consciousness'? It becomes clear from Gillian Beer's account, though she fails to confront the issue, that when we look at life unfolding what we see is 'achievement', and this is quite clear from Darwin's own observation of the marvels of the world. How did scientific theory come to settle for an explanation which strictly denies 'achievement', even if it eschews teleology? The 'truncated teleology' of Darwinism often surfaces in Gillian Beer's book:

> Man selects for his own good: Nature only for that of the being which she tends. Every selected character is fully exercised by her; and the being is placed under well-suited conditions of life.
>
> (Darwin)

Later Darwin tried to 'deconstruct' this mythological person Nature: but the point is not so much the use of personification, as the thinking which betrays a purposive force at work, which is then disallowed by strict Darwinian theory. As Gillian Beer says, 'he had to put something in the space left by God'. He is determined not to use a creationist language: but the baby goes out with the bath-water, and so there develops an implicit denial of quite clearly observable aspects of the world-its gradients, its tendencies towards order and complexity-while even in the impulse 'to survive' there is a forward-seeking dynamic which evolutionary theory on Darwin's terms cannot account for.

Gillian Beer, oddly, says that Darwinism was so unpalateable to the Victorians that they responded to it with a 'physical shudder': yet she sees also that it was very much a product of the thought of the time. 'Darwin's crucial insight into the mechanism of evolutionary change derived directly from his reading of Malthus'. Besides relating it to Freud's views, she shows that Marx also delighted in the blow given to teleology, though he also saw it as a version of Hobbes' *bellum omnium contra omnes*. What Gillian Beer fail to note is the reductive impulse behind much of this thinking–not least in the pronouncement quoted above, that there is no difference between man and the animals. Behind this in turn is an approach to the natural sciences which fails to find consciousness and culture–which is so strictly 'objective' that it fails to find the subjective sphere of reality: fails to find unique being. As Gillian Beer says, Darwin avoids the problem of man, but does not see how this means that he avoids the problem of mind. She fails to see that something is seriously wrong with a biologist who can write in his notes:

> People often talk of the wonderful event of intellectual man appearing – the appearance of insects with other sense is more wonderful. Its mind more different

probably and introduction of man nothing compared to the first thinking being, although hard to draw the line. (B207-8)

In 1838 Darwin had noted, 'Origin of man now proved :- Metaphysics must flourish. –He who understands baboon would do more towards metaphysics than Locke' (M84). But this indicates the kind of problem discussed by Ernst Cassirer: we have had so many disciplines involved in the study of man, but since they have tried to understand him as a natural object, they have failed. Psychology in this paradigm has tried to understand man by understanding rats, and has not even studied rats in terms of the 'kind of consciousness' in them. And one whole dimension has been left out – that very dimension which Gillian Beer is concerned with, the metaphorical mode, the application of consciousness to the world through symbolism. Man is the *animal symbolicum*, in a new dimension.

Darwinian evolutionary theory belongs to a long tradition, which includes Locke and Hobbes, Malthus, Adam Smith, and Bentham (whom Gillian Beer never mentions). The tendency is essentially atomistic, and the object of attention is a thing without the essential properties of life. In Utilitarian social theory it was an advantage to look at the human world in these terms since the needs of being could be ignored. It could then be acctually believed, by proper mathematical calculations, that it was a natural law, and appropriate, for the weakest to go to the wall. In Herbert Spenser's philosophy, as Gillian Beer shows, there was a moral component. Even as Nature selects, in the struggle for existence, for the good of the species which she tends, so there may be seen in society an accord between moral fitness and the ability to survive. (I believe that D.H. Lawrence was influenced by this, as is manifest in his curious equation, in which those who serve 'industry', the machine and that kind of 'will' which is devoted to the machine, have 'bad' sex, and so are in his terms morally inferior). Thus, although, as Gillian Beer says, 'Darwin was not seeking a covertly metaphysical world' the metaphysical effects of his evolutionary theory contributed not only to an implicit nihilism, in which man was a mere accident, not only a central achievement of the cosmos, but possibly a mistake. In this there is a devastation denial of consciousness, of mind and culture, of the capacity to *know*, and to be responsible for that knowing – which, as I have suggested, must undermine science itself.

But, as I try to argue in this work why should we accept it? Gillian Beer never questions the veracity of Darwinian evolutionary theory. She takes it simply to be proven. She begins by calling it a 'determining fiction by which to read our world': but she is not content to reply upon it simply as a brilliant fiction which is 'part of a form of imaginative history'. She takes it a 'fact': or at least, she does not demur when Darwin uses the word 'fact' in a very doubtful way: 'for him facts are identified with laws, and 'fact' and theory converge'. (p82)

How do we know evolutionary theory to be a 'fact'? Evolution may be a fact. The problem is to explain it. On page 8 Gillian Beer declares that 'evolutionary theory cannot be experimentally demonstrated sufficiently in any present moment'. On page 51 she declares that 'it took a hundred years for Darwin's projection to be authenticated empirically:

> He believed that he has discovered the *mechanism* of evolution but he did not expect to encompass the whole process. Indeed his theory was necessarily hypothetical rather than traditionally inductive. It took a hundred years for Darwin's projections, his 'fictions' or theories, to be *thoroughly authenticated empirically*. But the accuracy and scope of his observations were such that they carried conviction as scientific explanation long before they could be *proved*.
>
> p.15 (my italics)

But if they cannot be 'authenticated empirically', how have they then been 'proved'? The answer of course is that they have not. Despite the marvellous air of convincing explanation, Darwin has not proved either how life originated, how species began and developed, or how evolution 'worked'. One crucial question which Gillian Beer refers to is the eye: but she does not raise the important dilemma. Such developments require overall coordination of a very complex kind: how, then, could they have come about by 'chance' mutations, making small changes which brought small advantages which contributed to survival? Perhaps Gillian Beer supposed that the discovery of DNA confirmed Darwin's theories? She says

> the discovery of DNA demonstrated the organism as a structural narrative programmed to enact itself through time.

p.8

But this is not true, since, although we know something of how DNA replicates itself, we do not know what governs the changes in DNA structure and the forms of life (as through e.g. adolescence), or, indeed, in what dimension to look for this morphological dynamic: as Polanyi has shown by argument, this overall 'systems' dynamic cannot be reduced to physics and chemistry (Gillian Beer quotes Polanyi from the relevant essay, but does not register that in this he makes a radical criticism of the assumption that DNA theory confirms Darwinian mechanism).

So Gillian Beer is seriously wrong in places, as when at this point she says, 'its eschewing of fore-ordained design (its dysteleology) allowed chance to figure as the only determinant allowed'. The word 'allowed' is wrong. It may be that it has been postulated that chance may be the only formative dynamic allowed, but this is so patently inadequate to explain the origins and development of life that many scientists are now rejecting such explanations. Mathematical calculation alone can dismiss it (there has not been time enough in the earth's history). To say that it is all a matter of 'minute random mutations with their uncontrollable consequences' may suit Darwinian orthodoxy, but it is nonsense: one cannot believe that the multiplicity of life, and complex organisms, came into being and developed by such 'roulette': it is now against all reason. So it is wrong to talk of the 'all-inclusiveness of the explanation' despite the desire of science to find such all-inclusive explanations. The mechanism of natural selection, as Darwin conceives it, is not capable of 'producing change', because it is not a positive formative force.

At least, there are now many good arguments which ought to make the critic of Darwin extremely cautious. Gillian Beer seems to see Darwin as a primarily positive force, and uses here knowledge of his theories to examine his influence on the novel in the nineteenth century. But there are those who believe that the implicit denial of life and mind in the theory has had serious and nihilistic metaphysical effects, on our attitudes to ourselves and the world, and these questions should surely have been touched on in this book? But, alas, Darwinism is now a theology of its own, and felt to be beyond criticism and rebuttal, which cannot be good for science or the Humanities.

Gillian Beer does not convince me that George Eliot's novels are profoundly influenced by Darwinian evolutionary theory, though, of course, George Eliot herself was intelligently aware of what was going on in scientific thought. I don't for example feel that Gwendolen Harleth's moments of paralysed inaction can be related to Darwinian determinism. Gwendolen is not at the mercy of her unconscious life, but rather enacts the drama of will: and George Eliot's novels have at their centre the creative consciousness and moral sense of human beings. The objection to Darwin is that he failed to find man in his special dimension, in the natural world, and tended to imply that there were impersonal and implacable forces at work which undermined man's responsibility for his being, as George Eliot never does.

With Thomas Hardy, Gillian Beer's account is more illuminating. It is true, as Gillian Beer says, that

> Like Darwin, an ambiguous anthropomorphism pervades his writing – an anthropomorphism which paradoxically denies human centrality and gives the human a fugitive and secondary role in his system of reference but not in his system of values.

p.252

But this surely is an indication of the destructive effect of Darwinian theory on creative art? Hardy tended towards pessimism, because his own marriage was such a disaster, and so railed against the way in which 'life' seemed full of promise and potentiality, but then became frustrated and bleak. It is true that often he relieves this attitude by gloatingly contemplating the natural world in fine detail, as Gillian Beer shows by some marvellous quotations. But she says that Hardy 'perceived the malign tautology latent in.. 'the struggle for life': and yet, if Darwin is as unsatisfactory as he seems to be, this was to take a bleak scientific point of view to justify a bleak philosophy of being – one which bitterly portrays in the art the impossibility of there ever being a positive outcome to 'striving'. Indeed, on the basis of a willful interpretation of tragedy, Hardy makes blame sure that none of the striving of his characters will ever come to fulfilment, which makes his work a travesty of life. The strangely wilful quality of Hardy's position may be examined in the last episodes of *Tess*, in which Tess can only find sexual fulfilment and joy when she is already condemned, virtually, to death – a development which tells us more about Hardy's attitude to woman than about his philosophy of existence.

Gillian Beer seems to see the Darwinian influence on Hardy as a benign one, just as she sees the influence of Darwinism as generally benign. Yet it is really perverse, fostering Hardy's bitter nihilism, as it has continued to foster nihilism in modern literature. It prompted Hardy to conclude that he was right to believe that things can't be as they seem to want to be, which again is a travesty of the marvellous sense we may have – and which Darwin must have had in his observations – that life is a multiplicity of marvellous achievements. Hardy's perplexed conclusion seemed to be that while the world of life looked so marvellous, as to Tess, it was really hostile, while chance or destiny was indifferent if not malignant, so that man's capacity to know and perceive and fulfil himself was blocked by hostile and alien dynamics. The Darwinian paradigm persuaded him that this was 'truth', and this, I believe, encouraged him to distort the nature of tragedy: for though Democritus may have looked on life in this way, Greek tragedy does not. Hardy's view could be taken as implying that we cannot ultimately be responsible for our fate: the attitude of Greek tragedy is surely that not only must we take our fate upon ourselves, but that all must bear the suffering and the responsibility, too. This is the crux of the problem of Darwinism and the Humanities.

16 Conclusions

I have here tried to make my way through one small corner of the philosophy of science – the impact of scientific theories of evolution on the Humanities, and on the general philosophy of life at large. No doubt in venturing into unfamiliar regions for which I have not been trained, I have made some catastrophic errors. Yet I believe the effort has been worth it, because it is clear that the way we feel daunted by the scientific view is not only based on misunderstanding, but arises in part from fallacies in scientific thinking itself. For one thing science too often oversteps its bounds, its own self-defined limitations, and urges on us a metaphysic, and often a moral system, on which questions it has absolutely no authority to pronounce. Moreover, (as with E.O. Wilson and Richard Dawkins) it even goes further, in its fallacious acceptance of the assumption that it is 'objective' and so more true and realistic, to claim that (in the light of the particulate analysis of life-stuff) it should take over the moral sciences, if not the Humanities in general.

Yet, at the heart of such issues, we find that the concept of life at the heart of these life-sciences is hopelessly inadequate. Despite their aggressive confidence these sciences explain little or nothing, even at the level of the processes of replication which are the focus of its disciplines. Fundamentally, these scientists display a *failure to find the vital*, and fail to examine a number of fallacies – such as their implicit assumption that only the non-living is real, and that the only valid explanation is the micro-reductionist one, based on minimals, and linear, step by step, particulate analysis.

The basic assumptions from which the outrageous claims of the micro-biologists to take over philosophy, ethics and the Humanities, are, moreover, on examination, mere postulates : not 'truths' at all. Yet these hypotheses are now seriously placed in doubt, as by computer calculations, and seem no longer tenable. The mechanistic account of the origin of life seems inadequate (some of its adherents in consequence resort to space fiction fantasies). The directiveness of life, the development from inanimate to animate, the gradients in nature - all observable realities - are circumvented by evasions or denials, yet according to the mechanist model, some of these processes could only happen by miracle. Though by some of the titles of scientific books (*The Eighth Day of Creation*) it would seem that some scientists feel they have created the world, their explanations explain nothing. Even such a distinguished biologist as Dobzhansky confuses issues and tries to turn natural selection into a power that it cannot, by its very nature, be. We have only to ask where the 'missing link' is to be found, and to point to the many misrepresentations in this area, to demolish, for example, the common assumption that man evolved from apes. Hoping that a theory is true is no guarantee of its veracity. Of nearly all the explanations offered by evolutionary theory, we have to say 'not proven' : and to point out that they do not explain the predominant facts of life, its origins or development.

Shall the Humanities, then, be daunted by such doubtful hypotheses, evasions, false extrapolations, naiveties, expressions of blind faith, and manifestations of hubris?

more widespread, significance. As I have tried to show, Darwin's 'horrid doubt' about whether man's mind may be trusted, has a profound significance, because it shows that, in its inability to find mind, modern science is in danger of undermining itself. In some way the reality of mind and the implications of its being in the universe besides matter in motion so vividly brought home by Marjorie Grene as a follower of Michael Polanyi, must be embraced by science, or it is likely to find itself an enemy of truth. This would be a disaster to society, of cataclysmic proportions, since the whole posture of our way of life, in Britain and America at least, rests upon the assumption that in science there is embodied the truimph of reason and the quest for truth. If we lose our confidence in this, we could lose our confidence in man himself.

This was Husserl's warning to us, and here I wish to turn to two important figures in the debate : Kierkegaard and that same Husserl. The whole existentialist movement has picked up from Kierkegaard an attention to the uniqueness of the human existence. This does not mean (as some scientists seem to believe, not being able to escape from the mechanistic positivism) that each human organism is different, but that each *being* has its own life-experience, its life-world, in space and time : and this is irreducible. This is appreciated by medical research, which recognises that it cannot experiment on a hundred humans as it experiments on a hundred mice. The human being has an intrinsic worth, as people assert when they say : 'It is *my* life'. The consequences of mechanistic materialism, as Dickens knew, are that this unique intrinsic value is menaced. It is this Malthusian numerical menace to the experiencing 'I' that is an undercurrent of Darwinism. Then, as Husserl made clear, the human being, because of the intrinsic worth of his 'I' experience, cannot live with the world-view presented by modern science, because it excludes just that from the universe it is prepared to recognise, and which it creates in our minds.

In this, Husserl makes clear, there is a betrayal of the original telos of Greek civilisation and thought, the amazing moment in the cradle of our civilisation, when the pursuit of truth, that transcendent human impulse, was born. This pursuit was originally of both objective, 'out there', truth, and subjective, inner, truth. Since the seventeenth century, and especially in the rapid growth of the physical sciences since Gallileo, Newton, Kepler and Descartes, there has developed a fatal development of attention to the objective realities, split off, and divided from, the subjective realities. Until, in the end, it is now an aspect of our lives, as here especially in Cambridge, that those whose disciplines are of the inner life, exist daily alongside deniers of this realm : who tell us that consciousness is only chemicals, that the brain is only a computer, and that we are only survival machines built by genes. That is, not only is subjective truth *spilt off* from science, but science turns aggressively againt this area of truth, and seeks to bury it altogether. In this assertive naturalism, as Polanyi makes plain, in *Knowing and Being*, there are deep political implications. Man's relationship with life and the world seems deeply disturbed, and it seems clear that the roots of this disturbance are in the Cartesian mentality. There is not only a failure to bring objective and subjective together, and to overcome the consequences of the division between *res mensa* and *res extensa*. There are signs of a psychopathological disturbance, displaying a sense of alienation from 'mother earth', often combined with a hatred and fear of existence, life and being – not least because these cannot be entirely subject to analysis, and to triumphant manipulation.

The most urgent issue of all, doubtless, is man's consequent misuse of nuclear power. The Physicians for Social Responsibility have warned that nuclear war could make the continued existence of human beings, and possibly of all life on earth, impossible. (See Richard Severo, 'Scientists Say Nuclear Blast in City would kill two million', *The New York Times*, Saturday, 27 September, 1980 Section B, Page 7. I am grateful to *Teachers' College Record* for this source).

But a handful of scientists in the United States are also warning of parallel consequences from genetic engineering. These include George Wadl, Jonathan King and Erwin Chargaff. Beyond the questions of disease, guidelines and control, writes Chargaff, who is a biochemist whose own research was crucial in the discovery of DNA :

> there arise a general problem of the greatest significance, namely the awesome irreversibility of what is being contemplated. You can stop splitting the atom; you can stop visiting the moon; you can stop using aerosols; you may even decide not to kill entire populations by the use of a few bombs. But you cannot recall a new

form of life. Once you have constructed a viable *E. Coli* cell carrying a plasmid DNA into which a piece of eukaryotic DNA has been spliced, it will survive you and your children. An irreversible attack on the biosphere is something so unheard of, so unthinkable to previous generations, that I could only wish mine had not been guilty of it... This world has been given to us on loan; we live and we go; and after a time we leave earth and air and water to others who come after us. My generation, or possibly the one preceding mine, has been the first to engage, under the leadership of the exact sciences, in a destructive colonial warfare against nature. The future will curse us for it... I am being assured by the experts that nothing will happen. How do they know? Have they watched the web of eternity opening and closing its infinite meshes?

Erwin Chargaff, *Heraclitean Fire : Sketches from a Life Before Nature*, New York, the Rockefeller University Press, 1978, pp.189-191.

So, while no doubt some of our crises arise from mis-use of science, there are forms of menace from within science itself. The confident mechanists have so reduced the world to interacting machines that can be manipulated that they have become blind to *being*. The are blind to life's striving. Of course they believe that 'nothing will happen' because their DNA is a mere collocation of atoms operating only by the laws of physics and chemistry : they speak to them only of 'codes'. But, as Polanyi has argued convincingly, their account is false and wrong : DNA could not work like that. That there must be 'something else', some other organising principle, some other mysterious morphogenetic life-power, something like intelligence in life, has been asserted by responsible biologists like Sheldrake, and has been hinted at by many from Bergson to A.N. Whitehead and Collingwood. So long as they stick to the blind perspectives of mechanistic reduction, of course, the mechanists cannot see it, any more than the inhabitants of the Antipodes could see Captain Cook's ships as they sailed into their waters. But the mystery of life is *there*, and may one day strike us startingly in the face with its reality - the danger is it may then be too late, and the web of eternity could close its infinite meshes for the last time on Faustian man – not because of his hubris, so much as because of his ignorance : *because he did not think enough, in awe, before the mystery of his own being.*

If this were to happen, it would be a *poetic failure*, a failure to be aware through the imagination, of new possibilities, of new glimpses of the truth. Kekulé arrived at the structure of Benzine rings by a dream of snakes with their tails in their mouths. What we need now is far better dreams, by which, through the exercise of symbolism in the realm of the imaginative power, scientists may penetrate further into the secrets of life : not so much into substance and structure, but into form, with a new sense of *time*, and of *wholeness*.

Some scientists are already on this path : besides those mentioned here, David Bohm, W.H. Thorpe, Charles Birch (see : *Wholeness and the Implicate Order*, David Bohm, Routledge, 1980; John B. Cobb and D.R. Griffin, (eds) *Mind in Nature : Essays on the Interface of Science and Philosophy*, University Press of America, 1977). The implications of humanist criticisms must be followed through into scientific enquiry itself. It is no good merely bringing up mere subjective idealism : the problem is one of tough philosophical discourse, and this need, I hope, explains my inadequate attempts to raise these issues at large, by which I hope to prompt others more capable.

On the side of the Humanities, or the subjective disciplines, we must persist, certainly, in asserting the validity of our concern for being. Not only must we resist the extrapolations from mechanistic science into regions where it has no business to be. We must accept that we deal with realities not caught in that net. As many have pointed out, from Viktor Frankl to Karl Stern, scientific method - whatever breath-taking results it may have obtained in our analysis and control of the outer world - cannot be applied to inner questions which we know to be real ones, as of the presence of hate and corruption and their co-existence in the world along with love and beauty. We know that there are problems of being, and they are real ones, as anyone knows who has read *King Lear* or Keats or D.H. Lawrence. Ours may be an age in which analytical thought has tried to get rid of such problems by denying their validity or trying to reduce them to problems of definition. Scientists and philosophers of the strictly positivist schools even deny these as 'pseudo-problems' - even though to man they have always been such burning questions that we cannot live unless we try to solve them. Analytical philosophy may lend to reject all the major statements of the world's greatest religious leaders as 'emotional nonsense', since they cannot be 'verified'. Yet the dogmatic energy with which such assertions are pursued suggest not only hidden absolutes (which the analytical discipline denies) but even the desire to become the one true source of truth, as Tomlin has pointed out. We need to insist on the validity of genuine philosophy, and we must not be diverted by scientists who want the world to be a blind machine, so that these questions of being are by-passed.

As Karl Stern says in *The Flight from Women*, the processes of being and becoming, because these are 'inward' questions, resist such objective analysis. So, too, the metaphysical problems persist, however much they are dismissed by analytical philosophy and science. Our considerations of the nature of mind and being are not settled by atomistic physics or mechanism, and cannot be settled on that basis. Over these issues a philosophical battle has been raging for over a hundred years.[1]

But because of the confusions here, 'scientific' pronouncements contradict common sense, and offer a view of the world not reconcilable with our normal sane experience. As I have said, the psychoanalyst, D. Harry Guntrip, has suggested that there are schizoid elements in the detached coldness of the scientific world view, while Michael Polanyi has spoken of the 'pathology' of science. Karl Stern accuses the scientific account of cosmogenesis of being crazy : 'And I do not mean crazy in the sense of slangy invective, but rather in the technical meaning of psychotic' (*The Flight from Woman*, p.290). He characterises this world picture thus :

> At a certain point of time the temperature of the Earth was such that it became most favourable for the aggregation of carbon atoms and oxygen with the nitrogen-hydrogen combination, and that from random occurences of large clusters of molecules occurred which were most favourably structured for the coming about of life, and from that point it went on through vast stretches of time, until through the processes of natural selection, a being finally occured which is capable of choosing love over hate and justice over injustice, of writing poetry like that of Dante, composing music like that of Mozart and making drawings like that of Leonardo.

We do not even have to go as far as this, says Stern : 'I have a certain number of years ago been a single cell, microscopically small, and now I sit at a desk writing. Millions of data from the cumulative sciences form a fearfully intricate net of causalities to tackle this mystery but my being and becoming are not caught in that net' (my italics). The danger is that, under the influence of such scientific theories as Darwinian mechanism, we may be led to feel that, whatever we feel, we cannot uphold our own valued unique being against the kind of explanation Stern declares insane!

As Stern points out, what we are imprisoned in is a paradigm : the idea that all living things are random points of arrestation in a blind mechanism of physical occurences governed by pragmatic advantages is the expression of the Cartesian objectification of living things. It is a product of a whole philosophical tradition which ultimately found only geometry to be real, together with the atomistic micro-mechanistic explanation of life, so that the concept of 'wholeness' has been lost, and with wholeness, the recognition that in any living organism the whole is more than the sum of its parts, not in any mathematical sense, but in terms of the collective behaviour of highly complex living beings. No living system can be really understood apart from its whole ecological context, while important aspects of its existence such as intentionality in lived time, are lost in mechanistic approaches that suppose it possible to explain all life by breaking it down atomistically. Exploring the question : 'What is Life?', Erwin Schrodinger came to the conclusion : 'I... that is to say every conscious mind which has ever said or felt 'I'... as the person, if any, who controls the 'motion of the atoms' according to the laws of nature'.[2] George Wald, a Harvard biochemist and Nobel Laureate, put it another

---

1. See the work of Polanyi and Marjorie Grene; E.W.F. Tomlin's *The Approach to Metaphysics*; Herbert Spiegelberg, *The Phenomenological Movement*; Husserl, op cit; *Beyond Reductionism*, ed Koestler A. and Smithies R. See also *Education and Philosophical Anthropology* by the present author.
2. Erwin Schrodinger : *What is Life*? CUP, 1945.

way when he said : 'It would be a poor thing to be an atom in a universe without physicists. And physicists are made of atoms. A physicist is the atom's way of knowing about atoms'.

This is really another way of drawing attention to the problem, which Marjorie Grene and Michael Polanyi have bought to the centre of discourse, of bringing *mind* back to the attention of those who are trying to give a satisfactory account of the universe. Mind is a manifestation of life, and life is a manifestation of the way matter organises itself. Here there arise many perplexities and mysteries, including the question of our conception of what knowing is. As E.W.F. Tomlin puts it : 'The problem of life is the key to the traditional problems of matter and mind : it is the most fundamental of problems' ('The Concept of Life', *Heythrop Journal*, July 1977, Vol. XVIII, No.37).

As Tomlin puts it, 'life and living must be interpreted not in terms of some irrational vital sap or current, but in terms of consciousness and value'. Indeed, life is consciousness - 'consciousness being understood in a primary sense as the immediate self-possession of an organism, and in the secondary sense as that which, mediated through the brain, proceeds to the ordering of the external world'. Every living being is a *subject* rather than an object, a true rather than a pseudo-form. A being Tomlin defines as 'a domain of togetherness in dynamic organization or equilibrium, in which a hierachy of sub-systems are 'dominated", by a subjective unit, a 'here and nowness'.

There are two important things to say about such complexities of being : they could not have come by 'chance', any more than consciousness could have come by chance : and this is a feature of life which scientists, and indeed all of us must face. And secondly, to study these aspects of the world requires modes and disciplines capable of finding their special nature. Speaking for myself as a teacher in Humanities, as will be seen, there are words that I want to use which science threatens to deny me : I want to speak of 'higher things', a 'gradient' in nature, *order*, *harmony*, *direction*, *primary consciousness*, *intelligence*, *striving*, *ingenuity*, *achievement* and *aims*. The upshot of any exploration of the debate will be, I hope, that these words and the thinking that goes with them, are perfectly legitimate.

Appendix A: School text-books and evolutionary theory

It yields quite a shock, to turn from reading leading scientific exponents on evolutionary theory, to school text-books on the subject. While I was writing this book I borrowed my son's biology books. The first was *A New Biology*, by K.G. Brocklehurst M.A. M.I.Biol., Senior Biology Master, Warwick School, Formerly Examiner to the Oxford and Cambridge Schools examinations Board and Helen Ward B.Sc., M.I. Biol., Formerly Head of the Science Department, North Manchester Grammar School for Girls.

After summarising Darwin's theories of Natural Selection these authors say,

> It says much for his clear insight into the life of whole populations that the main idea of his theory of natural selection has stood up to being tested experimentally on a variety of organisms.
>
> p.252

As Norman Macbeth says,

> it is impossible to test large areas of Darwinism by experiments, since the crucial events either happened in the past or would need thousands of years before a conclusion could be reached.
>
> p.93.

He quotes on this matter Edward S. Deevey, 'The reply : a letter from Bronam Wood', 1967 *Yale Review* 61, pp.631-640 ; David Lack, Darwin's *Finches*, 1947, p.118 ; Ernst May's, 'Cause and Effect in Biology', *Science*, 13, pp.1501-1506, 1961. G. Ledyard Stebbings, *Variation and Evolution in Plants*, 1950, p.106-7.

It would be interesting to know what experiments Brocklehurst and Ward refer to. The question arises in relation to natural selection, whether this theory can satisfy Karl Popper's requirements at all. Popper, as a professor of logic and scientific method, insists that 'A theory which is not refutable by any conceivable event is non-scientific. Irrefutability is not a virtue of a theory (as people often think) but a vice'. (*Conjectures and Refutations*, 1963, pp.33-37).

In discussions of natural selection scientists often take a stand, not on whether the theory is reputable or irreputable, but on it being the only theory offered, as Macbeth shows.

G.C. Simpson, for instance, says that :

> The origin of such an organ as the eye, for example, entirely at random seems almost infinitely improbable...

There must, he says, have been 'some additional fact or process' and this *must have been natural selection*! (*This View of Life*, 1969, pp.18-19, 188, 207-208). As Macbeth declares,

> he creates a vacuum, offers natural selection as the only remaining possibility, and regards this as a proof that natural selection can do anything. It is unnecessary for him to show what natural selection can do. A logician would call this begging the question.
>
> p.91.

The school text-book seems thus quite wrong in saying that Darwin's theories of natural selection have been confirmed by experiment. They cannot be. And many scientific arguments in favour of natural selection are based on debaters' tricks. Nature is full of marvels : these can only have been produced by a watchmaker or natural selection. A watchmaker has no standing in science, ergo it must have been natural selection. The greater the improbability, the greater glory of natural selection, which is the only candidate! (Based on Macbeth pp.90-91).

Later the school text-book declares :

> about two million years ago the first human types developed from an ape-like ancestor...
>
> p.283.

This is presented as a fact. The truth is (as Macbeth puts it) :

> The remains found in the earth's crust are not those of our ancestors, while the bodies of our ancestors have not been preserved at all... we have no idea who our common grandparents are...
>
> p.14.

No-one is sure about the ancestral forms (p.142). In this school text book a postulate is presented as a proven fact. A sentence or two later the statement is properly qualified.

> Man certainly cannot have evolved from any one of the four modern apes, but it is likely that he shares with them a common ancestor.
>
> p.253.

As well as the misleading diagram showing the evolution of the horse, and an evolutionary tree of doubtful nature, *Illustrated Biology*, by B.S. Beckett (OUP, 1978) also has a section on *Evidence of Evolution from Anatomy and Embryology.*

> Embryos of mammals, birds, reptiles, and amphibia are so similar to fish embryos that it is difficult to tell them apart. All of these, including human embryos, have gill slits, a fish-like heart (with only one atrium and ventricle), fish-like kidneys, and a muscular tail. Human embryos later develop an amphibian heart (with two atria and one ventricle).
>
> Then their gill slips disappear and they develop reptition kidneys. Finally they develop a four-chambered heart, lose their tail, and develop mammalian kidneys. In other words, during embryonic development humans go through some of the stages by which they evolved from fish into anphibia, reptiles, and finally mammals.
>
> p.76.

It was Ernst Maecell (1834-1919) who carried the Darwin banner in Germany and propounded the biogenetic law, which declared that the growth of the embryo was a recapitulation of the history of the species.

English biologists however are not so convinced. Sir Gavin de Beer says that it is misleading than helpful and should be rejected (See G. Ledyard Stebbins, *Variation and Evolution in Plants*, 1950, p.488). G.G. Simpson calls it 'the overgeneralised and much abused aphorism of the nineteenth century' (*The Meaning of Evolution*, 1949, p.218). Embryology provides plenty of hints, but offers no solutions : it does less than Haedel asserted and less than Darwin hoped, declares Macbeth.

Both embryology and comparative anatomy suggest that certain animals and plants are cousins : but they do no tell us where they come from. They do not give the phylogenies.

In the two biology text-books under discussion there are assertions of fact which ought, in honesty, to have been presented as theories : little more than postulates. There are diagrams which totally falsify the latest observations of the development of the horse (see Hardin quoted above p.21). There are diagrams of embryo development and of bone structure given to provide *evidence* for evolutionary theory on Darwinian principles. One has an evolutionary tree in which (without explanation as to why this should be so) some creatures (e.g. the lungfish) do not change over millions of years, while others alongside change from being reptiles to snakes, or feathered lizards to birds. As Macbeth says, this sort of thing is very hard to swallow. No doubt the same excuse would be given, as was given to Mr. Macbeth :

> the audience for which this book was intended has very little scientific background and it is for this reason that the material is handled as it was...
>
> p.45.

Macbeth finds this double standard improper and unwise : all the illustrations and diagrams provided in these books are doubtful. They are offered as if the problems had been solved or were, indeed, solved by the diagrams, which make it seem evolutionary theory *must be correct*

: is a fact. Yet there are as yet no such solutions in evolutionary theory, and there is a great diversity of views, and it is dishonest - and, if one may say so - most unscientific, to make it seem to innocent children that this is not so.

The issue has surfaced yet again in *The Times* Here is a letter written in reply to a supporter who offered the Darwinian account of the evolution of the horse as 'fact' :

> Perhaps Dr. Jim Dorreen should get his facts straight before launching into totally unproven statements regarding the alleged 'horse series', or he might give credence for the rising tide of suspicion that much fossil research in our institutions bears some comparison with Haeckel's forgeries, the Piltdown hoax and the Cyril Burt scandal.
>
> These fossil mammals are nether totally equinine nor a stratigraphical series : the first two examples commonly quoted (Eohippus/Hyracotherium & Mesohippus) were, by dentition, browsers (eg deer-like) and not grazers. Nowhere in the rocks are all the species directly in sequence above each other, starting with the smallest at the bottom.
>
> The text-book representations always show a small animal at the start and indicate that, with extended time, they have become larger : few students who read these books realize that the King's Troop of the Royal Horse Artillery have tiny Falabella (Argentinian) horse (27in high) in their stables. The miniature horse farm at Inman, South Carolina, USA, also rears very small breeds.
>
> The size of horses is therefore no indication of age or descent. All the species shown in the charts could have lived simultaneously.
>
> (B.W. Grantham-Hill, 13 May 1982)

A further letter put the matter succinctly, for those in education to study :

> How long are we expected to tolerate the confusion in the present debate over Darwinism?
>
> According to Sir Karl Popper, scientific theories are recognized, amongst other things, by their openness to falsification. The good scientist should be looking out for ways to refute rather than confirm theories.
>
> Popper's philosophy of science classifies Darwinism as a meta-physical (non-scientific) theory because its claims are not open to falsification. Those aspects of the theory that can be tested have given negative rather than positive results.
>
> Research in genetics and breeding has shown that natural selection takes place at what Humphrey Greenwood calls 'fine tuning' and consequently cannot be used to explain complex changes. No one, as yet, has produced a new species. The fossil records shows an absence of intermediate forms, which contradicts Darwin's idea of gradual change.
>
> If natural selection has little application when explaining complex changes, and the fossil record is inconsistent with Darwinism, then what does Darwinism scientifically explain?
>
> 'Belief' in evolution is a different matter. Beliefs can exist independent of evidence. May I recommend that scientists involved in the present debate about Darwinism become more familiar with ways of distinguishing science from non-science?
>
> Leslie Cunliffe.

These letters suggest a debate which young biologists would enjoy pursuing. and would learn much from. Why do so many Biology textbooks prefer an erroneous dogmatism?

Appendix B: M. Pitman's evolution? A science teacher's doubts

A science teacher in Cambridge, Mr. M. Pitman, has listed other objections to conventional evolutionary theory, in a paper entitled '*Evolution*?' (Since developed into a book, *Adam and Evolution*, Rider, 1985).

First, he lists six objections to the postulated primeval reducing atmosphere, and many to various 'primordial soup' theories, all on physical and chemical grounds e.g. 'There exists a thermo-dynamic barrier' to the spontaneous formation of complex molecules such as proteins or DNA. There is practically no tendency for such compounds to form, but, on the other hand, they readily disintegrate. Some specific mechanism is required to drive the bonding process 'uphill'.

Millions of dollars of grant-aid have been devoted to investigations of the origins of life, to which scientists have devoted their lives, on erroneous principles : Professor A.E. Wilder Smith is quoted as saying 'we cannot - and should not - on theoretical and experimental grounds, expect any code to arise by random mechanisms ... (it) flies in the face of all scientific sense' :

> ... randomness in any code sequence progressively destroys the code. In fact, code sequences and randomness are incompatible. Randomness destroys code and putting a code on to a randomly arranged thread of binomers ('building blocks') will destroy randomness. If randomness and code sequences then are so mutually destructive, how can we ever come to the ridiculous conclusion that randomness gave spontaneous birth to code sequences of the super-specifity of the genetic code?... The whole idea is one huge paradox.
>
> *The Creation of Life*

It has been estimated, says Pitman, that the odds against chance producing a specific L-form amino acid chain of 100 units - the molecule of a single protein - in five million years (the supposed age of the earth) is one in 10^{71}.

Turning to mutations, Pitman quotes Huxley, who, admitting an exceedingly generous rate for favourable mutations (one in a thousand) calculated that the odds against one million successive, successful mutations which might result in the evolution of a horse was 10^{299843} to 1. It would take three 500 page volumes to write down the zeros : the time required would be hundreds of times the estimated age of the earth.

Many questions remain to cast doubt on evolutionary theory. If natural selection is supposed to weed out unfit kinds, it should decrease and not increase the number of species of creature. If the criterion of fitness is reproductive capacity, then why haven't bacteria, herring or rabbits overrun the earth? Why should protozoa ever have 'evolved' out of their extremely successful rut?

Mutation and variation could never bring about all the simultaneous complex biochemical and structural changes needed for the postulated transformation of one creature into another. Genetical systems deny it. Observation has not once witnesses it (despite endless scientific breedings of fruit-flies, E. Coli, and laboratory animals). The fossil evidence for such gradual change, which should be prolific, is absent. No 'common' or generalised ancestors have been found. Nor are transitional forms found.

The whole story of evolution is not only packed with massive improbabilities, it is constructed of whole sequences of them. Flight, with all the structural alterations it entails, has evolved on at least four seperate occasions (insects, pterodactyls, birds and bats). Eyesight has evolved on three occasions - with insects, squids and vertebrates.

There are puzzles in the fossil record. While millions of fossils are found in Cambrian rocks (600,000,000 years ago), not a single undisputable multicellular fossil has ever been found in pre-Cambrian rocks (1,000,000,000 - 600,000,000 years ago). Where did the Cambrian fossils evolve - a process that (according to evolutionary theory) must have taken 1,500,000,000 years?

There are gaps in the fossil record. But one great puzzle is the absence of transitional forms. If, for instance, insects developed from worms, there should be intermediate forms. While evolutionists like to draw 'trees' showing unbroken lines of development, this is not in fact possible. In plants, for example, 'we have not been able to trace the phylogentic history of a single group of modern plants from its beginning to the present' (Professor C.A. Arnold).

Transitional forms are lacking in the area of the supposed evolution of vertibrates from invertibrates from invertebrates. How did the endo-skeleton develop from the exo-skeleton? For a worm to develop into a vertebrate, five major structural transformations, let alone the growth of the endo-skeleton, are required – and no fossil evidence exists to show that they occured.

Here there are many problems. For example, when fish developed into amphibians, the eye had to change. The eyes of a fish, being lidless and unlubricated, would dry up and become infected. Lids and tear-glands developed - but how? A way of secreting liquid fluid inside the eyeball in order to change the refractive system in air came about – can we conceive of such ingenuity coming into being simply by a lucky chance mutation being preserved by brute necessity?

Other evolutionary developments pose similar questions. Reptiles lay their eggs out of water but the infant needs an aquatic environment, and also a lot of food as the infants are hatched fully developed. So the reptilian egg has to contain a large mass of yolk for food and albumin for water. Neither yolk or egg-white would give 'selective advantage'. To avoid evaporation, the egg-white would need a vessel to contain it - the shell. In such a shell the embryo could not get rid of waste products, so it is provided with a kind of bladder – the allantois. It also needs a chipping tool, to develop at exactly the right moment and place, along with the appropriate intinct. Imagine providing all these forms of equipment for a space-man in a capsule : what imagination and planning would be needed. Can we imagine that such interlocking, complex, forward-looking provisions can have come about by small chance changes? None of the could have given small advantages, and it any part of the provision were not fully available, the embryo would die. There is nothing in Darwinian theory to account for such a marvel.

Did birds develop from reptiles? There is no evidence for this remarkable change, which would have required the development of feathers from scales (can we imagine the feather developing from some chance mutation?). Cold blood to warm blood, non-flight to flight. There is no fossil evidence. Birds' eggs have to be incubated at a certain temperature : how did the instinct for this evolve?

For the (mammalian) bat to develop, there had to evolve a special bone structure, wing development, nictitating membrane, musculature with special tendons for hanging upside down, modified diaphragm, special blood vessels and nerve fibres – besides the complex sonar system for navigation. Each of those developments by itself would give no immediate advantage : some would be deleterious. They all had to come into being in a coordinated way. If any came into being separately, surely 'natural selection' would have weeded out the half-organised variety? When fossil bats first appear, in the middle Eocene period, they have all the characteristics of their order. There are no hints of evolution in the rocks : and can we believe chance and selection could have produced them?

Similar problems arise over complex organs in many animals : the different ways of construction found in ears; the different ways in which jaws are made (The organ of Corti, says Pitman, in the inner ear of mammals, comprises some 3,000 arches placed side to side so as to form a tunnel). There are differences in the complex chemistry of the blood of mammals and reptiles - warm and cold systems. Mammal blood corpuscles are non-nucleated. If this is an evolutionary step to make for greater economy, why has it not happened in birds and reptiles? In mammals the main produce of excretion is urea : in reptiles it is uric acid : how could evolution produce such a metabolic ambivalence? How did hair come to replace scales? Since reptiles do not have diaphragms, if reptiles changed into mammals, how did they manage to breathe during the transition? Where did the diaphragm evolve? How did milk develop – that is both mammary glands and the instincts to suck and suckle?

To read of all these problems makes one feel dizzy, not least because Darwinian theory seems less and less plausible, despite the conviction with which it is taught and held. What is clearly seriously lacking is any evidence of half-completed evolution : there are no half-hearts, half-stomachs, half-wings. In the world of today's mammals, there are no intermediaries, no suggestions of 'new kinds'.

Pitman next discusses Geology, pointing out that the *Origin of Species* appeared soon after the uniformitarian understanding of geology was propounded by Sir Charles Lyell in the 1830s. In 1863, Lyell published *The Antiquity of Man*, while Darwin confessed that many of his ideas came out of Sir George Lyell's brain.

Fossil remains are often used to date rock formations : certain fossils are called 'index fossils' (the Paradoxides trilobite is an index for Cambrian rocks and so on). Great embarrassment has been caused when such index fossils have been discovered still alive today (like the coelacanth).

The arrangement of fossils in a supposed time sequence is called a geological column. The arrangement of this column *assumes* evolution. And here we have a tautology :

> The assumptions of evolution are the basis upon which fossils... are used to date the rocks and these same fossils are supposed to provide the main evidence for evolution... Although the fossil has been interpreted to teach evolution, the record record itself has been based on the assumption of evolution. The main evidence for evolution is, therefore, the assumption of evolution.
>
> Pitman, p.33.

Some scientists, like the catastrophists, challenge the established 'geological column', and there are clashes between time estimates by radioactivity and paleontology. There could have been catastrophic disasters which overwhelmed the earth, including volcanic and other upheavels, axis wobbles, asteroid collisions and changes in the earth's magnetic field.

Certain discoveries certainly present a challenge to orthodox theory – e.g. the apparent discovery of human and dinosaur footprints found in Texas in the 1960s.

In all, Pitman suggests, established geology and paleontology are coming under scrutiny, and it seems that the fossil record does not fully support the general transformist theory of evolution.

Pitman next discusses the development of the horse, problems over which are discussed above (p.208). He quotes Professor G.A. Kerkut in *The Implications of Evolution* : 'it is a matter of faith that the textbook pictures are true, or even that they are the best representations of the truth that is available to us'.

The problem when it comes to man is more acute. Primates are supposed to have evolved from an insectivorous ancestor, but no series of transitional forms connecting them with such a type has been found. The first stage is missing. And as for the various family trees put forward, Pitman dismisses these as science-fiction, products of imagination, and as evolutionary fervour.

Ever since Darwin the resemblances between man and ape has been stressed. The differences, however, are even more striking. The ape's brain is 90-685 ccs in capacity : man's is 790 - 2350 ccs. Man's brain is highly developed and he has intellectual, inventive and aesthetic impulses. He has language. Man's opposable thumb enables him to use tools.

He has a upright posture : the whole pelvic structure of the ape is here different (any change from ape to man could not, of course, come by 'adaptation' according to strict Darwinian theory, but only by chance genetic mutations). Man's jaw is parabolic, rather than U-shaped and does not contain massive canines like an ape's jaw. In the sexual organs, human females have a hymen, sperm and ovum differ and the chromosome count is different. There are differences in the blood, in milk and other factors. There are some 150 differences between man and ape : for an ape to turn into a man there would have to be huge changes, to the brain, skeletal structure, and so on, far beyond any developments made possible by chance mutations. But was there a common ancestor? This was supposed to have existed 15 - 25 million years ago. The oldest skulls ever found, however, 10,000,000 years old, are practically indistinguishable from those of a modern European (Castenedolo, Italy). Neanderthal man (dated 100,000 - 250,000 B.C.) has been presented in imaginative drawings of heavy-browed and moronic-looking. In truth, 'the cranial capacity of the Neanderthal race of Homo Sapiens was, on the average, equal to or even greater than that in modern men'. (Dobzhansky, also points out that this is no criterion of intelligence). He also grew flowers, painted pictures, fashioned tools and buried his dead – so was no 'missing link', and was very different from an ape.

In this area, as Pitman makes clear, there are many frauds, flights of fancy, and speculations : 'artists' impressions' from skull relics are common, and often very different from one another. Passions run high in this sphere : Teilhard de Chardin is suspected of a forgery : one scientist refused to chair a meeting about Pithecanthropus Erectus, declaring the bones to be those of a giant gibbon. Yet the photographs of models and reconstructions appear automatically in childrens' biology books and higher level texts. In this matter, says Pitman, 'scientific integrity' has 'severely degenerated'. Behind this falisification is the need to preserve evolutionary dogma at all costs.

Pitman summarises this area of science thus : no proof of a missing link between ape and man has ever been established, though this is absolutely necessary for evolution. (Perhaps apes and men were always distinct types ?). There are no complete and clear skeletons of an ape-man. There is no sign today of apes turning into man, nor any sign of any species turning into another. And all this, despite considerable efforts to prove this key connection in evolutionary theory.

Some people, says Pitman, equate religion with superstition, and science with fact. Over evolutionary theory, however, the reverse seems true.

Pitman next discusses homology : the similarity between animal structures. Yet, as Sir A. Hardy said, this cannot be explainesd in terms of present-day biological theory. The 'coded-information' for intance behind a vertebrate column is too complex to have arisen from chance : it could not be stored in the genetic code. It seems (as it does to Sheldrake) the product of a plan, rather than order derived by chance from chaos. As Tomlin puts it, a structure cannot come from chaos but only from another structure : but what formed it in the first place? It is this kind of question which it is supposed to be *unreasonable* to ask, in evolutionary theory.

At the same time, evolutionary dogma insists that a few principles must be employed to explain everything, even though such a reduction is impossible. As Waddington puts it :

> The phenomenon if individuation by which cells become arranged into organs with definite shapes and patterns (is such that) biologists have to confess they still have hardly any notion of how this is done. It certainly must involve something more than purely chemical processes. *Development starts from a more or less spherical egg and from this there develops an animal which is anything but spherical... One cannot account for this by any theory which confines itself to chemical statements, such as that genes control the synthesis of certain proteins.*

Pitman next discusses the theory put forward by Kaeckel who believed that 'ontogeny recapitulates phylogeny', that an embryo passes through all the previous evolutionary stages. This theory is now discounted. But so passionately did Kaeckel want to believe it that he faked some of the drawings and was prosecuted and convicted by a university court.

Study of embryology, suggests Pitman, suggests a 'basic plan of development in all creatures, while the amazing progression suggests that there must be 'intelligent, economical programme and design' (Tomlin here speaks of primary consciousness).

Pitman next returns to the problem of the eye. Sherrington says that : 'beyond the intricate mechanism of the human eye lie breath-taking glimpses of a Master Plan'.

The eye appears in the fossil record fully formed. Can we believe that it began with a freckle or a light sensitive pigment spot which was gradually transformed by chance, until it became a working, *purposeful*, mechanism?

Consider the eye :

> each retina, consisting of 130 million correctly made and positioned cells : these are the rods and cones. In the human bi-convex lenses are free from blood vessels, focussing apparatus is exquisitely refined and a lid is present to clean and protect the surface. Each eye, itself just one of thousands of items which have developed from an ovum which is not one ten thousandth part as large, sends an estimated 1,000,000,000 impulses per second to the brain. This electrical storm is continuously translated... into a mental picture.

'Man sees, not the brain' : and so, we have, in addition to this working complexity, all the behavioural realities of perception : of positive seeing, such as Erwin Straus discusses.

What is there in evolutionary theory that could account for such an achievement? It would be of no use until it was complete and functional, and until it was, 'natural selection' would have eliminated it, surely? What use would a hole in the front of the eye be if it were caused by genetic mutation, but if there were no cells to receive the light? What use would the retina image be if there were no nervous system to interpret it?

The eye is also directed to an *end* : it is teleological. Some fish for example (Anableps) have bi-focal vision : how do we account for that by 'natural selection' ?

Pitman discusses various other examples, to show that the general theory of evolution by random mutation and natural selection can offer no real explanation.

Despite all these objections, evolutionary theory persists in the present world-view of intellectuals in every field.

Ingenuity in nature is perhaps the feature most impossible to explain on the basis of evolutionary theory. Pitman discusses the electric ray; the fish Gymnarchus Niloticus which has an electromagnetic underwater radar system; the angler fish which hangs out lumninous bulbs to lure its prey; the ink-clouds of squids; the lights on other fishes *which they can switch on and off* ; the bombadier beetle which by instinct injects into a 'gun' in its tail 10% hydroquinines and 23% hydrogen peroxide stored in an inhibitor : this inhibitor is neutralised when the liquid reaches the combustion chamber. (By what stages could such a system be evolved, since at any intermediate stage it would harm its bearer?).

The metamorphosis of caterpillars is a process by which they go through a kind of resurrection. Nothing in evolution could begin to explain such a process. What, indeed, does natural selection have to say about the scales on a butterfly's wing :

> The scales overlap like tiles on a roof and their surfaces are covered with row upon row of microscopic, transparent, glass-like blocks, whose floors are only about one hundredth-thousandth of an inch apart. The distance between the reflecting floors is half the wavelength of the light to be reflected : the patterns are not due to pigment but to reflected light.

How could mere negative processes produce such a piece of precision engineering?

The female mosquito is equipped with saws, lancets, syringe and siphon, contained in its beak. The walnut shell cannot be improved upon as a piece of engineering. The methods of predators for catching and digesting their prey display fantastic ingenuity.

How did spiders learn to produced and weave silk? And to make that fantastic piece of engineering, a web? What suggested to balloon-spiders that they spin parachutes of silk which they use to transport themselves across fields? Such skill are passed down by instinct, without thought or instruction : in what does this reside? Glow-worms anaesthetise snails, and hunting wasps paralyse caterpillars, leaving them alive to be fresh food for their larvae. This might haunt Darwin, but it cannot be explained on his principles - such instinctual behaviour could not have come into existence by his theory.

Nests, courtship-customs, the behaviour of bees, the systems of plant pollination, plant carnivores, various forms of symbiosis – all of these, as Pitman shows, raise problems which cannot be answered by evolutionary theory, and which show it ot be inadequate as an explanation of development. In many instances, if 'natural selection' were the predominant force, as it is in Darwinism, it would have weeded out most of these significant kinds of ingenuity.

Pitman ends his fascinating essay by glancing at the work of Pierre-Paul Grassé and Norman Macbeth, pointing out that every objection simply brings from the evolutionists increasingly dogmatic statements :

There are philisophical and methodological objections to evolutionary theory... It is too difficult to imagine or envisage an evolutionary episode which could not be explained by the formulae of neo-Darwinism.

Sir P. Medawar.

The theory cannot be falsified (according to the requirements of Karl Popper for scientific hypotheses), and cannot be refuted by any possible observation : so, the evolutionist simply can retreat to the broad dogma of his belief.

The teaching, Pitman believes, is now *doctrinaire*, and *alien to the spirit of academic freedom*. What goes on at the moment is a kind of brain-washing, based on falsifications and the reiteration of dogma in a faith.

All evidence favourable and contradictory needs to be fairly presented. The general humanistic philosophy based on evolutionary theory needs to be re-examined.

But evolution has been observed by no witnesses, and cannot be put to the test of experiment : the newest attempts have yielded negative results. Sir Julian Huxley may say : 'Our present knowledge forces us to the view that the whole of reality is evolution – a single process of self-transformation'. ('Evolution and Genetics'). But mathematical calculations seem to menace such pronouncements. In 'Mathematical challenges to the neo-Darwinian interpretation of Evolution', P. Moorhead and M. Kaplan conclude that 4,500,000,000 years is far too short a time for such organic complexity as our world displays to have arisen by chance mutations (leaving aside the question of whether it ever could).

Pitman wants creationism and evolutionary theory to be examined side by side. For my part, I believe the former offers no useful explanation. However, it seems clear that the question must be left open. And, as Pitman makes clear, it is certanly an issue that extends beyond the science classroom 'into the areas of social, moral and religious studies, politics, philosophy and history'. Evolution is even invoked to endorse some general vague notion of 'progress', while one finds it working behind all kinds of theories about human nature : at the time of writing it becomes clear from an article in *The Times* that Darwinian ideas of 'survival' lurked behind Melanie Klein's theories in psychoanalysis of a 'life instinct' in contest with a 'death instinct'. Other connections with Freud and Marx are clear in terms of the 'functional' model of man : Darwinian mechanism is fundamental to Marxist materialism.

But on what grounds can a really radical general philosophy be based on a theory which can offer so little to explain the primary processes and features of life on earth?

Note : It has not been possible to deal fully with Michael Pitman's book which appeared as this was going to press. He makes the following important points, in his critique of Evolutionary theory : 'no nascent organ has ever been observed emerging, though their origin in pre-functional form is basic to evolutionary theory' (p.67); 'the origin of life at every level needs to be thought of in terms of the origin of whole organisms rather than of self-replicating DNA molecules...' (p.135). 'Even the primordial cell must have been a miniature 'superlab' in which the most sophisticated, critical and purposeful series of interrelated reactions was carried out. Where did the coded information and precise retrieval systems arise to guide and control these labryrinthine pathways for releasing energy? How could this information possibly have been marshalled *fore* the cell chemistry as a whole became functional?'

Mr Pitman's most telling objections concern the eye (pp.215ff). 'Not only is there a problem of complex coordination, but the eye must be near perfect before it works at all, while there is also the question of seeing. Of what survival value is a lens... if not linked to a nervous system... Or a nerve without a brain to interpret the data? How could a visual nervous system have evolved before there was an eye to give it information? So questions continue until all parts of the body are woven into a single whole, a web of mutual necessity'.

Bibliography

A Abercrombie, M., C.J. Hickman, and M.L. Johnson. *A Dictionary of Biology*, Penguin, 1966.

Agassiz, Louis. *Methods of Study in Natural History*, 1863.

Arber, Agnes. *The Mind and the Eye*, Cambridge, 1954.

Ardrey, Robert. *African Genesis*, Collins, 1961.

Ardrey, Robert. *The Territorial Imperative*, Collins, 1967.

Ayala, F.J., and Dobzhansky, T. (eds.) *Chance and Creativity in Evolution*, 1974.

Ayala, F.J., and Dobzhansky, T. (eds.) *Studies in the Philosophy of Biology*, Macmillan, 1974.

B Baldwin, J.M. *Development and Evolution*, Macmillan, 1902.

Barnett, S.A. *Instinct and Intelligence*, MacGibbon and Kee, 1967.

Barzun, Jacques. *Darwin, Marx, Wagner, Critique of a Heritage*, Doubleday, 1958.

Barzun, Jacques. *Science, the Glorious Entertainment*, Harper and Row, 1964.

Beloff, J. *The Existence of Mind*, MacGibbon and Kee, 1962.

Bergson, H. *The Creative Evolution*, Macmillan, 1922.

Bergson, H. *Matter and Memory*, Allen and Unwin, 1911.

Bernal, J.D. *The Origins of Life*, Weidenfeld and Nicholson, 1968.

Bertalanffy, Ludwig von *Modern Theories of Development*, Harper, 1962.

Beveridge, W.I.B. *The Art of Scientific Investigation*, Heinemann, 1950.

Bonner, J.T. *The Evolution of Development*, Cambridge, 1958.

Bonner, J.T. — *The Ideas of Biology*, Harper, 1962.

Bose, J.C. — *The Nervous Mechanism of Plants*, Longmans, 1926.

Brain, Russell — 'Body, Brain, Mind and Soul', in *The Humanist Frame*, Allen & Unwin, 1961.

Burgess, J.M. — *Experience and Conceptual Activity*, MIT, 1965.

Burtt, E.A. — *The Metaphysical Foundations of Modern Physical Science*, Allen & Unwin, 1962.

Burtt, E.A. — *In Search of Philosophic Understanding*, Allen & Unwin, 1967.

Buytendijk, F.J.J. — *Das Menschliche, Wege zu seinem Verständis*, Kohler, 1958.

Buytendijk, F.J.J. — *Mensch und Tier*, Rowohet, 1958.

Buytendijk, F.J.J. — *The Mind of the Dog*, Houghton Mifflin, 1936.

C

Capek, M. — *Physics*, van Nostrand, 1961.

Cassirer, Ernst. — *An Essay on Man*, Yale, 1944.

Chargaff, Erwin. — *Heraclitean Fire : Sketches from a Life Before Nature*, New York, 1978.

Collingwood, R.G. — *An Autobiography*, Oxford, 1939.

Collingwood, R.G. — *The Idea of Nature*, Oxford, 1945.

Crick, F.H.C. — *Of Molecules and Men*, Washington, 1967.

D

Darlington, C.D. — *Evolution of Genetic Systems*, Oliver and Boyd, 1958.

Darwin, Charles. — *The Power of Movement in Plants*, Murray, 1880.

Darwin, Charles. — *The Movements and Habits of Climbing Plants*, Murray, 1882.

Darwin, Charles. — *Life and Letters*, edited by Francis Darwin, Murray, 1887.

Darwin, Charles. — *The Origin of Species*, John Murray, 1959.

Darwin, Charles. — *The Descent of Man in Relation to Sex*, Murray, 1870.

Darwin, Charles. — *The Variation of Animals and Plants under Domestication*, Murray, 1875.

Darwin, Charles. — *More Letters*, edited by Francis Darwin, Murray, 1903.

Darwin, Charles. — *Autobiography* London, 1958.

Dawkins, Richard. — *The Selfish Gene*, Oxford, 1976.

De Beer, Sir Gavin R. — *Embryos and Ancestors*, Oxford, 1958.

De Beer, Sir Gavin R. — *Evolution*, British Museum Handbook, 1958.

De Beer, Sir Gavin R. — *Homology : An Unsolved Problem*, 1971.

Denton, Michael. — *Evolution : A Theory in Crisis*, Burnett Books, 1985.

Driesch, H. — *History and the Theory of Vitalism*, Macmillan, 1914.

Driesch, H. — *Science and the Philosophy of the Organism*, Macmillan, 1914.

Dreisch, H. — *Mind and Body*, Methuen, 1927.

Drury, M.O'C. — *The Danger of Words*, Routledge 1973.

Dobzhansky, Theodosius. — *Genetics and the Origin of Species*, Columbia, 1941.

Dobzhansky, Theodosius. — *Mankind Evolving*, Yale, 1952.

Dobzhansky, Theodosius. *The Biology of Ultimate Concern*, New American Library, 1967.

Dobzhansky, Theodosius. *The Biological Basis of Human Freedom*, Oxford, 1956.

Duncan, Ronald and Weston-Smith, Miranda (eds.) *Encyclopedia of Ignorance*, Pergamon, 1977.

E

Eigen, M. and Schuster, D. *The Hypercycle*, 1979.

Eiseley, Loren. *The Immense Journey*, Vintage, 1958.

Eiseley, Loren. *The Firmament of Time*, Atheneum, 1960.

Eiseley, Loren. *Darwin's Centenary*, 1961.

Elasser, W.M. *The Physical Foundations of Biology*, Pergamon, 1958.

Elasser, W.M. *Atom & Organism*, Princeton, 1966.

Elasser, W.M. *The Chief Abstractions of Biology*, North Holland, 1975.

Emmet, D. *Whitehead's Philosophy of Organism*, Macmillan, 1966.

F

Fisher, R.A. *Creative Aspects of Natural Law*, 1950.

Fisher, R.A. *Genetical Theory of Natural Selection*, Clarendon Press, 1930.

Frey-Wissling, Albert. *The Morphology of Organic Forms*, 1977.

G

Ghiselin, Michael T. *The Triumph of the Darwinian Method*, California, 1969.

Gillespie, C. *The Edge of Objectivity*, Princeton, 1960.

Goldschmidt, R. *The Material Basis of Evolution*, Yales, 1940.

Goldstein, Kurt. *The Organism*, American Books, 1939.

Gould, Gray and Asa. *Darwiniana*, Harvard, 1876.

Gould, Stephen J. *Ever Since Darwin*, New York, 1977.

Gould, Stephen J. *Hen's Feet and Horses' Toes* W.W. Norton, 1983

Gould, Stephen J. *The Panda's Thumb* W.W. Norton, 1980

Graham, L.A. *Science and Philosophy in the Soviet Union*, Knopf, 1972.

Grassé, Pierre-Paul. *L'Evolution des Etres Vivants*.

Greene, John C. *Darwin and the Modern World View*, Mentor, 1963.

Grene, Marjorie(ed.) *Dimensions of Darwinism:Themes and Counterthemes in Twentieth Century Evolutionary Theory*, Cambridge University Press, Editions de la Maison des Sciences de l'Homme, 1986.

Grene, Marjorie. *The Knower and the Known*, Faber, 1966.

Grene, Marjorie. *Approaches to a Philosophical Biology*, Basic Books, 1968.

Grene, Marjorie. *The Understanding of Nature*, D. Reidal, 1974.

Grene, Marjorie. *Philosophy In and Out of Europe*, California, 1976.

Guntrip, Harry. *Schizoid Phenomena, Object-relations and the Self*, Hogarth, 1968.

H

Hardy, A. *The Living Stream*, Collins, 1965.

Hardin, Garrett. *Nature and Man's Fate*, Mentor, 1961.

Harré, Rom. *Principles of Scientific Thinking*, 1970.

Hinde, R.A. *Animal Behaviour*, McGraw-Hill, 1966.

Himmelfarb. *The Darwinian Revolution*, Chatto, 1959.

Hofstadter, Richard. *Social Darwinism in American Thought*, Beacon, 1955.

Hoyle, Fred and Wickramsinghe, C. *Lifecloud*, Dent, 1978.
Hoyle, Fred and Wickramsinghe, C. *Evolution from Space*, Dent, 1981.
Hull, D.L. *Darwin and his Critics*, Harvard, 1973.
Husserl, Edmund. *The Crisis of European Sciences and Transcendental Phenomenology*, Northwestern, 1970.
Hutchinson, Peter. *Evolution Explained*, David and Charles, 1974.
Huxley, J. *Evolution and the Modern Synthesis*, Allen & Unwin, 1942.
Huxley, J. *The Uniqueness of Man*, Chatto, 1942.
Huxley , J. *The Humanist Frame*, 1961.
Huxley, J. Hardy, A.C. and Ford, E.B. (eds.) *Evolution as a Process*, 1954.
Huxley, T.H. *Darwiniana*, Appleton, 1893, 1901.
Huxley, T.H. *Life and Letters*, Macmillan, 1908.

J

Judson, Horace Freeland *The Eighth Day of Creation*, Cape, 1979.

K

Kammerer, P. *The Inheritance of Acquired Characteristics*, Boni & Liveright, New York, 1924.
Kellogg, Vernon. *Evolution, the Way of Man*, Appleton, 1925.
Kierkegaard, Soren. *Concluding Unscientific Postscript*, 1846.
Klopfer, P., and Hailman, J. *Animal Behaviour*, Prentice Hall, 1967.
Koestler, A. *The Ghost in the Machine*, Hutchinson, 1967.
Koestler, A. *The Case of the Midwife Toad*, Hutchinson, 1971.
Koestler, A. and Smythies, J.R. (eds.) *Beyond Reductionism*, Hutchinson, 1969.
Kohler, W. *The Mentality of Apes*, Harcourt Brace, 1925.
Krutch, Joseph Wood. *The Great Chain of Life*, Houghton Mifflin, 1956.
Kuhn, T.S. *The Structure of Scientific Revolutions*, 1962.

L

Lack, David. *Darwin's Finches*, Cambridge, 1947.
Lack, David. *Evolutionary Theory and Christian Belief : The Unresolved Conflict*, Methuen, 1967.
Lack, David. *The Life of the Robin*, Wetherby, 1946.
Leclerc, I. *The Nature of Physical Existence*, Allen & Unwin, 1972.
Le Gros Clark, W.E. *Man-apes or Ape-man?* Holt, Reinhart & Winston, 1967.
Le Gros Clark, W.E. and Leakey, L.S.B. *The Miocene Hominides of Africa, Fossil Mammals.* British Museum of Natural History, 1947.
Leith, Brian. *The Descent of Darwin*, Collins, 1983.
Lerner, I. Michael. *The Genetic Basis of Selection*, Wiley, 1958.
Lewis, John (ed.) *Beyond Chance and Necessity*, A symposium, Garnstone Press, 1974.
Lewis, John and Towers, Bernard. *Naked Ape - or Homo Sapiens?* Garnstone Press, 1969.

Leakey, Richard and Lewin, Roger. *Origins : What New Discoveries Reveal About the Emergence of Our Species and His Possible Future*, Macdonald and Jane, 1977.

Lorenz, Konrad. *On Aggression*, Macmillan, 1966.

Lorenz, Konrad. *King Solomon's Ring*, Crowell, 1952.

Lorenz, Konrad. *Physiological Mechanisms in Animal Behaviour*, Cambridge, 1950.

M

Macbeth, Norman. *Darwin Retried*, Garnstone Press, 1976.

Mackie, J.L. *The Cement of the Universe*, Oxford, 1974.

Mackinnon, D.C., and Hawes, R.S.J. *An Introduction to the Study of Protozoa*, Oxford, 1961.

Mayr, Ernst. *Animal Species and Evolution*, Harvard, 1963.

Mayer, Ernst. *Systematics and the Origin of Species*, Columbia, 1942.

Medawar, Sir Peter B. *The Art of the Soluble*, Methuen, 1968.

Medawar, Sir Peter B. *The Future of Man*, Basic Books, 1960.

Medvedev, Z.A. *The Rise and Fall of T.D. Lysenko*, Columbia, 1969.

Merleau-Ponty, Maurice. *The Phenomenology of Perception*, Routledge, 1962.

Midgeley, Mary. *Beast and Man*, Harvester, 1981.

Moorhead, P and Kaplan, M. *mathematical Challenges to the Neo-Darwininian Interpretation of Evolution*

Monod, Jacques. *Chance and Necessity*, Collins, 1972.

Montague, Ashley. *The Direction of Human Development*, Watts, 1957.

Morris, Desmond. *The Naked Ape*, Cape, 1967.

Mudford, Peter. *The Art of Celebration*, Faber, 1979.

N

Needham, Joseph. *Biochemistry and Morphogenesis*, Cambridge, 1942.

Needham, Joseph. *Integrative Levels : A Revaluation of the Idea of Progress*, Oxford, 1937.

Nicolis, G. and Prigogine, I. *Self-organisation in Non-equilibrium Systems*, Wiley Interscience, 1977.

P

Paley, William. *Natural Theology*, 1802.

Pantin, C.F.A. *Science and Education*, University of Wales, 1963.

Pantin, C.F.A. *The Relations Between the Sciences* Cambridge, 1968.

Parsons, P.A. *The Genetic Analysis of Behaviour*, Methuen, 1967.

Pattee, H.H. *Natural Automata and Useful Simulations*, Spartan Books, 1966.

Pitman, Michael. *Adam and Evolution*, Rider, 1984.

Plessner, Helmuth. *Die Einheit der Sihne*, Bouvier, 1965.

Plessner, Helmuth. *Die Stufen des Organischen und der Mensch*, de Gruyter, 1965.

(Plessner is translated in fragments in *Approaches to a Philosophical Biology*, Grene, M.

Polanyi, Michael. *Personal Knowledge*, Routledge, 1958.

Polanyi, Michael. *The Study of Man*, Chicago, 1958.

Polanyi, Michael. *Science, Faith and Society*, Chicago, 1964.

Polanyi, Michael. *The Tacit Dimension*, Routledge, 1969.

Polanyi, Michael. *Knowing and Being*, Routledge, 1969.

Polanyi, Michael and Prosch, Harry. *Meaning*, Chicago, 1975.

Poole, Roger. *Towards Deep Subjectivity*, Allen Lane, 1942.

Popper, Sir Karl R. *The Logic of Scientific Discovery*, Hutchinson, 1959.

Popper, Sir Karl R. *Conjectures and Refutations*, Routledge, 1963.

Popper, Sir Karl R., and Eccles, J.C. *The Self and Its Brain*, Springer, Berlin, 1977.

Portmann, Adolf. *Animals as Social Beings*, Hutchinson, 1961.

Portmann, Adolf. *Animal Camouflage*, Ann Arbor, 1959.

Portmann, Adolf. *New Paths in Biology*, Harper, 1964.

R

Rensch, B. *Evolution Above The Species Level*, Methuen, 1959.

Ricard, M. *The Mystery of Animal Migration*, Constable, 1969.

Riedl, R. *Order in Living Organisms*, Wiley Interscience, 1978.

Roubiczek, Paul. *Ethical Values in an Age of Science*, Cambridge, 1969.

Rudwick, M.J.S. *The Meaning of Fossils*, Neal Watson, 1972.

Russell, Bertrand. *An Enquiry into Meaning and Truth*, Allen and Unwin, 1940.

Russell, Bertrand. *Human Knowledge, Its Scope and Limits*, Allen & Unwin, 1948.

Russell, E.S. *The Directiveness of Organic Activities*, Cambridge, 1945.

Ryle, Gilbert. *The Concept of Mind*, Hutchinson, 1949.

S

Schindewolf, O. *Grundfragen der Palaentologie*, (Schweizerbart, 1950).

Schopf, T.J.M.(Ed). *Models in Paleobiology*, Freeman, 1973.

Schrödinger, E. *What Is Life? The Physical Aspect of the Living Cell*, Cambridge, 1945.

Sellers, P.W. *Evolutionary Naturalism*, Open Court, 1922.

Sherrington, C.S. *The Brain and Its Mechanism*, Cambridge, 1933.

Sheldrake, Robert. *A New Science of Life*, Blond & Briggs, 1981.

Simon, Michael A. *The Matter of Life*, 1971.

Simpson, George Gaylord. *Tempo and Mode in Evolution*, Columbia, 1944.

Simpson, George Gaylord. *The Meaning of Evolution*, Columbia, 1953.

Simpson, George Gaylord. *This View of Life*, Harcourt Brace, 1969.

Simpson, George Gaylord. *The Major Features of Evolution*, Columbia, 1953.

Smith, John Maynard. *The Theory of Evolution*, Penguin, 1958.

Spiegelberg, Herbert. *The Phenomenological Movement*, Nijhoff, 1865.

Standen, Anthony. *Science as a Sacred Cow*, Dutton, 1950.

Stanley, S. *Macroevolution*, W.H. Freeman, 1979.

Stebbins, G.L. *Flowering Plants : Evolution Above the Species Level*, Harvard, 1974.

Stern, Karl. *The Flight From Woman*, Allen & Unwin, 1966.

Straus, Erwin. *The Primary World of Senses*, Free Press of Glencoe, 1963.

Straus, Erwin. *Phenomenological Psychology*, Basic Books, 1966.

Sykes, Sylvia. *The Natural History of the African Elephant*, Weidenfeld, 1971.

T

Tax, Sol. *Evolution After Darwin*, Chicago, 1960. (Three volumes : *The Evolution of Life*, *The Evolution of Man*, *The Issues of Evolution*).

Taylor, Charles. *The Explanation of Behaviour*, Humanities Press, 1964.

Taylor, Gordon Rattray. *The Great Evolution Mystery*, Secker, 1983.

Thompson, D'Arcy. *On Growth and Form*, Cambridge, 1942.

Thorpe, W.H. *Learning and Instinct in Animals*, Methuen, 1963.

Thorpe, W.H. *Animals Nature and Human Nature*, Methuen, 1974.

Thorpe, W.H. *Biology and the Nature of Man*, Oxford, 1962.

Thorpe, W.H. *Purpose in the World of Chance*, Oxford, 1978.

Thorpe, W.H., and Zangwill, O. *Current Problems in Animal Behaviour*, Cambridge, 1961.

Tinbergen, N. *The Herring Gull's World*, Collins, 1953.

Tinbergen, N. *Social Behaviour in Animals*, Wiley, 1953.

Tomlin, E.W.F. *The Concept of Life*, (as yet unpublished).

Towers, Bernard. *Concerning Teilhard*, Collins, 1969.

V

Vandel, A. *L'homme et L'evolution*, Paris, 1938.

Von Uexkull, J.J. *Theoretical Biology*, Kegan Paul, 1926.

Von Bertalanffy, L. *Modern Theories of Development*, Oxford, 1933.

W

Waddington, C.H. *The Nature of Life*, Allen & Unwin, 1961.

Waddington, C.H. *The Strategy of the Genes*, Allen & Unwin, 1957.

Waddington, C.H. *Towards a Theoretical Biology*, Edinburgh, 1969.

Watson, J.D. *Molecular Biology of the Gene*, 1970.

Weiss, P. *Principles of Development*, Holt, 1939.

Whitehead, A.N. *Nature and Life*, in *Modes of Thought*, Macmillan, 1938.

Whitehead, A.N. *Science and the Modern World*, Cambridge, 1928.

Whitehead, A.N. *Symbolism - Its Meaning and Effect*, Cambridge, 1927.

White, L.A. *The Science of Culture*, Ferrar, Straus, 1949.

Williams, George. *Adaption and Natural Selection*, Princeton, 1966.

Whyte, L.L. *The Unitary Principle in Physics and Biology*, Cresset, 1949.

Williams, Leonard. *Challenge to Survival*, Allison & Busby, 1978.

Willis, J.C. *The Course of Evolution*, Cambridge, 1940.

Wilmer, E.O. *Cytology and Evolution*, Academic Press, 1970.

Wilson E.O. *Sociobiology : The New Synthesis*, Harvard, 1975.

Wisdom, John. *Philosophy and Psychonalysis*, Blackwell, 1957.

Woodger, J.H. *Biological Principles*, Kegan Paul, 1929.

Index